Green Information Technology

Green Information Technology

A Sustainable Approach

Edited by

Mohammad Dastbaz
Colin Pattinson
Babak Akhgar

AMSTERDAM • BOSTON • HEIDELBERG • LONDON
NEW YORK • OXFORD • PARIS • SAN DIEGO
SAN FRANCISCO • SINGAPORE • SYDNEY • TOKYO
Morgan Kaufmann is an imprint of Elsevier

Executive Editor: Steven Elliot
Editorial Project Manager: Amy Invernizzi
Project Manager: Punithavathy Govindaradjane
Designer: Mark Rogers

Morgan Kaufmann is an imprint of Elsevier
225 Wyman Street, Waltham, MA 02451, USA

ISBN: 978-0-12-801379-3

British Library Cataloguing-in-Publication Data
A catalogue record for this book is available from the British Library

Library of Congress Cataloging-in-Publication Data
A catalogue record for this book is available from the Library of Congress

For information on all MK publications
visit our website at www.mkp.com

Contents

Foreword xiii
Preface xv
About the Editors xvii
Contributor Biographies xix
Acknowledgments xxv

SECTION I *GREEN IT: EMERGING TECHNOLOGIES AND CHALLENGES*

Chapter 1: Green ICT: History, Agenda, and Challenges Ahead 3
Introduction 3
The Second Industrial Revolution—The Emergence of Information and Communication Technologies 4
The Integrated Circuit (IC) Revolution 4
New Age of Computer Technology 5
Global Mobile Computing and Its Environmental Impact 6
The Agenda and Challenges Ahead 7
References 9

Chapter 2: Emerging Technologies and Their Environmental Impact 11
Introduction 11
Number of Connected Devices 13
Increased Functionality 15
Increased Number of Separate Functions 16
Increased Demand for Speed and Reliability 17
Obsolescence—The Problem of Backward Compatibility 18
The Other Side of the Balance Sheet—Positive Environmental Impacts or the "Other 90%" 19
Videoconference as an Alternative to Business Travel 20
Dematerialization of Product Chain 20
Travel Advice/Road Traffic Control 21
Intelligent Energy Metering 22
Building Management Systems 23

Saving IT Resources—A Drop in the Ocean? 24
Conclusion 24
References 25

SECTION II *GREEN IT: LAW AND MEASUREMENT*

***Chapter 3: Measurements and Sustainability* 29**
Introduction 29
ICT Technical Measures 32
Introduction 32
Service of Data Processing 33
Service of Data Transport 34
Service of Data Storage 34
Multimedia Service 34
Conclusion 35
Ecological Measures and Ethical Consideration 35
Introduction 35
ICT Impact on Pollution 36
Resource Efficiency 40
Main Green Measures of Performances 42
Ethics in ICT 45
Conclusion 46
Systems Engineering for Designing Sustainable ICT-Based Architectures 47
Introduction 47
Stakeholder Requirements Definition 47
System Requirements Analysis 48
System Requirements Validation and Verification 50
ICT Expertise and Results 50
Traceability Matrix 52
Ecoefficiency Metrics 53
Conclusion 55
References 57

***Chapter 4: The Law of Green IT* 61**
General Remarks on Law and the Regulation of Environmental Behavior 61
Direct and Indirect Governance of "Green IT" 61
Norm Addressees and Efficient Regulation 62
The Mechanisms of EU and National Law—Basics 64
Primary Law: Principles of Supremacy and Market Freedoms 64
Secondary Law: Directives and Regulations and Direct Effect 65
Comitology and Implementing Measures: Delegated Legislation by the Commission 66
Sustainability in EU Law 67
Article 3 TEU: Principle of Sustainable Development and Green IT 67

Integrated Product Policy (IPP) 68
Birth of a "Law for the Conservation of Natural Resources" 68
Specific European Legal Instruments Relevant to Green Computing and Their Implementation 69
Public Procurement: Environmental Standards as Criteria for Tenders 70
Ecodesign Directive: Regulation of Manufacturing and Encouraging Green Innovation 72
Energy Labeling Directive and Voluntary Ecolabeling 76
Mandatory Indication of Energy Consumption by Retailers....................78
Restriction of the Use of Hazardous Substances and Conservation of Natural Resources 79
Recycling and Disposal....................80
Conclusions....................80
References....................81

Chapter 5: Quantitative and Systemic Methods for Modeling Sustainability 83
Introduction....................83
Complexity....................83
Modeling Approaches....................84
System Dynamics and Control....................85
Influence Diagrams and Knowledge Representation....................86
Influence Diagram vs. Decision Tree 87
Modeling Approaches and Decision Support Systems....................88
Criticisms 89
Conclusions....................90
References....................90

SECTION III *SUSTAINABLE COMPUTING, CLOUD AND BIG DATA*

Chapter 6: Sustainable Cloud Computing 95
Introduction....................95
Challenges in the Use of Cloud Computing As Green Technology 98
Cloud Computing and Sustainability 99
Sustainable Applications of Cloud Computing....................100
Technologies Associated With Sustainable Cloud Computing....................105
Future Prospects of Sustainable Cloud Computing 105
Reflections on Sustainable Cloud Computing Applications 106
Conclusions....................107
References....................107

Chapter 7: Sustainable Software Design.................... 111
Overview and Scope....................111
Evaluating Sustainability Effects 111
Sustainability and the Product Life Cycle 112

Direct Effects: Sustainability During Use ... 114
Runtime Energy Consumption Basics ... 115
Analyzing the Energy Consumption of an Application ... 115
Energy Consumption Reduction Using Physical Properties of Semiconductors ... 117
Optimizing the Energy Consumption of an Application: Compiler Techniques ... 118
Optimizing the Energy Consumption of an Application: Runtime Approaches ... 119
Optimizing the Energy Consumption of an Application: Probabilistic Approaches ... 120
Indirect Effects: Sustainability vs. Production ... 121
Conclusions and Outlook ... 124
References ... 125

Chapter 8: Achieving the Green Theme Through the Use of Traffic Characteristics in Data Centers ... 129

Introduction ... 129
Green IT and the Cloud ... 129
Virtualization Behavior ... 131
Chapter Coverage ... 132
Rationale ... 132
Relationship Between Infrastructure as a Service (IaaS) and Power ... 132
Network Processes and Power ... 133
Need for Thermal-Aware Virtualization ... 133
Understanding Sustainability on the Cloud ... 133
Current State of Affairs ... 133
Achieving Sustainability on the Cloud ... 135
Sustainability with VM Management ... 136
Green Cloud as a Network Management Problem ... 136
Importance of Virtualization Management ... 136
Relationship Between Networking and Power Consumption ... 137
Need for Traffic Characterization in Virtualized Environments ... 138
Role of Hypervisors in Traffic Characterization ... 139
SNMP for Green Cloud Traffic Characterization ... 139
SNMP Operation in Context of Green Clouds ... 139
Related Work ... 140
A Model for Network Management for Green Cloud ... 141
Model Outline ... 141
Gathering and Using Statistics ... 143
Conclusions and Future Work ... 146
References ... 147

SECTION IV *FUTURE SOLUTIONS*

Chapter 9: Energy Harvesting and the Internet of Things ... 151

Energy Harvesting: Intelligence and Efficiency ... 151
The IoT, "Hyped" and "Hidden": A Green Technology ... 154

Conclusions 160
Further Reading 160

Chapter 10: 3D Printing and Sustainable Product Development 161
Introduction 161
3D Printing Design Pipeline 162
The Underlying Printing Processes 166
Stereolithography Apparatus 166
Selective Laser Sintering 168
Solid Ground Curing 168
Laminated Object Manufacturing 170
Fused Deposition Modeling 171
Inkjet Technology: Powder-Based Printers 172
Hybrid Systems: Integrating 3D Printing with Subtractive Machining Technology 173
Environmental Considerations 175
The Factory 2.0 Philosophy 177
Agile Manufacturing 178
Recycling 178
Organic Materials 179
Factory of the Future: The Next Steps 181
Conclusions 182
References 182

SECTION V *CASE STUDIES*

Chapter 11: Automated Demand Response, Smart Grid Technologies, and Sustainable Energy Solutions 187
Background 188
The Challenge of Maintaining Equilibrium in the UK Electricity System 188
The Opportunity 194
The Solution 194
ADR Value Streams 195
ADR Solution Platform 197
ADR Solution Overview 197
ADR Solution Description 198
Summary of ADR Project Goals and ADR Project Phases 200
Designing Reliable Load-Shed Strategies 201
Turnkey Implementation 202
Solution Elements 203
Demand Forecasting 207
DR Event Optimization 207
Intuitive Web Interface 208
Demand Response Automation Server 208
DRAS Baselining 209

Case Study 1: Thames Valley Vision Project, United Kingdom (Auto DR Element) ... 211
Why is ADR Important? ... 213
How Does ADR Work? ... 215
What are the Benefits? ... 216
SSEPD TVV Project SDRC Report ... 216
Early DSR Adopters ... 217
Trial of Promotion Success ... 217
Case Study 2: US Utility-Driven ADR Programs ... 219
CenterPoint Energy, Houston, Texas, United States ... 219
Consolidated Edison, New York, United States ... 220
CPS Energy, San Antonio, Texas, United States ... 221

Chapter 12: Critical Issues for Data Center Energy Efficiency ... 223
Introduction ... 223
Aim and Objectives ... 224
Literature Survey ... 225
Green ICT ... 225
Data Centers ... 226
Data Center Efficiency ... 226
Data Center Efficiency Measurements and Metrics ... 229
Methodology ... 230
Implementation ... 237
Operation of the Experiment ... 238
Assumptions ... 238
Results and Discussion ... 238
Results ... 238
PUE Analysis ... 238
Effect of Set Point Temperature ... 241
Effect of a Change in the Cooling System ... 241
Immediate Impact ... 244
Future Impact ... 245
Conclusions ... 246
Implications for the Future ... 246
Acknowledgments ... 246
References ... 246

Chapter 13: Communitywide Area Network and Mobile ISP ... 249
Introduction ... 249
Context of Application ... 249
Prototype Mobile Learning Environment ... 250
Architecture of Prototype ... 250
General Requirements ... 251
Infrastructure ... 253
Hardware Stack ... 253

Software Stack 256
Electrical Topology 257
Prototype 260
Prototype Specifications 260
Configuration 260
Langdale Pilot Study 262
Requirements 262
Core Equipment 262
Client Equipment 263
Measuring Cellular Backhaul Signal Strength 263
Measuring WAN Bandwidth/Latency 268
WAN Backhaul Conclusion 270
WLAN Planning 270
WLAN Propagation Losses 271
Squid Proxy—Control Measures 271
Squid Proxy—Policy 274
Feedback on the Mobile Learning System 274
Feedback (by Module Leader) 274
Feedback (Tutors) 274
Conclusions 275
References 276

Chapter 14: Thin-Client and Energy Efficiency 279
Introduction 279
Aims and Objectives 280
Literature Review 281
Green IT 281
Comparison of Thick and Thin Client Power Consumption 283
Project Reports on the Performance of Thin-Client Systems 284
Methodology 285
Overview 285
Variations of Thick and Thin Clients 285
Implementation 287
Operation of the Experiment 287
Assumptions 288
Results 288
Immediate Impact 288
Future Impact 290
Conclusions 291
Recommendations 291
Implications for the future 291
Acknowledgments 292
Appendix A: User Feedback Comments 292
References 293

Chapter 15: Cloud Computing, Sustainability, and Risk *295*
Introduction 295
Cloud Architecture and Risk Preferences 297
Green Cloud Computing and Risk Management 301
Risk Appetite and Tolerance 302
Risk Target and Optimization Model 305
Case Study: Petrogas Jahan Co 307
Conclusion 310
References 310

Index *313*

Foreword

This book delivers a comprehensive perspective of the challenges brought by our increasing demand for digital services and raises the perennial question as to whether we can sustain the growth of these services to meet that rising demand, furthering the prospect that we may overload the Internet with unforeseen consequences on our lives.

I would add three reflections to the book's perspective, returning to the three pillars of sustainability. First there is the impact of economics; as supply outstrips demand, up go the prices. Would so-called frivolous use of the Internet then be priced to provide capacity for those who can pay and are willing to do so for digital services that are important in their lives?

The social reflection is that growth in digital services is critical if we are all to live together successfully on this planet as our numbers grow to 9 billion and the resources we require to live increase exponentially beyond the immediate capacity of the planet. I believe that, without digital services and the Internet, we would not have the means to work together across communities, cities, nations, and the world to tackle the huge and complex problems that growth is bringing. Digital services will "extend" the resources available to us by enabling smarter, more closed-loop approaches that reduce our "take" from the finite new sources of minerals and metals still left in the earth. Perhaps we now have a new world war on our hands—a war between us and our appetites and behaviors. We will have to return to a mentality of "waste not, want not," of being frugal so our neighbors can survive, and, above all, to recover and further develop our humanity from the digital silos into which it is sinking.

Finally, the third reflection returns to environmental concerns. Digital services will and are delivering elsewhere significant reductions of the resources we take from the planet in living our lives. Through the "Internet of things," we will be able to improve the monitoring and management of our planet's resources and behaviors, and our own, to achieve greater efficiencies and the means to live within our "means."

I therefore remain optimistic that the developments and opportunities set out herein will enable the growth in digital services to be sustained and thereby enable us to survive and prosper, albeit living in ways that would be alien to the founders of this great Digital Age.

Bob Crooks, MBE
Chair of the British Computer Society Green Specialist Group
Defra Lead for Innovative and Sustainable ICT

Preface

We are living in an era of amazing technological changes. With more than 2 billion PCs across the world and 48 billion Web pages indexed by some 900,000 servers, the era of "Big Data" and how we use it on a daily basis, and the environmental impact of storing and safeguarding this data have become a critical issues facing system developers and researchers across the globe. *Green Information Technology: A Sustainable Approach* offers, in a single volume, a broad collection of practical techniques and methodologies for designing, building, and implementing a green technology strategy in any large enterprise environment, which up until now have been scattered in difficult-to-find scholarly resources. Included here is the latest information on emerging technologies and their environmental impact, how to effectively measure sustainability, and discussions on sustainable hardware and software design, as well as how to use big data and cloud computing to drive efficiencies and establish a framework for sustainability in the information technology infrastructure.

This book is arranged in five sections. Section 1 provides an introduction to emerging technologies and the challenges that lie ahead. Section 2 addresses some of the most important topics concerning how we measure "sustainability" as far as "Green Information Technology" is concerned and the available methodologies and their gaps. It also deals with the critical issue of the current legal framework (or lack of it) in regulating the development of Information Technology. Section 3 takes a look at "cloud computing and big data," providing a thought provoking discussion around some of the challenges in these areas. Section 4 discusses some future solutions, discussing the "Internet of Things" and "Energy Harvesting," as well as "Factory 2" concepts. The final section of this book, Section 5, is devoted to a number of state-of-the art case studies covering a range of research and their important findings.

About the Editors

Professor Mohammad Dastbaz is Pro-Vice Chancellor and Dean of Faculty of Arts Environment and Technology at Leeds Beckett University. Professor Dastbaz main research work in recent years has been focused on the use and impact of emerging technologies in society, particularly learning, training, issues of privacy and cyber security and the development of "e-Government." Mohammad has led EU- and UK-based funded research projects and has been the symposium chair of Multimedia Systems in IEEE's Information Visualization (IV) conference since 2002. He has more than 50 refereed publications, including numerous journal articles, conference papers, book chapters, and books on e-learning and e-government as well as design and development of multimedia systems. Professor Dastbaz is a Fellow of the British Computer Society and UK's Higher Education Academy as well as a professional member of ACM and IEEE's computer society.

Professor Colin Pattinson is the Head of School Computing, Creative Technologies and Engineering at Leeds Beckett University. Professor Pattinson's research and teaching have reflected the massive changes in computer and communications technologies beginning with very basic computer-to-computer connections through the development of the Internet to current developments in smart phones and cloud computing. Professor Pattinson continues to be an active researcher in these areas with a particular interest in measuring and understanding the performance of IT systems. This interest in performance measurement has led to current research interests in sustainability issues within IT on which he has supervised research projects in both the environmental impact of IT systems and the use of IT to enhance the sustainability of other aspects of human activity. He led a pan-European project to promote sustainability among data center professionals and is currently involved in two EU projects to develop and deliver sustainable IT curriculum for university-level courses. He is a committee member of the Green IT Specialist Group of BCS: The Chartered Institute for IT.

Babak Akhgar is Professor of Informatics and Director of the Centre of Excellence in Terrorism, Resilience, Intelligence, and Organized Crime Research (CENTRIC) at Sheffield Hallam University (UK) and a Fellow of the British Computer Society. He has more than 100 refereed publications in international journals and conferences on information systems with specific focus on knowledge management (KM). He is a member of the editorial boards of several international journals and has acted as chair and program committee member for numerous international conferences. He has extensive hands-on experience in the development, management, and execution of KM projects and large international security initiatives (e.g., the application of social media in crisis management; intelligence-based combating of terrorism and organized crime, gun crime, cybercrime, and cyberterrorism; and cross-cultural ideology polarization). In addition to this, he is the technical lead of two EU Security projects: "Courage" on Cyber-Crime and Cyber-Terrorism and Athena on the application of social media and mobile devices in crisis management. He has coedited several books on intelligence management. His recent books are *Strategic Intelligence Management* (*National Security Imperatives and Information and Communications Technologies*), *Knowledge Driven Frameworks for Combating Terrorism and Organised Crime*, and *Emerging Trends in ICT Security*. Professor Akhgar is a member of the academic advisory board of SAS UK.

Contributor Biographies

Dr. Christian DeFeo has worked in the information technology industry since 1994, when he was a management intern with International Computers Limited. He has since had roles as a webmaster, developer, and project manager; he was the Site Producer for Ebookers.com, the Web Development Manager for the Marine Trader Media branch of Trader Media Group, and had the opportunity to work with leading Digital Economy thinkers while working as a Collaboration Manager at the University of Southampton. He presently works as the Global Community Supplier Manager for the element14 community; he is responsible for working with leading electronics manufacturers, including Texas Instruments, Cisco, and Würth Elektronik and finding new means to increase user engagement and generate innovation. Among his successful projects were extensive community marketing and education programs involving webinars, product road tests, and competitions to publicize the advent of new Wireless Power technologies (http://www.element14.com/beyondthephone), Energy Harvesting technologies (http://www.element14.com/community/groups/energy-harvesting-solutions), Smarter Homes technologies (http://www.element14.com/smarterlife), and the Internet of Things (http://www.element14.com/forgetmenot). He is also the program manager for a unique "crowdsourced" project to develop a mobile application for the Bath Institute for Medical Engineering (http://www.element14.com/project-nocturne).

Dr. Azad Camyab has held senior business development and project management roles in the energy sector for over 25 years. He started his career with CEGB and National Power in the UK where he was involved in the development and construction of a number of CCGTs in the early 1990s and developing and managing IPP (Independent Power Producer) projects globally.

Azad is currently the CEO of Pearlstone Energy Limited and an Associate of the Laing O'Rourke Centre for the Masters course in Construction Engineering & Technology, and the Masters programme in Sustainability Leadership, at the University of Cambridge. He is also a Visiting Professor at the London Met Business School.

Azad is a Fellow of The Institute of Engineering and Technology (FIET), a member of the Renewable Energy World Europe (Power-Gen Europe) Executive Advisory Board, and a Fellow of the Leeds Sustainability Institute Advisory Board.

Kiran Voderhobli is a Senior Lecturer at Leeds Beckett University who specializes in teaching Network Management and Network Security, among other areas related to networking systems. He is the course leader for Masters in Networking and Masters in Computer Science at Leeds Beckett. Before becoming an academic, Mr. Voderhobli also worked in industry, developing secure commercial VoIP solutions. His research areas include network security and sustainable computing. He received the MPhil degree from Leeds Beckett University after undertaking research into ubiquitous paradigms for network security. He is currently working toward a PhD in the area of sustainable networks and green ICT. He is an active researcher in the sustainability research group at Leeds Beckett University.

Dr. Nick Cope is currently Associate Dean for Enterprise and Employability in the Faculty of Arts, Environment and Technology, Leeds Beckett University. Nick holds a BSc in Bio-Medical Electronics from the University of Salford and a PhD in Human Interface/Assistive Technology from the Department of Electronics at Southampton University. Dr. Cope's research interests include 3-D computer graphics, the application of 3-D print technologies, computer simulation, computer game technologies, augmented reality, 3-D motion capture, 3-D scanning technologies, and digital applications in medical technology. Previously Dr. Cope had a Post-Doctoral Research Fellowship at Aston University, Department of Mechanical Engineering (Manufacturing Systems and Factory Simulation); he was also Senior Research and Development Software Engineer, Ferranti Computer Systems, Human Interface Technology Group.

Dr. Stephen Wilkinson worked in industry in the design office, designing very large machine tools, before entering academia. He has been a lecturer for more than 32 years, having taught a range of subjects from Robotics and Automation to 3-D Visualization in both undergraduate and postgraduate courses. He used this experience while obtaining his PhD on 3-D simulation of flexible manufacturing systems and in his research on augmented reality for the 3-D simulation of hand operations. This work has enabled him to co-author two books in the areas of manufacturing technology and e-manufacture. His current research and course development have concentrated on eco engineering using 3-D printing as an advanced manufacturing technology. His other interests include the development of new engineering courses.

Professor Eric Rondeau is a full professor at the University of Lorraine, France. His research domain is Networked Control Systems (NCS) and green ICT. He was the coordinator of the FP6 NeCST STREP project. He is the coordinator of Erasmus Mundus Joint Master Degree in PERCCOM (Pervasive Computing and Communications for Sustainable Development). He is participating in an Ecotech ANR (Research National Agency) project on indoor pollution in designing a new smart formaldehyde sensor. He is a member of IFAC TC 1.5 on Networked Systems. He has supervised 10 doctoral students, and he is co-author of more than 100 conference or journal papers.

Professor Francis Lepage is a full professor at Lorraine University in Nancy, France. His research interest previously focused on time-driven systems, especially critical real-time controlled systems. Now he studies time-constrained communication networks and large wireless sensor networks. He has supervised 25 PhD theses, and he is author or co-author of six books and about 100 publications.

Dr. Jean-Philippe Georges received his PhD in network engineering from the University of Lorraine (France) in 2005. In 2006, he worked as a researcher at Aalto University, Finland. He is currently an associate professor with the Research Centre for Automatic Control of Nancy at the University of Lorraine. He has conducted research in areas such as ethernet-based real-time networks and performance evaluation of embedded networks, especially for spatial launchers. He has published more than 50 journal and conference papers. His current research interests include performance evaluation, dependability and sustainability of wired and wireless networks with quality of service, and green IT metrics. He is also involved in the Complex Systems Engineering master program and in the Erasmus Mundus Master PERCCOM (Pervasive Computing and Communications for Sustainable Development).

Professor Gérard Morel is a full professor at the University of Lorraine. He has supervised about 30 PhD theses and Accreditations to Supervise Research and published more than 150 articles in the area of systems and automation engineering. He has held scientific positions in national and international research networks and served in several positions in IFAC (International Federation of Automatic Control), as well as working as the journal editor for *Engineering Applications of Artificial Intelligence* and for *Journal of Intelligent Manufacturing*. He has also held evaluator positions for the European Commission and for the French Agency for the Evaluation of Research and Higher Education. He has also served as vice-chairman of AFIS, the French chapter of the International Council on Systems Engineering.

Professor Hamid Jahankhani recently joined the Department of Digital Technology and Computing, Faculty of Social Sciences, Law and Technology, GSM-London.

He obtained his PhD from Queen Mary College, University of London. In 2000 he moved to the University of East London, and he became the first Professor of Information Security and Cyber Criminology at the university in 2010.

Over the last 10 years Hamid has also been involved in developing new and innovative programs and introducing the "block mode" delivery approach at UEL, including MSc Information Security and Computer Forensics (block mode delivery), Professional Doctorate Information Security.

Hamid's principal research area for a number of years has been in the field of information security and digital forensics. In partnership with key industrial sectors, he has examined and established several innovative research projects that are of direct relevance to the needs of UK and European information security, digital forensics industries, Critical National Infrastructure (CNI), and law enforcement agencies.

Professor Jahankhani is the Editor-in-Chief of the *International Journal of Electronic Security and Digital Forensics* published by Inderscience, www.inderscience.com/ijesdf, and general chair of the annual International Conference on Global Security, Safety and Sustainability (ICGS3). Professor Jahankhani has edited and contributed to more than 10 books and has more than 100 conference and journal publications.

Dr. Michael Engel is currently a Senior Lecturer for Computer Systems Engineering at the School of Computing, Creative Technologies and Engineering at Leeds Beckett University. Before this, he was Assistant Professor at the Faculty of Computer Science at TU Dortmund, Germany. In 2006/2007, he was interim Professor for Operating Systems at TU Chemnitz, Germany. He received his doctoral degree in 2005 from the University of Marburg and holds degrees in Computer Engineering and Applied Mathematics from the University of Siegen, Germany. Dr. Engel has published more than 50 papers on topics including dependability, operating systems, embedded systems, and software engineering. His work on probabilistic computing received the Best Paper Award of the ARCS conference in 2012. Dr. Engel's research interests lie at the intersection of hardware and software, especially for embedded systems. His current research concentrates on improving embedded system dependability under the influence of errors, probabilistic and variability effects to enable reductions in energy consumption, cost, runtime, and memory consumption. Previous projects included the application of software engineering techniques to operating systems as well as large-scale simulation environments for wireless systems. Dr. Engel's research efforts have been supported by federal and private foundations as well as industry. He was Principal Investigator of the "FEHLER" project in the context of the German National Priority Research Program SPP1500 "Dependable Embedded Systems," funded by the German Research Foundation (DFG). Overall, he has been a key participant in projects attracting more than $1 million of research funding, with his contributions valued at $500,000.

Ashkan Tafaghodi is a PhD candidate in the field of Information Technology Management at the Department of Management, Tehran University. He is also an MSc graduate in the same field of study and holds a bachelor's degree in Applied Mathematics. His major fields of interest are cloud computing, business intelligence, and big data. He is currently involved in a couple of cloud projects on a national scale.

Dr. Konstantinos Domdouzis is a researcher at the Centre of Excellence in Terrorism, Resilience, Intelligence & Organised Crime Research (CENTRIC), Sheffield Hallam University. He is a computer scientist with extensive experience in the applications of computing to different scientific and technical fields. He has applied computational techniques in Civil & Building Engineering, Agricultural & Biological Engineering, and Healthcare Modelling. Dr. Domdouzis holds a BSc (Hons) in Computer Science from the University of

Luton, United Kingdom, and an MSc in Computer Networks and Communications from the University of Westminster, United Kingdom. Dr. Domdouzis successfully completed his doctoral thesis, entitled "Applications of Wireless Sensor Technologies in Construction," at the Department of Civil & Building Engineering of Loughborough University, United Kingdom. Following the completion of his PhD, he worked as a postdoctoral research associate in the field of Agricultural & Biological Engineering at the University of Illinois at Urbana-Champaign, United States, and as a Knowledge Transfer Partnership (KTP) Associate for Whole Systems Partnership, a UK-based healthcare modeling consultancy in cooperation with Brunel University, United Kingdom. His main research interests lie in the fields of distributed systems, pervasive computing, and intelligent computation.

Dr. Ah-Lian Kor is a Course Leader for MSc Sustainable Computing at Leeds Beckett University and a Leeds Sustainability Institute Strategic Committee Member. She is a Senior Lecturer specializing in web applications, software development, and green ICT. She is active in AI as well as sustainable ICT research and has published numerous papers in these areas. Currently, she is an Editorial Advisory Board Member for *International Journal on Advances in Intelligent Systems, International Journal on Advances in Security,* and *IBIMA Publishing Journal*. She sits on many international conference program/technical committees (e.g., IEEE Cloud Computing, ICAART, CeDEM, and others). She is involved in EU projects relating to pervasive computing and communications for sustainable development and green computing and communications. She has co-authored online learning resources for a cross-boundary EU masters degree program on green sustainable data centers. She has also offered consultancy services relating to food freshness technologies to a global company through the Faraday Retail Institute at Leeds Beckett University. She is a member of the BCS Specialist Group for Data Centers.

Dr. Eva Julia Lohse is a postdoctoral research and teaching fellow in European administrative and constitutional law at the Friedrich-Alexander-Universität, Erlangen-Nürnberg. Her doctoral thesis was on school administrative law and the right to education. She is now working on a book about harmonization mechanisms in the European Union. Her LLM in European and Comparative Law is from the University of Kent, Canterbury.

Dr. Amin Hosseinian-Far is currently a Senior Lecturer in the Faculty of Arts, Environment & Technology at Leeds Beckett University. He received his MSc in Satellite Communication and Space Systems from University of Sussex, and he received a PhD from the University of East London after completing his doctoral thesis, titled "A Systemic Approach to an Enhanced Model for Sustainability." He has held lecturing and research positions at the University of East London and at a number of private HE institutions and strategy research firms in London. He has served as a program committee and steering committee member for international conferences, symposiums, and workshops. He is also a reviewer for a number of international journals. Dr. Hosseinian-Far is a Member of the Institution of Engineering and Technology (IET) and a Fellow of the Higher Education Academy (HEA) and the Royal Society of Arts (RSA).

Roland Cross graduated with a degree in Biophysics from the University of Leeds in 1988 and immediately created an IT business that he ran for ten years, during which time he experienced the PC boom. In 1999, he joined Leeds Metropolitan University, and he currently holds the post of Technical Projects Consultant. His research interests include green IT governance, and he is currently researching methods to improving data center efficiency, and linking business systems to building management systems with the objective of reducing organization carbon footprints.

Richard Braddock graduated from Leeds Metropolitan University in 2009 with a first-class degree in in Computer Communications. He started out working in embedded systems, carrying over those lessons learned from low-power x86 and ARM architectures into real-world contactless systems by helping to launch successful initiatives for Orange and O2. He currently works for Etihad Airways, taking those same lessons in efficiency and translating them into high-performance guest experiences. His areas of expertise include web performance, infrastructure architecture, and scaling Enterprise Application Support structures.

Acknowledgments

We would like to thank all the contributing authors for their tireless efforts in providing us with their valuable research work that has made this book possible. We thank Steve Elliot and the Elsevier team for working closely with us and providing advice about how to develop the structure and content of the book.

Special thanks go to Carole Smith for providing valuable administrative support, being very paitent, sending numerous e-mails as deadlines loomed, and gathering information and material for us.

We want to acknowledge Scott Petersen, business development director of Smart Grid Solutions Europe and North Africa for Honeywell Building Solutions. He stepped in and provided us with valuable advice and help on the book's case study on Smart Grid Automated Demand Response and Sustainable Energy Solutions.

We would also like to thank the Centre of Excellence in Terrorism, Resilience, Intelligence, and Organised Crime Research (CENTRIC) at Sheffield Hallam University for its support.

Finally, our thanks go to all our colleagues at Leeds Beckett University for supporting us and making this book possible.

Mohammad Dastbaz
Colin Pattinson
Babak Akhgar

Green IT: Emerging Technologies and Challenges

CHAPTER 1

Green ICT: History, Agenda, and Challenges Ahead

Mohammad Dastbaz
Leeds Beckett University, Leeds, UK

Introduction

The dawn of the Industrial Revolution and the growth of machine-based industries changed the face of our planet for good. While Charles Dickens darkly depicted poverty, disease-ridden cities, and squalid living conditions in *Great Expectations*, the replacement of the farming/cottage-type production industry with large factories changed Britain's skylines in the 18th century and signaled the beginnings of monumental change and innovation in addition to immense scientific discoveries.

The Industrial Revolution also fundamentally changed Earth's ecology and humans' relationship with their environment. One of the most immediate and drastic repercussions of the Industrial Revolution was the explosive growth of the world's population. According to Eric McLamb (2011), with the dawn of the Industrial Revolution in the mid-1700s, the world's population grew by about 57% to 700 million, would reach 1 billion in 1800, and within another 100 years, it would finally grow to around 1.6 billion. A hundred years later, the human population would surpass the 6 billion mark. This phenomenal growth in population put enormous pressure on the planet, forcing it to cope with a continuously expanding deficit of resources.

The transformation from cottage industry and agricultural production to mass factory-based production led to the depletion of certain natural resources, large-scale deforestation, depletion of gas and oil reserves, and the ever-growing problem of carbon emissions—mainly the result of our reckless use of fossil fuels and secondary products. The pollution problem following the Industrial Revolution led to the atmospheric damage of our planet's ozone layer as well as air, land, and water pollution.

The two world wars in the early twentieth century brought with them catastrophic human and natural disasters as well as rapid development of military technologies. These developments laid the groundwork for the emergence of what some would like to call the Second Industrial Revolution.

The Second Industrial Revolution—The Emergence of Information and Communication Technologies

In a visionary paper entitled "As We May Think" (published in *Atlantic Monthly* July 1945), Vannevar Bush, the scientific adviser to President Theodore Roosevelt, predicted a "bold" and exciting future world, based on "memory extended" machines. He wrote:

> Consider a future device for individual use, which is a sort of mechanized private file and library. It needs a name, and, to coin one at random, "memex" will do. A memex is a device in which an individual stores all his books, records, and communications, and which is mechanized so that it may be consulted with exceeding speed and flexibility. It is an enlarged intimate supplement to his memory...

It was not long after this that the first large-scale technological innovations started to appear, marking the beginning of a new dawn in information and communication technologies, changing almost every aspect of our lives.

In 1946, the first new generation of computers emerged from US military research. Financed by the United States Army, Ordnance Corps, and Research and Development Command, the Electronic Numerical Integrator and Computer (ENIAC) was announced and nicknamed the "giant brain" by the press. In reality, ENIAC, as compared to the smartphones of today, had very limited functionality and capabilities. According to Martin Weik (December 1955), ENIAC contained 17,468 vacuum tubes, 7200 crystal diodes, 1500 relays, 70,000 resistors, 10,000 capacitors, and around 5 million hand-soldered joints. It weighed more than 27 tons, was roughly $8 \times 3 \times 100$ ft (2.4 m $\times 0.9$ m $\times 30$ m), took up 1800 ft^2 (167 m^2), and consumed 150 kW of power. This led to the rumor that whenever the computer was switched on, the lights in Philadelphia dimmed.

In fact, all ENIAC could do was to "discriminate the sign of a number, compare quantities for equality, add, subtract, multiply, divide, and extract square roots. ENIAC stored a maximum of twenty 10-digit decimal numbers. Its accumulators combined the functions of an adding machine and storage unit. No central memory unit existed *per se*. Storage was localized within the functioning units of the computer" (Weik, 1961).

The Integrated Circuit (IC) Revolution

Although there was much excitement about the development of ENIAC, its sheer size and physical requirements, coupled with the limited functionality it offered for such a huge cost, used such systems limited to large mainly military-based research labs. At the same time that ENIAC was being developed and launched, there were other significant developments in the field of information technology.

Werner Jacobi, a German scientist working with Siemens AG, fielded a patent for developing cheap hearing aids based on an "integrated-circuit-like semiconductor amplifying device." Further development on a similar idea was pioneered by Geoffrey Drummer, working for the British Royal Rader. Drummer used the symposium on Progress in Quality Electronic Components in 1952 to discuss his revolutionary idea but was unfortunately unsuccessful in realizing his vision and building such a circuit.

It was not until 1958 that a young scientist working for Texas Instruments, Jack Kilby, came up with a solution for a truly IC and was able to successfully demonstrate a working version of it on September 12, 1958. The idea of being able to create much smaller systems, coupled with computer applications developed to offer greater functionality, soon changed the face of computing and led to the march of "machines." Jack Kilby won the 2000 Nobel Prize in Physics for his part in the invention of the IC, and his revolutionary work on the development of IC was named as an IEEE Milestone in 2009.

New Age of Computer Technology

According to the Historical Museum of computers online, there were only around 250 computers in use in the world in the 1950s. The development of IC and the microprocessor manufacturing industry and the emergence of corporations such as Intel meant that the world of computer technology had changed significantly. Intel's founder, Gordon Moore, observed that as the hardware industry grew rapidly, the number of transistors in a dense IC doubled approximately every two years—this estimation proved to be accurate and by the 1980s, more than 1 million computers were in use throughout the world. According to Gartner Inc., the number of installed PCs worldwide has surpassed 1 billion units, and it is estimated that the worldwide installed base of PCs is growing just under 12% annually. At this pace, it has been estimated that the number of installed PCs will surpass 2 billion units by early 2014. To this staggering figure, one must add the number of tablet and smartphone devices to understand the scale of this new technological age and its potential environmental cost.

It is important to note that it is not just the scale of the hardware development and emergence of more and more powerful PCs that have changed our way of lives but also the phenomenal development of applications available on our PCs and mobile devices.

I recall that in early 1992, I had my first taste of using the Internet and managed to log in to the Library of Congress information page, which showed me its opening and closing times. This was incredibly impressive, and I was thrilled that I could connect to the other side of the Atlantic at such speed; nevertheless, sitting in Kingston University's research lab, I also wondered how useful this information was for me if I could not access the rich content that the library had to offer. But it was only a matter of time before this changed, and the doors were opened to the "information super highway." The clumsy text-based interface that I had used was replaced by a much more interesting and exciting Mosaic (graphic capable) interface and then the wonderful World Wide Web (WWW) arrived, and the rest, as they say, is history.

I do not think many of us living through the rapid development of this exciting yet unknown technology could have imagined that in a couple of decades our lives would be so dominated by it with frightening and somewhat crippling effects.

In reality, we have reached a stage where estimating the size and growth of the Internet and the WWW is extremely difficult, if not impossible. Nevertheless, one can find some terrifying figures online: according to www.factshunt.com, the size of the Internet and the WWW is something like:

- 48 billion—webpages indexed by Google Inc.
- 14 billion—webpages indexed by Microsoft's Bing
- 672 exabytes—672,000,000,000 gigabytes of accessible data
- 43,639 petabytes—total worldwide Internet traffic (2013)
- More than 900,000 servers—owned by Google Inc. (largest in the world)
- More than 1 yottabytes—total data stored on the Internet (includes almost everything; 1 yotta = 1000^8)

Global Mobile Computing and Its Environmental Impact

Clearly, the next question one has to ask is what the overall cost and the energy footprint of the new digital age are. Although there have been a number of attempts to identify a definitive cost model for the new digital age, clearly the scale and complexity of such calculations and the constantly changing data set have meant that at best we can come up with are intelligent guesses rather than definitive answers.

To answer the question of what the energy footprint of the new technology age requires, we must consider the electricity consumed by the world's laptops, desktops, tablets, smartphones, servers, routers, and other networking equipment (Gills, 2011). The energy required to manufacture these machines also needs to be included, as more importantly, does the energy required to keep such a wide range of devices and systems operational (such as the fact that smartphones typically require charging on daily basis).

Another key issue considered by researchers is the enormous volume of data transferred across our global mobile network of devices and the cost associated with maintaining and transferring these data.

Coroama et al. (2013), discussing the "electricity intensity of the Internet and its data consumption," point out that studies have already explored the electricity intensity of Internet data transfers (such as Koomey et al., 2004; Baliga et al., 2007, 2008, 2009; Taylor and Koomey, 2008; Weber et al., 2010; Hinton et al., 2011; Kilper et al., 2011; Lanzisera et al.,

2012). Comparing their estimates is difficult because of inconsistent boundaries, data uncertainties, and the range of methodologies used.

In an interesting report sponsored by the National Mining Association American Coalition for Clean Coal Electricity) produced in August 2013, Mark Mills (CEO of Digital Power) states, "The information economy is a blue whale economy with its energy uses mostly out of sight. Based on a mid-range estimate, the world's Information Communications Technologies (ICT) ecosystem uses about 1500 TWh of electricity annually, equal to all the electric generation of Japan and Germany combined as much electricity as was used for global illumination in 1985. The ICT ecosystem now approaches 10% of world electricity generation. Or in other energy terms the zettabyte (1000^7) era already uses about 50% more energy than global aviation."

Referring briefly to the beginning of this chapter, human society's rapid industrial development over the past three centuries has caused far more lasting damage to our environment than ever before. As we destroy our forests and plunder our natural resources, the promise of technological innovations in saving our planet from further damage has not yet been realized. The ever-growing demand for data as a commodity that rules our lives along with the concept of "never switching off" and "contact on demand" with the requirement to keep the devices necessary to entertain this "new digital age" has brought with it further serious challenges that require immediate attention. Cloud computing was an incredibly promising concept that was perhaps hyped beyond reasonable measures to convince mass consumer migration. The matter of cost was never discussed or scrutinized as a factor before multinational corporations "migrated" us into the cloud. Although the move toward cloud computing has provided us with new opportunities that cannot be overlooked, the era of "big data" and the computing power required to deal with them both in terms of the data's energy consumption and technical complexity is one of the key areas of urgent further research and development. The US Environmental Protection Agency estimates that centralized computing infrastructures (data centers) currently use 7 GW of electricity during peak periods. This translates to about 61 billion kilowatt hours of electricity used. As already indicated, we are moving to the age of zetta and yotta data (1000^7 and 1000^8), but pressure from environmentally conscious consumers has forced computer and Internet giants such as Microsoft, Google, and Yahoo to build their new data centers on the Columbia River, where there is access to both hydroelectric power and a ready-made source of cooling.

The Agenda and Challenges Ahead

So what are the key agenda items for the "Green Information and Communication Technology" debate, and what are the key challenges facing researchers and developers in this area?

- While there is focus on big infrastructure and big data computing, the debate quite often tends to overlook the large number of existing "legacy" systems (which are neither green nor efficient). Even our current laptops are viewed as "old legacy" systems. While the statistics show that there are around 2 billion PCs in operation currently, most of these systems suffer from "old" hardware (i.e., power hungry) designs. It is not surprising that manufacturers such as Intel have spent billions of dollars designing the next generation of microprocessors ("Haswell") and moving toward fanless, less power-hungry systems.

- Another key challenge is how we measure ICT performance and sustainability and what tools we can use to provide reliable data. It is clear that unless we have a reliable methodology to measure ICT suitability, we will be experiencing what environmental campaigners have been experiencing over the past decades: claims and counterclaims and data being dismissed for not being accurate or substantial enough to corroborate our arguments.
- One of the most critical challenges facing green Information Technology is the legal framework within which system developers and providers need to work. Although recently the European Union (January 2015) brought in regulation to European Union rules to oblige new devices such as modems and internet-connected televisions to switch themselves off when not in use, nevertheless, we are far from having robust sets of enforceable regulations for green IT on the national or international level.
- While cloud computing was hailed as the "green way" of moving away from device-dependent clunky power-hungry applications and data storage approaches, and although it could be claimed that the use of virtualized resources can save energy, a typical cloud data center still consumes an enormous amount of energy. Also, because the cloud is comprised of many hardware and software elements placed in a distributed fashion, it is very difficult to precisely identify one area of energy optimization.
- Some of the most interesting areas of research in green Information Technology concern "energy harvesting" or "energy scavenging," and the concept of the internet of things (IoT). Energy harvesting is exploring how we can take advantage of various ambient power sources scattered all around us. The concept of the internet of things examines a series of technologies that enable machine-to-machine communication and machine-to-human interaction via internet protocols. Being able to use the World Wide Web and the global connectivity of billions of machines, smartphones, and tablets to monitor and control energy usage can have a tremendous impact on developing a much more environmentally friendly computing world that is sustainable.

To predict the future and particularly the future of green Information Technology is a futile exercise that one should not entertain. The pace of changes over the past two decades has proven to us that even the most visionary thinkers of our time struggled to provide us with a vision and a road map to follow. Perhaps the best way to conclude these introductory remarks is quote the visionary who gave us the World Wide Web and these days is very worried about how large multinational corporations and indiscriminate government surveillance can ruin a tool that has provided such amazing opportunities for humankind. In the words of Sir Tim

Berners-Lee, "One way to think about the magnitude of the changes to come is to think about how you went about your business before powerful Web search engines. You probably wouldn't have imagined that a world of answers would be available to you in under a second. The next set of advances will have a different effect, but similar in magnitude."

References

Baliga, J., Ayre, R., Hinton, K., Sorin, W.V., Tucker, R.S., 2009. Energy consumption in optical IP networks. J. Lightwave Technol. 27 (13), 2391–2403.

Baliga, J., Ayre, R., Sorin, W.V., Hinton, K., Tucker, R.S., 2008. Energy consumption in access networks. In: Optical Fiber Communication Conference and The National Fiber Optic Engineers Conference (OFC/NFOEC). Optical Society of America, San Diego, CA, USA.

Baliga, J., Hinton, K., Tucker, R.S., 2007. Energy consumption of the internet. Paper presented at Joint International Conferences on Optical Internet, and the 32nd Australian Conference on Optical Fibre Technology, COIN-ACOFT, 24–27 June, Melbourne, VIC, Australia.

Berners-Lee, T., 2015. BrainyQuote.com. Xplore Inc. http://www.brainyquote.com/quotes/quotes/t/timberners444499.html (accessed 02.02.15.). Read more at http://www.brainyquote.com/citation/quotes/quotes/t/timberners444499.html#cwclkW9U8U6sRqvV.99.

Bush, V., July 1945. As we may think. Atlantic Monthly J. http://www.theatlantic.com/magazine/archive/1945/07/as-we-may-think/303881/?single_page=true.

Coroama, V., Hilty, L.M., Heiri, E., Horn, F., 2013. The direct energy demand of internet data flows. J. Ind. Ecol. 17 (5), 680–688. http://dx.doi.org/10.1111/jiec 12048.

Gills, J., 2011. Internet responsible for 2 per cent of global energy usage. New Scientist http://www.newscientist.com/blogs/onepercent/2011/10/307-gw-the-maximum-energy-the.html.

Hinton, K., Baliga, J., Feng, M., Ayre, R., Tucker, R.S., 2011. Power consumption and energy efficiency in the internet. IEEE Netw. 25 (2), 6–12.

Jacobi, W/SIEMENS AG, 1952. "Halbleiterverstärker" priority filing on 14 April 1949, published on 15 May 1952.

Kilper, D.C., Atkinson, G., Korotky, S.K., Goyal, S., Vetter, P., Suvakovic, D., Blume, O., 2011. Power trends in communication networks. IEEE J. Sel. Top. Quant. Electron. 17 (2), 275–284.

Koomey, J., Chong, H., Loh, W., Nordman, B., Blazek, M., 2004. Network electricity use associated with wireless personal digital assistants. J. Infrastruct. Syst. 10 (3), 131–137.

Lanzisera, S., Nordman, B., Brown, R.E., 2012. Data network equipment energy use and savings potential in buildings. Energy Efficiency 5 (2), 149–162.

McLamb, E., 2011. The Ecological Impact of the Industrial Revolution Ecology Global Network | News and Information for Planet Earth. http://www.ecology.com/2011/09/18/ecological-impact-industrial-revolution/.

Taylor, C., Koomey, J., 2008. Estimating Energy Use and Greenhouse Gas Emissions of Internet Advertising. http://energy.lbl.gov/EA/emills/commentary/docs/carbonemissions.pdf.

Weber, C.L., Koomey, J.G., Matthews, H.S., 2010. The energy and climate change implications of different music delivery methods. J. Ind. Ecol. 14 (5), 754–769.

Weik, M.H., December 1955. Ballistic Research Laboratories Report No 971 - A Survey of Domestic Electronic Digital Computing Systems - page 41. US Department of Commerce. Retrieved 2009-04-16.

Weik, M.H., 1961. Ordnance Ballistic Research Laboratories. Aberdeen Proving Ground, MD.

CHAPTER 2

Emerging Technologies and Their Environmental Impact

Colin Pattinson
Leeds Beckett University, Leeds, UK

Introduction

That we live in an age of rapid technical development is, if anything, a massive understatement of the actual situation. It is remarkable that so many of these developments have resulted in devices that have become almost indispensable to everyday activity; it is impossible to envisage life without things such as mobile phones, digital TV on demand, or computer programs and applications such as Google, eBay, and Facebook. It is also remarkable that these technologies have become such an integral part of everyday life (at least in "developed nations") over such a short period. In spite of (or maybe because of) the rapid pace of development during the past 20 years, there seems to be no letup in the expected pace of future innovation: technologies such as 3-D printing are only beginning to be exploited, and devices such as wearable computers are beginning to emerge. Even without these newly emerging uses of existing technology and of the radical new technologies that are yet to take shape, we are in the grip of one very clearly predictable consequence of technological advancement: the fact that more users are making more use of more devices to do more things. In addition, it is usually the case that each of these new activities requires more resources as it becomes more complex.

The result of the connection between easier (cheaper, more rapid) production and a wider range of uses for the product has been seen before in many situations: eighteenth-century improvements in iron manufacture led not to making the same limited set of products in a shorter time using fewer resources but to a much wider range of iron products. The mechanization of textile (wool and cotton cloth) manufacture in nineteenth-century Great Britain meant that more cloth was available to be made into a wider range of clothing. Mass production techniques such as those associated with Henry Ford led to producing more cars rather than building the same number of vehicles as before but at lower cost, and more quickly (in days rather than weeks). The most commonly quoted example of this phenomenon is the *Jevons paradox* (Alcott, 2005). The Victorian economic scientist W.S. Jevons noted that improvements in the efficiency of the steam engine had led not to the same amount of work being done with fewer machines but instead to an increased number of steam engines accompanied by a growth in the scale and scope of their application. He could have been predicting the development of information technology (IT) and computing power. Instead of running the same set of application software in an efficient manner, the availability of more processing power, greater data storage, and quicker, more reliable data transfer have allowed the creation of a vast range of applications including the examples mentioned.

A full discussion of capitalism and market economics is not appropriate here, but suffice it to note that the cost of investing in improved production techniques demands that the production system make more units of product and sell them at the demanded profit and, in turn, that new outlets and uses be found for them.

In some of these cases, the improvement process is continuous, interrupted only by sudden changes (the replacement of human labor by robots in much car manufacturing is one such example). In others, new products overtake the old: many former cast metal products are now made in plastic (children's toys and garden furniture being two such cases). Sometimes the development process itself reaches a point at which no significant further development is possible. In the domain of IT, the most obvious case is that of Moore's law. There are physical limits to the miniaturization of the transistor-based integrated circuit devices governed by immutable characteristics such as the properties of light and the size of the electron. However, this does not mean that development is likely to stop the demand for new technologies driven by the applications and services of those technologies but seems set to continue beyond the limit of Moore's law—even without the prospect of many more years of improvement in silicon-based IT and even if new developments such as quantum computing fail to materialize.

The number of devices provides an additional multiplying factor. In common with many "developed" countries, some years ago the United Kingdom passed the mark of having more registered mobile phones than it has citizens. There is similar growth worldwide, not by any means limited to "first-world" countries. The billionth PC was shipped in 2008, and the data—in its many different formats—that this growing collection of devices generates and processes also grows year on year. The volume of stored data is actually growing more quickly as the twin factors of easier production and greater precision make it simpler to produce more while the declining cost per unit of storage reduces the need to be selective about what is kept.

The final factor to be considered is obsolescence. In some cases, obsolescence is physical, caused as technology "wears out": each time an on/off switch is used, wear (metal fatigue) occurs; memory read/write operations can be carried out reliably only a limited number of times before the magnetic characteristics of the device become unreliable. In other cases, obsolescence is brought about by external factors: the UK government's decision to reallocate the spectrum formerly used by analog TV transmission meant that TV sets and associated recording devices designed to receive analog transmission no longer have a signal to receive. In other situations, obsolescence arises as the result of newer devices that are more desirable because either they possess greater functionality or they are perceived as fashion items. Smartphones offer examples of both these cases: some are replaced because a newer version offers more or better quality user experience, others because they are no longer the current "must-have" devices. Lastly, devices become unusable through damage and are replaced by newer ones.

The purpose of this chapter is to show that all these changes tend to increase the amount of computing processing and memory needed and that delivering that processing and managing that memory lead to an increase in energy requirements. It seems likely that this increase may far outstrip any "savings" made in efficiency improvements *within the technology itself*: therefore, if we are to maintain or improve the overall environmental balance sheet for emerging technologies, those technologies must deliver significant positive environmental benefits in their own right. We will therefore conclude by considering these benefits and placing them into context.

Number of Connected Devices

The continued development of existing technologies, as opposed to the emergence of new ones, is likely to provide the majority of growth in the number of devices, although there are likely to be new technologies (wearable computers, the "internet" of things) that will also have significant impact on this growth.

As mentioned in the introduction to this chapter, the growth in Internet-connected devices is clear and some of the numbers that capture this bear repetition:

- Of the world's population, 50% possesses a mobile phone (World Bank, 2014).
- In many countries, there is already more than one registered mobile phone per capita (Stonington and Wong, 2011) although the data do not show how many of these remain active, and it is likely that a significant proportion is unused, having been superseded but not disposed of.
- By the end of 2013, 1.78 billion PCs and 1.82 billion units of browser-equipped mobile devices were in use (Gartner, 2013).
- In the United Kingdom in 2012, there were an estimated 10 million office PCs, and it was estimated that 50% of the working population used a PC in their daily work; this is expected to increase to 70% by 2020 (1E, n.d.).
- The number of smartphones in use is predicted to pass the number of PCs in use during 2014 (Blodget, 2013).
- *Sales* of tablet devices are likely to exceed those of PCs during 2015 (Blodget).
- By 2020, there are predicted to be 50 billion Internet-connected devices; much of this will be a consequence of the Internet of things explored in subsequent paragraphs, but almost all of the devices (phones, PCs, and tablets) mentioned possess some form of Internet connectivity (Chiu et al., 2010).

Most of these devices and most of the emerging applications are not stand alone; they rely on technologies such as cloud computing, device connectivity, and information sharing to deliver their functionality. This makes it pertinent to include the growth in networking and storage within the consideration of the environmental impact of emerging technologies.

The increase in the numbers of devices, the features each one possesses, and inbuilt complexity all tend to increase the volume of data. In addition to the obvious fact that more devices are likely to produce more data, registering, tracking, interconnecting, monitoring, and securing large numbers of devices are all data-generating activities. Additional functionality also tends to result in more data: a one-color document (on screen or printed) is a rare sight in most offices; people who have digital cameras in their smartphones use them to capture events that would have gone unrecorded by the previous generation of analog camera users. The widespread adoption of cloud storage means that the need to be selective about which data to retain—a need already reduced by growth of on-board storage capacity—has now been pushed even further into the distance. Clouds offer almost limitless data storage at low additional cost, so there is little incentive for "housekeeping" to remove redundant, duplicate, or out-of-date information. The inevitable result of all this has been an increase in the overall volume of stored data: in 2011, it was estimated that a zettabyte (10^{21} bytes) of digital data existed in storage systems (Wendt, 2011); by the time this book is completed, it is expected that this will have quadrupled.

Consideration of data storage leads us to data centers, whose power consumption doubled over the four years between 2007 and 2011; in the single year of 2012 (a year of major investment in cloud technologies), this increased by an additional 63% (Venkataraman, 2013). Note that these figures relate to power consumption, not the actual number of data centers or the units installed within them. If we accept that efficiency has improved over that time, we are forced to conclude that the power consumption is an underestimate of the installed units because each unit should require less power in 2012 than its equivalent in 2007. Since then, growth has continued, albeit at a lower rate, but still in excess of 10% per annum. Not surprisingly in view of this, data centers are responsible for a significant portion of the total emissions for the IT sector (14% of the total in 2007). It is estimated that this will rise to 18% by 2020.

Greenpeace's 2011 estimate (Greenpeace, 2013) that, considered as a country, "the Internet" would be in fifth place, ahead of Russia and below Japan in a report of electricity use, is a compelling illustration of the energy consumption required to support the world's demand for IT. The future seems likely to be less one of new technologies and more of the ever-increasing use of existing technologies to create more data. Cisco has dubbed the five years ending in 2018 as "the zettabyte era," a recognition of the fact that global networks will carry that amount of traffic in the calendar year 2016, increasing to 1.6 ZB in 2018 (Cisco, 2014). In light of this, it is interesting to estimate the likely electricity requirement of the global telecommunications system of 2016.

In a study of conventional wired communications technology (mixed fiber and copper) Coroama et al. (2013) offer a "pessimistic" estimate that overall Internet *transmission* uses 0.2 kWh of electricity per gigabyte (GB) of data. "Pessimistic" means that the estimate is a worst-case scenario, and improvements in energy efficiency will probably reduce the estimated use over time. However, even if we assume that this pessimism has doubled the actual figure to move 1 ZB at 0.1 kWh/GB will require 100 TWh of electricity per year. An estimate by Raghavan and Ma (2014) uses a different definition of the "Internet" that includes end devices and a full life cycle costing (both excluded by Cororama et al. whose focus is solely on the communication links). As a result, their estimate is significantly higher at 8 kWh/GB; applying this figure to Cisco's "zettabyte Internet" would mean an energy requirement a factor of 100 higher than the preceding figure. For comparison, the total domestic energy consumption of the UK for 2010 was 112,856 Wh/GB (Rose and Rouse, 2012).

Additional growth in what might be considered *hidden connectivity* will occur. The internet of things is a concept in which almost any object can or will possess Internet capability. Much of this could create very positive forces for good by delivering reductions in energy use without requiring operator (human) intervention. Obvious examples are the washing machine, which is able to determine the optimum combination of time, heat, and water use for its current load or the in-vehicle systems that will provide the most fuel-efficient route and driving patterns for current road and weather conditions. The very need to provide such connectivity means that each device effectively becomes a computer in its own right. Providing devices in this quantity will require the creation of more IT with consequential increases in the demand for raw materials and the energy needed in operation as well as the need for disposal. Allowing these devices to communicate with each other, whether wired or wirelessly, will make a significant contribution to the growth in traffic predicted by Cisco.

Increased Functionality

Added functionality is widely used as a selling point for new products. The development and upgrade cycle of consumer electronic products is the most apparent illustration, for example, the way in which improvements in camera resolution, creating higher-quality images, have been a selling point for new devices. A higher resolution camera will generate more data (each pixel is 3 bits, so a 12 megapixel camera at the lowest end of the current market requires 36 Mb of uncompressed data per image before any compression. Higher resolution calls for one or more pixels, information per pixel, and less compression).

Another rapidly growing consumer product is in-vehicle navigation. Here the same factors are apparent. New units are marketed on their performance, accuracy, and reliability: a more precise geolocation system will require more processing to calculate with greater precision (the basic in-vehicle geolocation systems typically transfer less than 2.5 MB per day in constant use and offer a positioning accuracy of 10-15 m, higher levels of precision, an accuracy call for increased sampling rates, and more processing, for example, the use of "least squares" to minimize misfits between modeled and real location [Bilich, 2006]).

In both cases the result is the same: the amount of resource (energy or hardware) is larger: more storage and transfer capacity are required because the volume of data generated is larger, and/or there is more processing of this additional data to provide the higher levels of accuracy and precision.

Increased Number of Separate Functions

Multifunction devices (MFDs) are now commonplace, whether in the guise of the phone that also acts as a camera, personal organizer, Internet access device, entertainment center, mapping and location system, and so on, or the office printer that also provides copying, scanning, and image processing. On public transport, a single device can issue and check tickets, provide a live timetable and journey planning information, ATM functionality, and the generation of data relating to cash sales. Less obviously, single devices provide multiple functions in networking, connecting both wired and wireless devices, and a single vehicle system can support both engine maintenance and management and route control. Initiatives in the field of wearable computing expand this use with devices that allow health monitoring to be added to the existing functions of the mobile phone.

This could be viewed as a good thing because it should reduce the number of separate devices needed by any one person, but the reality is rather less distinct. We argue that the increased number of functions also contributes to the negative environmental impact of technology for the following reasons.

- *A proportion (often significant) of this added functionality is unnecessary*: Often many of the added functions available are rarely used if at al, but they still require support, which is rarely energy free. Simply running a background process on a device consumes CPU cycles and hence power. Unless the user chooses otherwise, it is quite likely that the application will be monitored and upgraded, regardless of whether it is actually used.
- *Unsuitability for multiple purposes*: It is often difficult to combine the demands of different user interfaces into the design of a single device while retaining the comfortable and efficient (in human terms) use of any one function. This can often lead to compromise, resulting in dissatisfaction. Although it might be almost acceptable to take holiday photographs with a regular tablet computer, using it as a "normal" telephone (without the addition of a hands-free earphone and microphone) would not be acceptable to most people. How this dissatisfaction is then manifest may have a further environmental impact: the user may discard the device in an attempt to find a "better" multifunctional device or purchase more than one device, using each for the subset of roles the user finds most satisfactory (and then adding to the communication network load by

synchronizing them); or other technology may be added in an attempt to make the use more comfortable. As an example of the latter case, consider the growing market in detachable keyboards for tablet computers to meet the need for a more "user-friendly" keyboard for sustained operator.

- *Separation of ownership/primary use*: A rush hour journey on public transport will quickly reveal that many users have more than one example of what is ostensibly the "same" device. This typically is one mobile phone for work and one for personal use. This practice may be driven by individuals' wish to maintain a separation between their public and private lives or by security-driven management policies to avoid misuse of company-provided equipment.

Increased Demand for Speed and Reliability

Describing the communications between units of the British Army in South Africa during the military campaigns of 1899-1902, Arthur Conan Doyle gives us what was then clearly considered a remarkable example of speedy and reliable communication: "it is worthy of record that . . . at a distance of thirty miles [48 km], they succeeded in preserving a telephonic connection, seventeen minutes being the average time taken over question and reply" (Doyle, 1902). In the 112 years since then, military (and more peaceful) operations in both sound and vision are controlled remotely over thousands of kilometers in real time.

The use of the word *controlled* in the modern-day description is in deliberate contrast to *question and reply* in the 1902 example: whereas local commanders in the army of 1902 would be expected to make local decisions based on information received, the *remote control* of an operation now calls for instructions to be issued and received with 100% reliability and accuracy.

Emerging trends in a number of fields are the requirements for increases in the use of such systems. Smart city initiatives call for monitoring environmental characteristics, traffic patterns, power, water and sewage flows, traffic systems, water and power grids, and building heating and ventilation units. The mapping and location requirements of navigation systems (in vehicle or handheld) that will guide users to specific locations will need real-time information about current location, and automated transport (exemplified by the recent announcements of driverless cars) will call for ever more precise and accurate location. At the moment, there is no suggestion that driverless cars be remotely controlled, although many mass transit systems do operate in a highly automated mode (e.g., the Docklands Light Railway in London), and this will be extended whether through entirely new systems or additions or modifications to existing ones. In those applications in which remote control of systems whose incorrect functioning can present a risk to life and property, the speed and reliability of information and commands are paramount.

Similar demand for speed, reliability, and accuracy now exist in commonplace activities. Although they may not be essential, these device have become desirable or expected. The consequence of this is the need for systems that deliver these three requirements. In turn, these demands are met by the user of higher performance devices that typically require more memory and processing power. The standard approach to providing increased reliability is through *redundancy* with stand-by units both hot (ready to go instantly) or cold (able to be powered up at short notice); by load sharing to avoid overloading any one component; and by system designs that seek to eliminate any single point of failure. It hardly needs to be stated that the overall result of attempts to meet the demands for speed, reliability, and accuracy in any single activity is often to increase the volume of resource needed to support that activity.

Obsolescence—The Problem of Backward Compatibility

The IT industry is one in which change and innovation are rapid and often fundamental. Consider the change from mainframe computers to minicomputers to stand-alone PCs and then to networked PCs, the client-server era, wireless networks, and the current trend to use cloud and mobile devices. These step changes—occurring roughly once per decade—have meant that otherwise fully operational and functional systems have been rendered obsolete before their designed lifespan has expired. *Backward compatibility*—allowing newer and older software and equipment to work together—may be achievable for a time but is not sustainable in the longer term without constricting development. The need to support a variety of different hardware types (old and new) also adds to the complexity of the software development process for applications, operating systems, and device drivers. After some time, the cost and complexity of maintaining backward compatibility will become too high and will be dropped. The majority of users typically will have already moved on to the newer product as part of normal update/renewal processes. Those who are left then must decide whether to continue with (now unsupported) equipment or to seek alternative applications that will run on their current equipment and are supported by the developer. Note, however, that this approach brings the added cost of learning the new system and possibly the need to convert data formats.

The speed of development and upgrading of technology attenuates this process, giving rise to the well-known situation of fully functional equipment being discarded because of obsolescence rather than being worn out.

Within the PC environment, Microsoft's recent decision to cease support for the XP operating system is a classic example of a step change, and user response is a classic example of what happens as a consequence (Covert, 2014). Although it is possible to install other operating systems, it is likely that many of the current XP devices will go out of use in relatively short order as application support, development, and (probably more significantly) security updating are discontinued. The difficulty—real or perceived—of transiting to a newer operating system will add to the likelihood of accelerated obsolescence.

Another example of accelerated obsolescence, already mentioned in the introduction to this chapter, is the UK's decision to no longer provide analog TV signals. Here the driving force was government policy rather than manufacturer decisions. Although newer equipment was already being adopted by a significant proportion of users as a consequence of demand (or desire) for the additional features offered by cable and satellite provision, it is clear that the government's decision to reallocate the frequency bands used by analog TV in order to auction them to mobile operators created a step change in this process. The provision of set-top converter units offered the option to retain existing analog units, but the added functionality of new TV sets—particularly HD—has led to an increase in sales of the newer ones. Whether the sets rendered redundant by this are stored or disposed of is unclear.

The Other Side of the Balance Sheet—Positive Environmental Impacts or the "Other 90%"

We have already seen that the likely result of continued development in the use, range, and application of IT will be an increased demand for the resources required to create, maintain, and dispose of the various hardware items in use. The use of *resources* here refers to everything from the raw materials (including rare earth metals) and other chemicals (acids for etching, water for cleaning) and the energy used in the manufacturing process to electricity for powering the devices and the energy needed to handle them at the end of their lives. A wider definition would also include factors such as the energy required to transport both new and discarded raw materials and finished products.

The case is frequently made that much of the focus has been on the greening *of* IT systems; that is, to make the technology more energy efficient, driving down the energy consumption of data centers, and so on. However, if we take the widely quoted figure that IT is responsible for 10% of the world's energy consumption, this leaves the question of the "other 90%." In addressing this, we consider the opportunities for greening *by* IT (i.e., using the technology to reduce the energy use of other aspects of human activity). Most prominent among these opportunities are transport and building heating and lighting. They are simple because they are the largest consumers of energy and therefore even a small proportional reduction in them would be significant.

One possible approach to create a balance sheet is to consider a particular activity that is performed in a particular way that allows a calculation of the resource required to support that activity. Then consider how the activity would be undertaken using IT as an enabler, calculate the resource requirements for *that* mode of operation, and compare the two. One activity—replacing the intercontinental business trip by videoconference—will become almost the de facto comparative factor.

Constructing a balance sheet for an activity such as a trip involves determining the energy used by a single air traveler on a return journey to conduct a face-to-face meeting and then working out the energy required to support a videoconference of the same time length. Such a calculation makes a number of assumptions, including that a videoconference is as productive as a face-to-face meeting and that removing a number of *passenger* trips would necessarily lower the number of *airplane* trips as opposed to the same number of planes traveling at reduced occupancy.

Attempts to calculate the actual energy (or resource) used for a particular transaction results in very widely variable outcomes; there are significant variations in the determination of which devices are considered to support a transaction. The problem is further complicated by the task of apportioning the resource use of any shared device that actually handles a specific transaction and by the discussion of whether the device in question would have used any less resource had the specific transaction not taken place (e.g., if a particular call did not take place, would a router that was not required to route the particular call have used any less energy or resource?).

Videoconference as an Alternative to Business Travel

Perhaps unsurprisingly, most calculations of this form consider the use stage alone; there is no attempt to include the embedded energy of manufacturing a passenger airliner and then apportioning a part of the embedded energy to a given passenger on a particular journey. Nor should it be noted whether these calculations include any consideration of the energy costs of building and operating airports or of the cost of operating the communications infrastructure necessary to direct air traffic. Even without these added embedded energy costs, the typical energy balance sheet is heavily in favor of the videoconference solution. For example, Raghavan and Ma (2014) calculate that replacing 25% of the current 1.8 billion air journeys by videoconference would "save about as much power as the entire Internet [consumes]." An added benefit is that the energy saved is energy that would be generated by burning aviation fuel.

Dematerialization of Product Chain

Perhaps a more interesting and exciting opportunity is offered by dematerialization of the supply chain for a particular product item. Consider the situation in which some mechanical device (e.g., an office air-conditioning unit) fails. Currently, the sequence of operations is typically as follows: a maintenance engineer is called to the site, diagnoses that the problem requires replacement of a particular component, which is then delivered to the site, and the engineer then returns to perform the repair. Alternatively, the engineer may obtain the component or, very rarely (or so it seems), has the required part on hand, but this is only at the cost of maintaining a stock of such parts, resulting in a larger vehicle to carry more weight.

A 3-D printer on site would allow the replacement part to be produced locally and fitted in one visit. As an additional IT-related benefit, remote diagnostics (also supported by IT) would provide an additional enhancement: the engineer could arrive knowing that the task was to replace a particular component and find it waiting upon arrival, thus saving time.

The potential for 3-D printing is enormous. Although the spare part example is one of the more trivial ones, it is easy to envisage other products being "delivered" in the same way; examples include clothing, food, and buildings, both terrestrial (BBC, 2014) and extraterrestrial. NASA and the European Space Agency are currently considering the possibility of constructing a moon base using 3-D printers that use lunar soil as the raw material. The cost of transporting the 3-D printers would be significantly less than that of transporting building components from earth (NASA, 2014). Although it is still necessary for the raw materials to be available for the printer to use on site and the limitations of choice and quality (particularly for 3-D food) may leave something to be desired with current technology, 3-D printing could clearly offer potential energy savings in reducing the transportation and packaging of finished goods.

What are arguably more trivial examples of dematerialization are already with us: the decline in printed newspaper, magazine, and book volumes brought about by their replacement with e-readers is beginning to have a noticeable impact on sales and production methods. The change to online consumption of music and home entertainment has already led to significant declines in the supply of physical media such as CDs and DVDs with consequences in the decline of the high street outlets for these products.

The last example is only one part of an even more significant disruptive use of IT: the online shopping boom and its impact on shopping habits. We do not believe that there has been any attempt to calculate the cost (in energy and resources) of the typical online shopping spree in the run-up to Christmas 2014. Nor has there been an attempt to compare the cost with the energy and resources that would have been necessary to collect the same set of goods by traditional shopping. It seems likely, however, that savings quoted for videoconferencing coupled with the economies of scale and the logistical improvements achievable by the major online retailers and the opportunities for planning fuel-efficient delivery would occur for the online activity. If products such as clothing, toys, kitchen goods and even food items were to be 3-D printed, the cost savings would be even higher. Whether we are prepared to accept the societal impacts of this (the inevitable disappearance of actual shops in towns and cities) or the loss of the physical and mental activity of the shopping trip on us as humans adds extra complications to any such calculation.

Travel Advice/Road Traffic Control

Although the replacement of journeys and physical movement of goods described here offer the potential to decrease some travel, the human need (or desire) to move around in vehicles of some form apparently is likely to continue. Travel in more populates areas can be met by a mixture of private and public transport, and IT can make a positive difference to overall energy use. We already have implemented a number of these: modern automobiles are fitted with engine management systems and other devices to ensure optimum use of fuel; communications systems allow information on traffic flows and road conditions to be made available so that drivers can consider changing travel plans to avoid congestion. Public transport vehicles also benefit from such devices, and a well-planned information system allows control of the whole network while keeping customers informed and allowing them to plan their journeys. Finally, citywide traffic control systems provide traffic management across a city so that local changes to traffic flows, such as sequencing traffic lights and rerouting traffic, will reduce delay, thus offering the option to use less fuel. It is likely that these initiatives will further develop and expand to incorporate data such as weather forecasting (cold and rainy weather encourages the use of cars and buses rather than walking or cycling for essential journeys and reduces the number of nonessential journeys—both of which have predictable effects on transport patterns and demand).

Other uses of ICT in traffic and travel management are more subtle: charging fees that to persuade leisure users not to travel at peak times can be facilitated by tracking technologies such as automatic vehicle license plate recognition; requiring changes of public transport mode (e.g., bus to train to bus) is made easier by ticketing and using an integrated timetable.

Fuller integration of traffic management, journey planning, and timetable integration, together with longer-term planning and modeling of city infrastructure (to allow easier access to places of work and leisure), bring us to the reality of smart cities, the potential of which is now beginning to be exploited.

Intelligent Energy Metering

In the United Kingdom and other countries, most domestic energy use (gas and electricity) was until recently recorded and paid for using a labor-intense process. A "meter reader" would visit each household to physically make a reading (by reading a meter's dial and writing down the number on it). This reading then generates a bill, which is probably the first indication of their energy use that most customers have. On receipt of the bill, some users might take action to reduce use, but without a positive feedback loop and accurate (and timely) information, this is often a piecemeal activity. The location of meters in difficult to access locations was another disincentive to proactive energy awareness.

The concept of so-called smart meters can provide easier access to information about energy consumption in real time, allowing more obvious changes to be made to current energy consumption. A web-enabled meter can communicate real-time usage to an end user device (a smartphone or tablet) without requiring the user to locate and check the meter. Among other benefits, the time-consuming round of meter reading can be removed; if the data can be sent to the user's smartphone, it can also be communicated to the energy company.

Emerging applications of the potential for this technology will include energy suppliers' ability to negotiate and smart meters to balance demand, allowing customers to choose between suppliers using a number of criteria. Moving the negotiation processes further into the domestic arena, customers can determine the pattern of their energy costs for their household devices—especially washing machines and dishwashers—and can use them during periods of lower demand.

More controversially, there is the prospect that the energy supplier could control operation and behavior more directly by restricting supply. We are not aware of this actually taking place, but it is indicative of the challenges resulting from the opportunities for more control by more complex technology. The issue of the extent to which we are willing to allow technology to control our behavior and lifestyle is a topic worthy of a full research study in its own right but is beyond the scope of this book.

Building Management Systems

We are familiar with systems that control the temperature of buildings in which we live and work from domestic central heating controllers to office air-conditioning units. In large buildings, such as office blocks, hospitals, and university and colleges, these control systems are being merged with sensors and other input systems in the form of a building management system (BMS).

A complex BMS has the potential to have very detailed control and operation not just of the heating, ventilation, and air conditioning (HVAC) system but also lighting and security systems; they can also provide room ambience (mood music) appropriate to the occasion. By including inputs from weather forecasting systems, the BMS can adjust HVAC settings to prepare a building for coming weather conditions—for example, providing some heating while energy is at a cheap rate because the outside temperature is predicted to fall over a three- to five-day window rather than waiting until the temperature falls below a set level after two days of cooler weather, which requires using more expensive energy to provide required heating.

Other BMS-related opportunities include a linkage between room use and HVAC settings and modification to heat needs for different numbers of occupants or for different levels of activity; more active occupants would need less heat (or more cooling) with the less active the needing the opposite. Security could be enhanced via an intelligent BMS: movement detected in a room that is scheduled to be unoccupied would trigger an alarm; the ability to count the number of people entering a given room could be compared with the room's safe occupancy level, allowing action to be taken as necessary.

As with energy metering, it is likely that the potential range of uses to which this technology is put will be limited as much by human factors as by any limitations in the technology itself. The degree to which we are prepared to accept the "intrusion" that many of these applications bring with them—in both monitoring behavior and adjusting the environment—remains to be seen. Once data are available relating to an individual's presence in a given location (or to the intensity with which the individual is moving within that location), legal and regulatory action is needed to prevent the use of that information for other purposes—such as monitoring active working hours.

Saving IT Resources—A Drop in the Ocean?

It is clear that the effect of individuals making the changes that are available will have negligible (a drop in the ocean) impact on the overall energy consumption of IT—whether current or emerging. One person deciding to cull unwanted photographs stored "somewhere in the cloud" is not going to change the resource requirements of the cloud servers or the network that connects them. However, a number of individuals involved would probably be sufficient to have a measureable impact *if they all did the same thing*. The likelihood of this is, of course, increasingly small although workplace campaigns to raise awareness can have a limited but measureable effect. Campaigns to encourage office staff to turn off their computer systems on weekends typically have some impact: the problem is that the impact of the campaign and hence the level of response diminishes over a relatively short time scale. To make a major, lasting difference, legislators or providers need to take action; the most effective tool is financial—making it more expensive to produce and retain more data. Providers could also introduce a "delete by default" form of operation in which stored photographs are automatically deleted after a set time period unless the owner explicitly tags them for retention, thereby taking advantage of the inertia that affects the majority of people.

Another "drop in the ocean" response to the call for greater efficiency in the IT sector relates to the relative size of its contribution to the environment (10% of energy, 2% of greenhouse gas) compared to other sectors, not only the airline industry but also heavy manufacturing of such items as steel and chemical works. This is in addition to the underlying inefficiency of much fossil fuel electricity generation. Although this is undoubtedly true, a counterargument is that if some proportion of that 2% (or 10%) can be reduced, the opportunity should be taken. With energy demands at the level discussed earlier for the zettabyte network, even a small relative change is significant in absolute terms.

Conclusion

Our capacity for innovation and for acquiring (or coveting) innovative products has often seemed to be unlimited. Even the exhaustion of a particular resource often fails to have a major effect: we appear to be able to innovate to find a replacement either for the resource or the product that used it.

In the short to medium term, there is little reason to expect a reduction in the increased demands for IT products and services discussed in this chapter. Globalization means that those areas of the world that have yet to feel the benefit of increased IT will be more and more exposed to it. Market economics will cause more products to be made and create new needs to exploit them. Improvements in efficiency (through customer demand, energy cost, and legislation) will continue but are likely to be outstripped by the overall growth in the market. Similarly, the "greening by IT" developments discussed here will have an effect on overall resource use, but it is likely that the best that can be achieved is to slow the rate of increase of demand, not to achieve an absolute reduction.

Whether this will continue into the future remains to be seen. The pessimistic view is that we will continue in a business-as-usual fashion and will be unable to change our behavior until it is too late. A more optimistic alternative to this scenario relies on our capacity for innovation to stop the problem. The true optimist hopes that a combination of our ingenuity and our realization of the risks will bring about changes before the consequences become unbearable (although noting that this point has been reached for some parts of the world, affecting the plant and animal species that live there).

References

1E, n.d. The PC energy report. http://www.1e.com/energycampaign/downloads/1E_reportFINAL.pdf.

Alcott, B., 2005. Jevons paradox. Ecol. Econ. 54 (1), 9–21.

BBC, 2014. How Dutch Team is 3D-Printing a Full-Sized House. http://www.bbc.co.uk/news/technology-27221199 (03.05.14).

Bilich, A., 2006. Improving the Precision and Accuracy of Geodetic GPS: Applications to Multipath and Seismology (Ph.D. Thesis). University of Colorado.

Blodget, H., 2013. The Number of Smartphones in Use is About to Pass the Number of PCs. http://www.businessinsider.com/number-of-smartphones-tablets-pcs-2013-12#ixzz3CdoVKxAX (11.12.13).

Chiu, M., Loffer, M., Roberts, R., 2010. The Internet of Things. http://www.mckinsey.com/insights/high_tech_telecoms_internet/the_internet_of_things (March 2010).

Cisco, 2014. http://www.cisco.com/c/en/us/solutions/collateral/service-provider/ip-ngn-ip-next-generation-network/white_paper_c11-481360.pdf.

Coroama, V., Hilty, L., Heiri, E., Horn, F., 2013. The direct energy demand of internet data flows. J. Ind. Ecol. 17 (5), 680–688.

Covert, A., 2014. Microsoft is About to Take Windows XP Off Life Support. http://money.cnn.com/2014/01/29/technology/enterprise/windows-xp/index.html (January 2014).

Doyle, A., 1902. The Great Boer War. Smith, Elder & Co, London.

Gartner, 2013. Gartner Highlights Key Predictions for IT Organizations and Users in 2010 and Beyond. http://www.gartner.com/newsroom/id/1278413 (13.01.10).

Greenpeace, 2013. How Dirty is Your Data? http://www.greenpeace.org/international/Global/international/publications/climate/2011/Cool%20IT/dirty-data-report-greenpeace.pdf (April 2011).

NASA, 2014. Building a Lunar Base with 3D Printing. http://sservi.nasa.gov/articles/building-a-lunar-base-with-3d-printing/2014.

Raghavan, B., Ma, J., 2014. The Energy and Emergy of the Internet. http://www1.icsi.berkeley.edu/~barath/papers/emergy-hotnets11.pdf.

Rose, W., Rouse, T., 2012. Sub-National Electricity Consumption Statistics and Household Energy Distribution Analysis for 2010. https://www.gov.uk/government/uploads/system/uploads/attachment_data/file/65933/4782-subnat-electricity-cons-stats-article.pdf (December 2012).

Stonington, J., Wong, V., 2011. The 20 Countries with the Highest Per Capita Cell-Phone Use, Business Week. http://images.businessweek.com/slideshows/20110213/the-20-countries-with-the-highest-per-capita-cell-phone-use (14.02.11).

Venkataraman, A., 2013. Global Census Shows Datacentre Power Demand Grew 63% in 2012. http://www.computerweekly.com/news/2240164589/Datacentre-power-demand-grew-63-in-2012-Global-datacentre-census (08.10.12).

Wendt, J., 2011. A Zettabyte of Data Puts New Premium on Scale-out Storage Solutions. http://www.dcig.com/2011/07/a-zettabyte-of-data-puts-new-premium-on-scaleout.html (20.06.11).

World Bank, 2014. Digital Finance: Empowering the Poor via New Technologies. http://www.worldbank.org/en/news/feature/2014/04/10/digital-finance-empowering-poor-new-technologies (10.04.14).

Green IT: Law and Measurement

CHAPTER 3

Measurements and Sustainability

Eric Rondeau, Francis Lepage, Jean-Philippe Georges, Gérard Morel
Université de Lorraine, Vandoeuvre-lès-Nancy, France

Introduction

Information and communications technology (ICT) engineers are developing and creating a virtual world offering new services and new applications to help people both in their work and their daily lives. Nevertheless, ICT engineers are almost always constrained in their project development by the intrinsic hardware performances of calculators, storage systems, and communication systems. The monitoring and measurement of physical ICT system performances are crucial to assess the computer processing unit (CPU) load, the available memory, the used bandwidth, and so on to guarantee the ICT-based services correctly work regarding their expected use. The famous Moore's law states that the number of transistors on a chip tends to double every 18 months; this enabled a rapid growth of digital system performances. However, these technical performances do not provide direct information about the level of quality of the offered services. In the International Telecommunication Union-Telecommunication (ITU-T) standardization sector's E.800 recommendation (ITU-E800, 1994), quality of service (QoS) is defined as "the collective effect of service performances, which determine the degree of satisfaction of a user of the service."

The satisfaction of users is then the key of ICT business and must be specified in a service-level agreement (SLA) signed by ICT experts and their customers. By definition, SLA is not a technical document and must be understandable by all stakeholders who are not necessarily aware of ICT terms. Two major issues should be analyzed during the SLA specification: the identification of relationships between the performance indicators defined by the user's application and ICT technical performances and one related to the monitoring of these indicators to be sure that the contract is fulfilled.

The purpose of translation between ICT and the user's application performances covers many types of problems. The translation can be direct such as having the application response time correspond to the time for ICT experts to process and transport the applicative request. In this case, the main problem is to clearly define the context of this requirement in terms of number of users, opening hours, and so on and to characterize the type of delay (worst case, average, confidence interval, etc.). However, the translation can be more complex when the user's requirements are expressed in specific professional terms. For example, the control of an industrial process is assessed by analyzing the stability of its behavior around a set point to be reached. In the research on networked control systems, the identification of impact of ICT performances (and especially network) on industrial process stability requires complex preliminary studies and development of new approaches (Vatanski et al., 2009). One other barrier in the SLA definition is the specification of the user's requirements with qualitative, not quantitative information. The user's perception of the quality of a phone call, TV broadcast, Web site, and so on is subjective and complicated to analyze and to associate with quantitative ICT parameters (delay, jitter, etc.). Usually, the user's perception is transformed in a metric based on mean opinion score using a scale between 0 (no service) and 5 (perfect service) to guide the ICT experts in their technical configurations.

Monitoring indicators specified in SLA is essential to identify the border between the ICT systems and the application itself. The goal is to be able to understand and identify the cause of malfunctions and to determine the stakeholders' responsibilities. ICT systems must be continually supervised to analyze their performances in order to detect, anticipate, and recover faults. The assessment of ICT performances can be based on measures or models or a combination of both.

Basically, the measurement requires probes, a monitoring system, and standardized protocols to access the whole metrics. Usually, a management information base (MIB) is implemented for each piece of equipment (computer, printer, switch, router, and so on) and supports in a standardized hierarchical structure all the equipment properties (name, OS, version, storage capacities, bandwidth, etc.). The monitoring systems (such as Nagios, Centreon, etc.) can then collect or modify the information stored in MIB by using simple network management protocol (SNMP). This approach is apparently easy to implement, but in practice the selection of pertinent information (regarding the SLA contract) defined in MIB is a complex process. Instead of measuring the parameters of equipment, another solution is to directly analyze the performances of the application or service defined in SLA. For that, robots are developed and simulate the user's behavior using the application. In all cases, one inherent issue of measurement is its intrusiveness with two consequences: (1) each request for a monitoring ICT system consumes CPU and bandwidth and has an impact on its performance and (2) the response time of a monitoring request depends on the performance of the ICT infrastructure.

Another approach to assess ICT performances is to use mathematical theories from one of two methods, constructive and black box. The constructive methods are based on the assembly of elementary components with specific properties; their combination can be used to estimate average delays, average buffer occupation (queuing theory) or bounded delays, and backlog bounds (network calculus theory) (Georges et al., 2005). In the second method, the ICT system or a part of it is considered as a black box and its behavior (output) is analyzed regarding the changes of ICT parameters (input). From this analysis (i.e., experiment design method), a model of the ICT system can be defined. The black box method is less generic than the constructive method, but it is better correlated with real ICT properties.

With a monitoring system, it is very interesting to couple the measures and models. The models represent the expected behavior of an ICT system and the measures its real behavior. A difference between models and measures can be used to detect anomalies and to anticipate faults according to a trend analysis. This combination of models and measures is one way to develop a successful monitoring system.

Since the birth of computer science and telecommunications, their performance evaluation mainly focused on technical and cost indicators. But, as explained in the beginning of this introduction, the final goal of ICT is to facilitate people's lives without undesirably affecting either their health or quality of life. Therefore, these effects also should be assessed during the full life cycle of an ICT product or ICT-based solutions in considering its manufacturing step, its use step, and its next life. An overall assessment of pollution must be performed to determine the ICT carbon footprint, the toxic material rate used in ICT devices, and so on, which impacts the health of people. Moreover, the Earth's resources used for ICT must be continually decreased to preserve the quality of life of current and future generations. The preservation of the Earth's resources includes recycling and extensive use of renewable energy. On another issue, quality of life is also related to ethical questions for both employees in ICT companies and ICT users with general considerations such as the salary of employees, the gender balance, and so on, or questions more specific to the ICT area such as the protection of privacy and personnel data.

In summary, ICT must be assessed using the three Ps or pillars of sustainable development (Figure 3.1) during the engineering process of the target system as a whole by balancing people, planet, and profit requirements with the final objective to design green ICT solutions.

ICT engineering should be systemic in order to analyze the negative or positive effects among the three pillars. For example, the mitigation of data center energy consumption is interesting both in terms of the environment and profit. But increasing data center capacity to grow

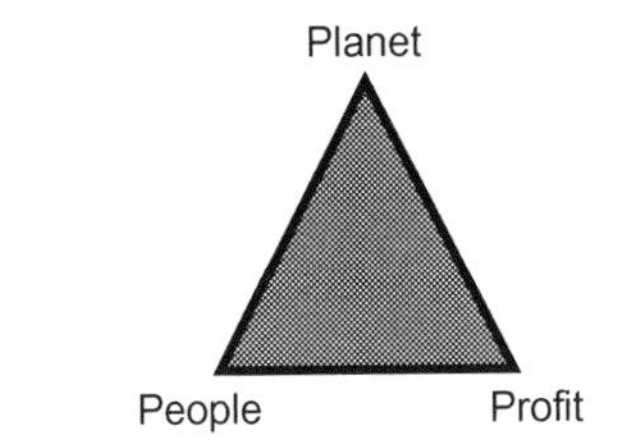

Figure 3.1
The three pillars of sustainable development.

business activities consumes more energy, generating a negative impact on the planet. The well-known analysis using a C2C fractal tile tool (Donough and Braungart, 2002) is a new challenge and obligates ICT engineers to study the solution spaces not only on business and technical performances but also in associating new metrics coming from ecology and ethics. Thus, the SLA specifications are more complex in integrating additional requirements. This complexity can be managed in using systems engineering to guarantee that the development of ICT products, services, and ICT-based solutions is sustainable and measurable for validation and verification of all technical solutions. The sustainability is not only expressed in term of longevity of solutions but also must include properties of modularity, flexibility, scalability, and recyclability.

The objective of this chapter is to show the different aspects of ICT metrics. The chapter is then organized as follows: Section "ICT Technical Measures" briefly explains the traditional metrics used to assess intrinsic ICT performances. Section "Ecological Measures And Ethical Consideration" focuses on the performance indicators specific to the environment and ethics. From a simple ICT architecture, Section "Systems Engineering for Designing Sustainable ICT-Based Architectures" presents systems engineering approach to specify and to consider the measures in Green ICT-based solutions. Section "Conclusion" concludes the chapter.

ICT Technical Measures

Introduction

The measurement of ICT systems can be compared with the measurement applied to any natural or artificial system of the world. However, the main difference with other systems is that ICT measures are related to data performance, which is an abstract thing. In general way, ICT functional model can be described as in Figure 3.2a in considering an ICT system as a black box. The input and the output are data crossing the ICT system. Controllers can be implemented over the system to manage the ICT QoS such as scheduler, smother, load balancer, and so on in order to meet the user's requirement. Finally, energy is the resource required for supplying an ICT system. An ICT system offers three elementary types of services:

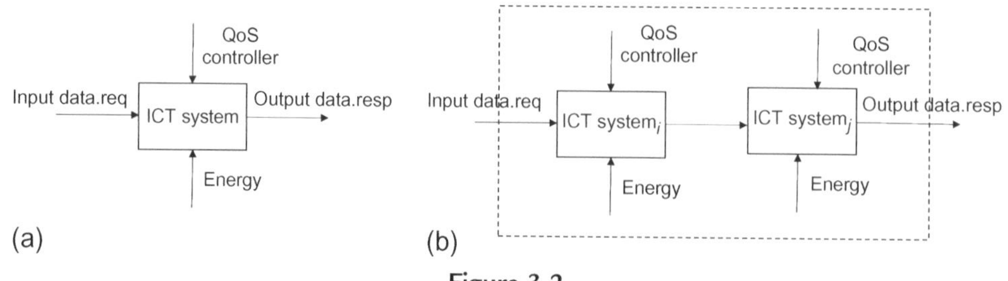

Figure 3.2
ICT system functional models.

data processing, data transport, and data storage. Moreover, a set of ICT systems can provide global services such as multimedia. Traditionally, ICT system performances are assessed by observing the deviation between the requests of service sent by the user containing the input data and the corresponding response of service supporting the output data. This deviation characterizes the service offered by ICT system to the user. Finally, until very recently, energy has not been analyzed and is considered an infinite resource with no impact on cost and environment.

The main attributes used in the data measurement are quantity, throughput, quality, availability, and security. However, in general, the measure of these attributes is not directly useful in assessing ICT systems. *Metric* is a concept defined as a measure of an attribute or a combination of attribute measures that provides pertinent information. One example of an ICT metric is the available communication bandwidth. It is expressed in bits per second, and its value is obtained by using models and several measurements.

Moreover, the assessment of a complex system composed of subsystems (Figure 3.2b) requires derived metrics that can be developed from a combination of basic metrics relative to each subsystem. The three classes of operation to obtain derived metrics are additive, multiplicative, and concave. The additive metric is calculated by adding the basic metric of each part of the system. The one-way delay of a path calculated by summing the one-way delay on each link of the path illustrates the principle of additive metric. The multiplicative metric is calculated by multiplying metrics of each part of the system. The metric of packet loss ratio can use the multiplicative operation to assess a complex architecture. The concave metric is determined by selecting the minimum value of each part of the system. For example, the estimation of end-to-end bandwidth is constrained by the link with the lowest bandwidth.

A common metric applicable to all ICT systems is *response time* corresponding to the delay between service request arrival time in the system (input) and service response available (output). It depends on both type of service and the data amount in the request. The response time is sometimes called *delay* or *latency*. The delay is calculated from two measures obtained from different devices and then requires synchronizing their clocks or using the same clock. In the next sections, the different types of services are described.

Service of Data Processing

Data processing is characterized by two specific metrics: the processing capacity and the processing quality.

The *processing capacity* is the higher number of requests per time unit that could be sent (without loss of request) to ICT system input in order to be processed. Service tests (mathematically complex functions) are defined to compare the processing capacity of different services.

The *processing quality* is determined by the accuracy of the results. This metric is very important in scientific computation.

Service of Data Transport

The data transport service uses three specific metrics: the bandwidth, the jitter, and the packet loss rate (Michaut and Lepage, 2005). There are two types of bandwidth. The first type is the *total path bandwidth*, also called *path capacity*. It expresses the maximum total throughput accepted by the path. The second type is the *available bandwidth* expressing the maximum throughput offered to the user. Thus, the available bandwidth is the path capacity minus the used traffic (concurrent traffic). The bandwidth measure is based on interval packet times.

Jitter is the delay variation between the same successive requests. It is not a measured value but is obtained by subtracting the delays of two successive requests.

The reliability of a communication network path is expressed by the *packet loss rate*. This metric is equal to the number of packets not received divided by the total number of packets sent. It should be noted that erroneous packets are not generally considered outside the lost packets in computer networks because most applications require data integrity. Indeed, the frame reception driver discards each erroneous packet that becomes lost. However, this is not a general rule, especially in audio or video networks, because applications can accept a low error rate. The measurement is achieved by using a sequence number in each packet and counting the missing numbers.

Service of Data Storage

Data storage includes two metrics: the storage capacity and the data throughput. The *storage capacity* is difficult to measure using the service black box approach. Fortunately, specific requests provide the total or remaining capacity of the storage system.

Data throughput is sometimes different for read and write operations. For a write operation, the measure is achieved by counting the maximum amount of data that the service accepts per time unit. For a read operation, the measure is similar but is applied to the service output. The time unit must be large enough to integrate gaps between storage units.

Multimedia Service

A multimedia service must allow the user to play multimedia data from voice or video acquisition crossing a chain of ICT systems. Two methods are used, the full reference and the no reference.

The *full reference (FR)* method consists in comparing the quality of reference signal from the source (perfect signal) with the received signal (degraded signal). FR measures deliver the highest accuracy and repeatability. However, FR measures can be applied to dedicated tests only in live networks.

The *no reference (NR)* method uses a degraded signal without information about the reference signal. From the NR method, some characteristics of a conversation, such as voice, gender, background noise, and so on, can be identified.

Conclusion

The fundamental activity of an ICT engineer is to configure ICT systems to offer its customers the best or optimized services regarding the main ICT metrics described in this section. However, the recent high increase in energy prices and the new environmental awareness by politicians and people require an ICT engineer to consider energy in SLA, which has been renamed *green SLA*. For example, Laszewski and Wang (2010) and Makela and Luukkainen (2013) present new metrics on energy, especially in the context of green SLA for data center. Moreover, power over ethernet (PoE) technology is a new network service enabling the transport of energy on data link to remotely supply ICT products and finally to control the status of ICT devices by using different sleeping modes. Thus, energy becomes an integral part of ICT metrics. However, limiting green SLA to only an energy metric is very restrictive. The goal of the next section is to describe all the facets of green ICT.

Ecological Measures and Ethical Consideration

Introduction

The consideration of the three pillars of sustainable development in the design of ICT-based solutions is complex because many new parameters should be added during the design process and because the limit of study space is always difficult to determine. Indeed, an enterprise involved in a sustainable development approach must analyze the environmental impact of its design process itself (project system) and of the complete life cycle of designed systems (system of interest). The issue is to obtain a global result in integrating and estimating not only the enterprise's environmental footprint but also in including all stakeholders (subcontractors, suppliers, sellers, users, recycler, etc.) involved in the enterprise activity. The goal is to prevent the enterprise from creating a marketing communication about its green and ethical virtues in keeping its environmentally friendly activities and in outsourcing the others. In this context, standardization is a major aspect to successfully reach this objective to ensure seamless environmental monitoring between the different stakeholders. The International Organization for Standardization (ISO) 14001 standard provides a framework for an enterprise to follow to set up an effective environmental management system. However, the purpose of ISO 14001 is not to estimate environmental performances and cannot explicitly show whether the enterprise pollutes.

The impact of an enterprise on the environment should be determined by using a list of key performance indicators (KPIs). A KPI is a metric or measure used to quantify and evaluate an organization's performance in relation to meeting of targets and objectives. Many standardization organizations or working groups have proposed environmental KPIs. The ITU report "General Specifications and KPIs" (ITU-T report, 2012) referred to an impressive list of initiatives specifying KPI. There are global initiatives such as the Carbon Disclosure Project, Dow Jones Sustainability Index, GHG Protocol corporate standard, Global Reporting Initiative, ISO 14031, ISO 14064-1 and initiatives from ICT sectors such as European Telecommunications Network Operators Association, European Telecommunication Standards Institute, the Green Grid, GSM Association, International Electrotechnical Commission (IEC), International Telecommunication Union—Study Group 5 (ITU-T SG5). In conclusion, the 2012 ITU-T report does not propose a KPI harmonization but suggests a general process to define environmental KPIs. This process has three steps: defining the needs, listing relevant KPIs, and verifying third-party data collected for all environmental indicators. This method fully respects the basis of systems engineering developed in Section "Systems Engineering for Designing Sustainable ICT-Based Architectures" of this chapter.

The standardization requirement concerns not only the definition of environmental KPI but also the specification of protocols enabling the collection of KPI and control of ICT-based applications regarding the KPI status. The IEEE P1888, "Ubiquitous Green Community Control Network Working Protocol" (2011) is an example illustrating the standardization of remote control architecture for digital communities, intelligent building groups, and digital metropolitan networks in terms of the energy, environment, and security domains.

The intention of this section is not to describe the list of environmental KPIs defined in the different working groups (mentioned previously) but to classify KPIs according to three categories—pollution, Earth's resources, and ethics—to present all facets of sustainable development.

ICT Impact on Pollution

By definition (Chapman, 2007), *contamination* is simply the presence of a substance where it should not be or concentrations above specified levels. That is, pollution is a contamination that results in or can result in adverse biological effects to communities. It means that all pollutants are contaminants, but not all contaminants are pollutants. Various forms of pollution can affect the air, soil, light, noise, heat, vision, and water. All these forms of pollution can be generated by the ICT sector. Therefore, it is essential to identify pollution causes to mitigate or to eliminate their impact on environment.

ICT products are composed of many chemical substances that are toxic for the environment and people. For example, beryllium used in relays is dangerous for workers manufacturing this electronic equipment, bromated flame retardant used in mobile phones is neurotoxic, cadmium used in the rechargeable computer batteries is toxic for kidneys and bones, mercury in the flat screens can affect the brain and central nervous system, and so on. The management of these chemical substances during the life cycle of electronic products is essential in the green context and must include informing manufacturers, users, and recyclers about the presence of toxic elements. Especially, in the end-of-life product cycle, electronic wastes including hazardous products pollute soil, atmosphere, water, and the beauty of landscape. Incineration releases heavy metals and ashes into the air. Obviously, there are two ways to limit this pollution. One consists of mitigating the use of hazardous elements in manufacturing electronic equipment, and the other pertains to the proper management of dismantling and recycling steps. The GS1 EPC global standard in the consumer electronics supply chain (GS1, 2010) is an interesting solution to ensure the traceability of electronic equipment by using radio frequency identification (RFID) technologies, which provides storage for a cartography of an electronic product's components. A through knowledge of electronic product composition is essential for improving the efficiency of hazardous materials management and for preventing health issues of employers working on recycling processes and have direct contact with electronic wastes. Kubler et al. (2014) propose a similar concept to embed RFID in material for improving the monitoring of the life cycle of materials.

"If the cloud were a country, it would have the fifth largest electricity demand in the world" (Green Peace International Report, 2012). The Smart 2020 report (Smart 2020, 2008) provided quantitative information on carbon emission produced by ICT by considering the complete electronic product life cycle including the ICT use step and the manufacturing and dismantling steps (embodied carbon). The energy consumed by the ICT sector significantly adds carbon emissions to the atmosphere, causing air and thermal pollution.

The energy consumed by ICT use corresponds to the power supply to feed ICT equipment (computers, networks, etc.), cooling systems, and energy transport. The installation of a base transceiver station (BTS) in the desert is a typical example that illustrates clearly the logistical chain to be considered in the energy consumption estimation. It includes fuel to produce electricity from generators. This electricity is used both for BTS and cooling systems. Finally, it is also necessary to manage a crew of tank trucks consuming fuel to transport it to BTS generators. Another interesting example is the study of data center energy consumption. Emerson Network Power (2009) defines two categories of energy use: the demand side and the supply side. The demand-side systems are the servers and storage, communication, and other ICT systems that support a business. The supply side systems include uninterruptible power supplies (UPS), power distribution, cooling, lighting, and building switchgear. The power usage effectiveness (PUE) proposed by Green Grid (2012) to assess data center efficiency is the ratio of facilities energy (supply side) to IT equipment energy (demand side) and is expressed as follows:

$$\text{PUE} = \frac{\text{Total Facility Energy}}{\text{IT Equipment Energy}}$$

However, as Ascierto (2011) mentioned, PUE is a simple metric that does not reflect the entire complexity of data centers regarding its global efficiency. Especially, PUE does not provide performance on its use (bandwidth, bytes, etc.).

Nevertheless, the interest in identifying all energy consumption sources is not only to control the global energy consumption of data center but also to develop a strategy to mitigate the energy consumption. Indeed, Emerson Network Power (2009) shows the cascade of effects between component chain in explaining that the energy reduction of a server has an impact on the power supply, which has an impact on power distribution, which has an impact on UPS and on cooling, and so on. Emerson Network Power estimates 1 W saved at the processor saves approximately 2.84 W of total consumption.

A thorough knowledge of these interactions is crucial in decreasing energy consumption. In this process, two criteria concerning energy and carbon emission must be clearly identified. The mitigation of energy in ICT is mainly a business objective corresponding to the profit pillar of sustainable development. As mentioned in (Emerson Network Power, 2009), the global electricity prices increased 56% between 2002 and 2006 and continue to growth. Thus, ICT sectors must lessen their energy consumption to limit the impact of energy costs on their business.

Carbon emission participates in the increasing of greenhouse gas (e.g., as methane) and is related to the planet pillar of sustainable development. Even if carbon emission is correlated to the energy consumption, its decrease regarding the planet criterion and the profit criterion can provide two opposite strategies. Indeed, the selection of a clean energy to reduce pollution can increase the electricity prices. Moreover, the decrease of carbon emission is complex, and its complexity will increase with smart grid development because of the variation of carbon rate in the energy production. The carbon emission algorithms should integrate these variations. The French Company of Energy Transport (RTE) provides such information in real time that can be used to develop optimal strategies in the usage management of ICT equipment in an enterprise. For example, the observation of a peak on carbon to produce electricity could automatically configure the laptops in battery mode.

The Smart 2020 report also analyzes the embodied carbon in carbon emissions during both the manufacturing of ICT components and their end-of-life treatment. In general, RICS QS & Construction Standards (2012) explain that the extraction, manufacture, transportation, assembly, replacement, and deconstruction of construction materials or products involve embodied carbon.

The knowledge of embodied carbon and the carbon emitted during the usage phase enables the determination of the sustainability of an ICT-based solution or product and then an estimation of its optimum obsolescence. From this information, the global energy consumption of an active ICT product can be compared with a new generation of ICT products using more energy-efficient components to anticipate the ICT installation renewal. The approach in extending the ICT product life at all costs may be in opposition to the sustainability objective of promoting environmental-friendly solutions. The concept of point or band of optimum obsolescence was explained by Tuppen (2013) to define the right time to change a car, a notebook, and so on.

Electrosmog is the electromagnetic radiation emitted by electronic equipment (computers, mobile phones, etc.) and is a hot topic in the scientific community to determine its real effect on health and then to determine whether electrosmog is a form of pollution or a contamination. The health effects of radio waves were intensively studied by national and international organizations including the International Commission on Non-Ionizing Radiation Protection, the Institute of Electrical and Electronics Engineers (IEEE), and the World Health Organization (WHO). The effects of electromagnetic fields on the human body depend not only on their field level but also on their frequency and energy. All study results indicated that the unique noncontroversial effect of nonionizing electromagneting field (EMF) is thermal (WHO, 2006, 2011). To avoid it, all organizations have determined the maximum acceptable values for EMF (IEEE, 2005) associated with the frequency range of the radio frequency (RF) channel. Based on these recommendations, the government of each country or state has defined the legal maximum level of EMF generated by wireless network antennas and their maximum specific absorption rate (SAR) value. SAR is a measure of the maximum energy absorbed per unit of mass of the head of a person using a mobile phone. These values are sometime lower than the maximum. It is not disputed that electromagnetic fields above certain levels can trigger biological effects. For example, weak electromagnetic transmitters with a frequency spectrum between 0:1-10 MHz can affect animal behavior, particularly the orientation of migrating birds (Ritz et al., 2004). But biological effects do not necessarily cause health hazards. However, researches are actively continuing to confirm that low level, long-term exposure to radio frequency fields do not generate adverse health effects.

We now explain many proposals to limit electrosmog. One way is to use cognitive radio-enabling automatic detection of the available channels in a wireless spectrum and accordingly to change their transmission and reception parameters (Stevenson et al., 2009). Another way is the smart antenna technique, which uses spatial multiplexing and coding for removing interferences (Cui et al., 2005). Other solutions consist of implementing base stations and mobile phones using low power transmissions. However, this approach tends to multiply (for an equivalent level of QoS) the number of base stations and then to increase the energy consumption and CO_2 pollution while spoiling the esthetic environment. Based on this idea, Cerri and De Leo (2004) propose to reduce electromagnetic pollution of mobile communication systems in optimizing radio base station locations. Finally, a good improvement in wireless communication is to put antennas in fake chimneys or fake trees. This is a good way to avoid the electromagnetic hypersensitivity effect (Seitz et al., 2005) and to minimize the visual pollution.

Two additional aspects related to pollution induced by ICT, but not really considered in green ICT, are light and noise pollution. A recommendation has been made to limit paper documents to preserve the environment. The systematic use of electronic documents is often recommended. However, computer screens generate a light pollution affecting health. The computer vision syndrome defined by the American Optometric Association includes sore, dry, irritated, or watery eyes; headaches; sleep disorders; and so on of people using computers. A specific information on that could be mentioned in each proposed ICT solution.

The noise pollution generated by cell phone ringtones and conversations in public area has also impacted quality of life. This problem is related to ethical behavior in the use of ICT and is more complex to measure and assess.

Resource Efficiency

Janine Benyus explains in the last chapter of her book *Biomimicry* (Benyus, 2002) "how will we conduct business," in an ecosystem that is a complex of living organisms, their physical environment, and all their interrelationships within it in a particular unit of space (Encyclopedia Britannica, 2012). The evolution of ecosystems generally occurs in two phases: the developing stage and the mature stage (Allenby and Cooper, 1994). The developing stage involves few species and short food chains. At this stage, the ecosystem is unstable but highly productive, in the sense that it builds organic matter faster than it breaks it down. The mature ecosystem is more complex, more diversified, and more stable. Currently, the business model used in our society, especially in the ICT sector, is the developing stage. The challenge is to move to the mature stage.

Businesses based on the developing stage consider natural resources to be infinite. However, different studies have estimated that the earth reserves of a number of materials (Diederen, 2009; Cohen, 2007; Kesler, 2007). For example, the reserve of Indium is between 5 and 10 years, of copper 20-50 years, and of gallium 5 years. Indium, copper, and gallium are components used in semiconductors and there are two actions to be able to manufacture ICT products in long term. The first one consists of substituting other materials having the same properties; the second one is to generalize the recycling of ICT products. "A tonne of gold ore yields 5 g of gold, compared to a staggering 400 g yielded from a tonne of used mobile phones" (ITU-T report, 2012). In this context, the interest in recycling is both ecological and economical. The EuPs report (2007) provides much interesting information about materials inside computers and screens.

Recycling in the ICT sector is a big issue because of the rapid obsolescence of hardware and software, which continually offer new functionality with better performance, lead to premature equipment renewal, and produce electronic waste. The different recycling aspects include the reuse of old equipment in other applications and in equipment dismantling. For example, reuse might involve replacing an old mobile phone with a new one and identifying a market for selling the old one to people with different needs. In the book *Cradle to Cradle*, Donough and Braungart (2002) propose that products be designed with their raw materials separated into biological nutrients and technical nutrients. The interest is in avoiding the design of "monstrous hybrids" and in facilitating the recycling step by having one part dedicated to biological metabolisms and another part dedicated to technical metabolisms (enabling the recovery of rare materials, for example). This aspect is very important in the context of Internet of things when electronics and batteries are more and more mixed with the natural environment and its dismantlement should be carefully analyzed during its design.

Standardization and legislation have a primary role in the success of recycling. For example, the Waste Electrical and Electronic Equipment Directive (WEEE, 2003) in the European Community (2002/96/EC) imposes responsibility for the disposal of electrical waste and electronic equipment on the "manufacturers." Other examples for limiting e-waste are the recent ITU recommendations (ITU, 2011; ITU-T L.1001, 2012), proposed to specify universal power adapter and charger solutions for both mobile and network devices enabling consumers to reuse them when they buy new mobile phones or other electronic equipment.

The limitation of natural reserves (Kesler, 2007) also concerns energy (oil, gas, etc.) and suggests the use of renewable energies (solar, wind, water, etc.) to preserve the earth's resources. Moreover, this method can be also applied to energy and materials. Electronic equipment consumes energy and dissipates heat. This heat can be reused as a resource for other applications. For example, the heat emitted by a data center can provide heat for buildings (Val-Europe, 2012).

"Water is life, sustaining ecosystems and regulating our climate. But it's a finite resource, and less than 1% of the world's fresh water is accessible for direct human use" (European Commission, 2010). The water scarcity defined as the lack of sufficient available water resources to meet the demands of water use within a region is another large challenge to the green ICT. As Lewis (2013) explained, the ICT sector has a strong demand for energy; data centers and the energy sector consume large amounts of water. Moreover, the air cooling in data centers has recently replaced by a water cooling system because water conducts more heat than air, and warm water can be reused more easily to heat buildings and swimming pools near data centers. For measuring the water consumption in data centers, the Green Grid uses the metric water usage effectiveness (WUE) corresponding to the annual water use divided by IT equipment energy; it is expressed in liters/kilowatt-hour. This metric is applied to the site consumption WUE_{site} and to the source consumption WUE_{source}. Nevertheless, the manufacturing, transport, and recycling of water consumed during the life cycle of ICT products should be assessed.

Main Green Measures of Performances

From the previous analysis, the objective now is to define three main metrics to measure of performance (MoP) in the context of system engineering, enabling the development of green ICT regarding recyclability, energy consumption, and carbon emission. Obviously, these MoP are first proposals (Drouant et al., 2014) and should be refined and completed by others in order to take into account of all kinds of pollution and resource scarcity issues. Table 3.1 provides all notations used in the different equations.

MoP on Recyclability

This metric in Equation (3.1) involves the set X_r of equipment that will never be repackaged or refurbished for use in another architecture. In the worst case, it corresponds to $|X|$ the total number of items of ICT equipment used in the ICT-based architecture plus X_s equipment used to substitute for existing ICT equipment. Items of ICT equipment might be replaced because of failure, upgrades, or even extensions to the architecture. It is important to note that the list of equipment in Table 3.1 must be considered in terms of the entire architecture life cycle.

The recyclability rate of an item of equipment (ρ_i) ranges from 0 to 1 (1 corresponds to 100% recyclability and means that the item is fully reusable as a resource in other applications). From

Table 3.1 Notations

Symbol	Description
X	Set of equipment for the architecture
$\|X\|$	Number of items of equipment in the architecture
X_s	Set of substitute equipment
X_r	Set of repackaged equipment: $X_r \in (X \cup X_s)$
$\overline{X}_r$	Set of used equipment never repackaged: $\overline{X}_r = (X \cup X_s) - X_r$
ρ_i	Recyclability rate for an item of equipment i
Γ	Recyclability rate of the whole architecture
E	Energy consumed by the whole architecture
E_m	Energy consumed by the equipment's manufacturing process
E_u	Energy consumed by the whole architecture during its use phase
P_u	Power consumption of the whole architecture during its use phase
E_d	Energy consumed during the dismantling phase of $\overline{X}_r$
f	Factor related to the environment (air, feed efficiency)
h	Number of operating hours per year
δ	Power consumption gain (relative to the traffic profile)
ω_i	Traffic load for a switch port i
ε_i	Power consumption of an item of equipment i
$\emptyset$	Power consumption of a switch in the idle state (no traffic)
σ	Power consumption of a busy switch port (full load)
Φ	CO_2 emission for the whole architecture
α	Multiplicative gain between energy consumed and CO_2 emission by country
τ	Multiplicative gain caused by energy transportation losses

Equation (3.1), the management of metal can be taken into account in ρ in using the recyclability rate of metals defined in Graedel et al. (2011), for example, and is referred to as the *end-of-life recycling rate* (EOF-RR), including recycling as a pure metal (e.g., copper) and as an alloy (e.g., brass).

$$\Gamma = \frac{\sum_{i \in \overline{X}_r} \rho_i}{\left|\overline{X}_r\right|}, \quad \overline{X}_r = (X \cup X_s) - X_r \tag{3.1}$$

To optimize this metric, it is necessary to select highly recyclable ICT equipment, to anticipate any repackaging, and to limit the number of items of ICT equipment to be implemented except for those that are 100% recyclable ($\Gamma = 1$). Even if it does not mean zero energy consumption corresponding to the MoP on energy consumption, 100% recyclability is an ideal objective for a circular economy (Donough and Braungart, 2002).

MoP on Energy Consumption

In the estimation of energy consumption in Equation (3.2), there are three stages, namely the manufacturing of the ICT equipment (E_m), the use of the ICT-based architecture (E_u), and the dismantling of ICT-based architecture (E_d). It is important to note that the objective of green ICT is to reduce the energy consumption. With Equation (3.2), the dismantling step has a negative impact on this MoP to limit the effort of recycling. For this, Williams and Sasaki (2003) propose a similar equation with a negative term for (E_d) to indicate the positive gain of recycling. However, Equation (3.2) should be globally analyzed. A company developing an efficient management of its electronic waste will actually consume energy for the dismantling but will mitigate the energy consumption for manufacturing its new ICT products in avoiding the extraction of new earth resources. Thus, the optimization is not based on one product life cycle but on the manufacturing of multiple products always using the same earth resources. Another interpretation of Equation (3.2) is to recommend the reuse of the obsolete products in other applications or for other users before recycling.

$$E = E_m + E_u + E_d = E_m + \int_{t=0}^{\text{end of life cyle}} P_u(t)\mathrm{d}t + E_d \tag{3.2}$$

The energy used by the transport of ICT products from the manufacturer to the customer is included in (E_m). In the same way, the energy consumed to transport an ICT product for recycling is associated with (E_d).

P_u represents the power consumption by the ICT-based architecture during its use phase. Many research works propose expressions for modeling the energy consumed by ICT equipment. There are two main classes of proposal. One involves high-level modeling in which the interest is in approximating the energy consumption of general network architectures. The other is specific to network technologies and provides precise outcomes.

Two examples in the network domain illustrate these two classes.

Firstly, Foll (2008) considers a high-level model without specifying a networking technology. The objective is to develop a management tool for the Orange Company that offers higher visibility for the current and future consumption of its network. In this context, the following macroscopic model is defined per year:

$$\int_{t}^{\text{one year}} P_{\text{u}}(t)\mathrm{d}t = \sum_{i \varepsilon X} \varepsilon_i \times f \times h \tag{3.3}$$

Secondly, Reviriego et al. (2012) propose a more specific equation for estimating the energy consumption of a network architecture based on switched-ethernet technology. The model for the energy-efficient ethernet switch is:

$$P_{\text{u}} = \varnothing + \sigma \sum_{\text{Port}\, i} \min(1, \delta\omega_i) \tag{3.4}$$

In Equation (3.4), σ represents the difference between the power consumption at full load and the traffic-free power consumption divided by the number of ports in the ethernet switch.

P_{u} can be also approximated by a simple measure of ICT product or by using information provided by the Energy Star label regarding the efficient energy consumption of computers, displays, and imaging equipment. However, the understanding of the relationship between both the network activities and configuration and its energy consumption is open and is very important regarding two issues. Firstly, a formal relationship will enable proposing ICT-optimized strategies to mitigate the energy consumption. Secondly, a formal relationship could be compared with the current monitoring of ICT infrastructure in order to develop smart sensors able to observe deviations with expected results and to automatically detect possible anomalies.

MoP on Carbon Emission

Carbon emission is estimated from the energy used during the entire life cycle of an ICT product or ICT-based architecture and from the information in Table 3.2 matching rate between energy consumption and CO_2 emission in electricity production for some European countries.

Because a piece of equipment might be manufactured in one country, used in another, and dismantled in a third, the gain α is related to the stage of the life cycle (α_{m} will hence correspond to the factor during the manufacturing stage). The estimate of CO_2 pollution is then obtained by:

$$\Phi = \Phi_{\text{m}} + \Phi_{\text{u}} + \Phi_{\text{d}} = \alpha_{\text{m}} E_{\text{m}} + \alpha_{\text{u}} E_{\text{u}} + \alpha_{\text{d}} E_{\text{d}} \tag{3.5}$$

Table 3.2 Factors for kg of CO_2 per kWh

Country	CO_2 Factor
Sweden	0.04
France	0.09
Finland	0.24
Italy	0.59
Germany	0.60
Ireland	0.70
Luxembourg	1.08

Source: IEA

Moreover, Equation (3.5) addresses the location of energy production in order to use energy produced locally, thereby limiting energy transport losses. To integrate this parameter, it is necessary to make visible that part of the CO_2 emission caused by energy transportation. Therefore, a factor τ_s is added to Equation (3.5), giving:

$$\Phi = \sum_{s=m,u,d} \frac{\alpha_s}{\tau_s} \times E_s \tag{3.6}$$

The factor τ_s represents the additional rate for transporting energy at a stage s; for example, in France, the average energy lost during transport is 5% (i.e., $\tau_s = 0.95$).

Moreover Equation (3.6) can be extended to include other local sources of energy, such as oil-burning power generators (to power GSM antennas in an emergency) and solar panels installed locally and dedicated to the network. The impact depends on the nature of the energy (electricity, oil) so that the general metric for CO_2 emission becomes:

$$\Phi = \sum_{s=m,u,d} \sum_{\text{energy source } i} \frac{\alpha_{s,i}}{\tau_{s,i}} \times E_{s,i} \tag{3.7}$$

where $E_{s,i}$ represents energy-source component i consumed during the manufacturing, use, or dismantling stages.

Equation (3.7) is suitable in the context of smart microgrids (Perea et al., 2008) whose the objective is to encourage the consumption of renewable energy produced locally.

Ethics in ICT

The design of green ICT solutions should integrate all pillars of sustainable development: planet, profit, and people. Mainly, the papers on green ICT focus on the environment and neglect to consider human and ethical aspects. Nevertheless, the basis of ICT development is to help the citizens in their daily life and in their working activities. Moreover, Herold (2006) explains that "the consideration of computer ethics fundamentally emerged with the birth of computers." The use of the people pillar is to cover many areas and employ subjective metrics, making the assessment difficult. In general, each ICT company should analyze its social responsibilities to its customers, employees, and community (Uddin et al., 2008).

Social responsibilities toward the customers: The ICT company must develop safe and durable ICT products and services, offer efficient after-sale services, and be prompt, reliable, and courteous in dealing with queries, complaints, and so on (Uddin et al., 2008). The company must not exaggerate the performance of proposed solutions, but should propose alternative solutions by clearly explaining its reasons for offering multiple choices to customers, and so on (Grupe et al., 2002). Rules regarding privacy and anonymity should be clearly defined to prevent the installation of cookies to obtain customer (or personal) data to sell to other companies (Herold, 2006). The company must also inform customers about the content of applications and any restrictions in regard to the law (for example, parental controls for videogames).

Social responsibilities toward employees: Ethics must be analyzed during the design process of each ICT project. The company must define a general code of conduct for the management of employees (salary, gender, productivity, etc.) and for the selection of subcontractors according to the consideration of the working conditions of their employees (exploitation of children, no salary, etc.). Moreover, during the design process, the company must respect simple and evident rules related to good practices in ICT such as not using unlicensed software to develop new software; it also must respect the private lives of employees based on e-mail exchanges and so on (Grupe et al., 2002).

Social responsibilities to the community: Ethics must also be considered before an ICT project is started. A preliminary study should analyze the impact of the newly developed system on society and individuals and whether it increases the digital divide and the social divide. According to Herold (2006), the issues regarding computers in the workplace require assessing the impact of ICT on the elimination of jobs, loss of skills, health issues, and computer crime in installing spyware, hacking and so on. Finally, social responsibilities to the community include educating ICT users to respect the environment, develop good new habits in turning off equipment not being used, turn off mobile phones in meetings, recycle ink cartridges, so forth.

To summarize, ICT companies should define ethical guidelines in order to produce ICT-based solutions in compliance with the requirements of the people pillar of sustainable development. Implicitly, an ethical approach has a positive impact on the choices realized in the planet pillar (environmental ethic) and in the profit pillar in promoting the fair business.

Conclusion

The assessment of green ICT solutions must consider many performance indicators gathering both numerical and subjective information about specific equipment. Many initiatives specify metrics that are either general (outside ICT scope) or specific to one ICT domain such as data centers. This section described three concrete MoPs that necessitate refinement and completion with new ones to cover all aspects of sustainable development. In waiting for standardized metrics, the design of ICT-based solutions must follow systemic approaches such as systems engineering. The interest of systems engineering is to handle complex systems, to specify measurable requirements in order to translate them into MoPs and to ensure that the performances are satisfied during all the project life cycle.

Systems Engineering for Designing Sustainable ICT-Based Architectures

Introduction

The development of green ICT solutions is highly complex because they must be based on an analysis of the system as a whole including the effects on business, the environment, and the people in the design loop during its entire life cycle. Therefore, the design of green ICT systems requires formal or semiformal methodologies and tools to specify, verify, and validate measures of performances according to both the user's requirements and sustainable development. Moreover, to reach the global green ICT objectives, the assessment should focus not only on the ICT performance of a system of interest but also on its environmental impact. In this section, the goal is to present an example for studying ICT systems in using systems engineering (Pyster et al., 2011; INCOSE, 2010). This example of a student project in the Erasmus Mundus Master in Pervasive Computing and Communications (PERCCOM) for sustainable development (www.perccom.eu) highlights some major issues of the requirements specification process (Jin, 2006).

Stakeholder Requirements Definition

The will-to-be operational domain in this study is an ICT architecture composed of ten computers and three Cisco 3560 PoE-24 ports switches. Two more reliable alternative solutions are also analyzed (Figure 3.3). The first one implements 4 switches and the second one 5 switches. The ICT architecture is planned for five years (use step). The ICT products are manufactured in China and the architecture is installed in France.

The objective of the system engineering process (Figure 3.4) is to analyze the performances of these ICT architectures (GICT0) regarding the pillars of ecology (GICT0.1), ethics (GICT0.2), and economics (GICT0.3) from the three ICT architecture solutions. In the ecology pillar, the intermediate requirements are to estimate the recyclability of ICT devices (GICT0.1.0), the radio wave emission (GICT0.1.1) and the ICT carbon emission (GICT0.1.2). In the ethic pillar, the requirement is to control anonymity and privacy of ICT users with appropriate data access policy. In the economy pillar, the requirement is to calculate the energy cost of ICT architecture.

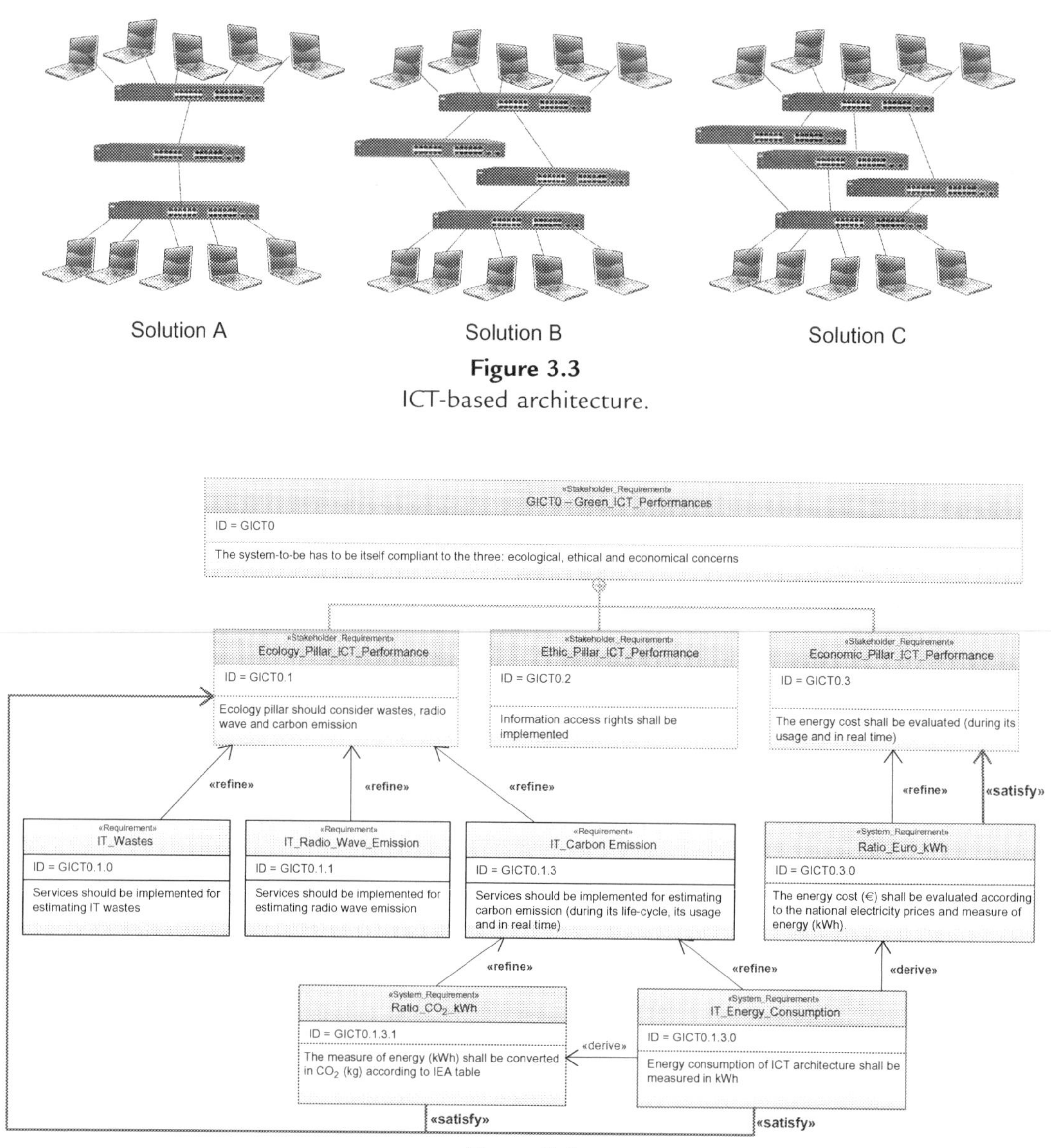

Figure 3.3
ICT-based architecture.

Figure 3.4
System requirements of SysML diagram.

System Requirements Analysis

This chapter analyzed only the estimations of carbon emission and energy cost. Thus, system requirements (Figure 3.4) defined by the expert in systems engineering are specified in order to satisfy the stakeholder requirements. In both cases, the ICT energy consumption should be measured (GICT0.1.3.0). The carbon emission is calculated from the ICT energy consumption and the conversion tables between CO_2 and energy (GICT0.1.3.1). The energy costs are calculated from ICT energy consumption and the electricity price (GICT0.3.0).

The expert in systems engineering has no competence in the ICT domain to propose ICT solutions regarding the system requirements. As stated by Bouffaron et al. (2014), the driver of any specification process (as a requirement analysis process) is a quest for knowledge as design property from any specialist engineers to the system engineer. The aim of this knowledge is to bring the gap between stakeholder requirements, system, and ICT architectural constraints. Thus, ICT experts are involved in the design process to provide their knowledge in applying the green ICT MoPs defined in Section "Main Green Measures of Performances." Figure 3.5 shows the interactions of all the knowledge domains of the study.

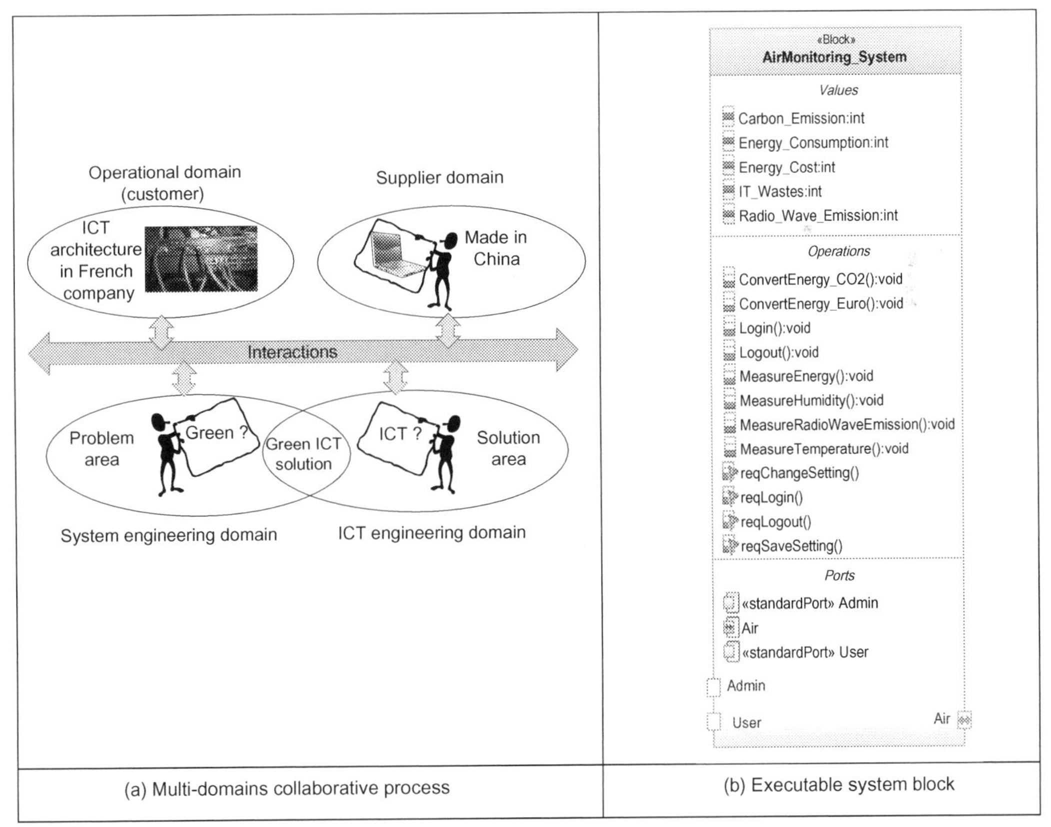

Figure 3.5
Systems specification process.

System Requirements Validation and Verification

The systems engineering project framework is based on systems thinking (Lawson et al., 2014) and on a model-based integrative approach aiming to check the "right-system requirements-right" from the early stages of a project. The goal is to explore in depth the problem to design first the overall required system in a concurrent, recursive, and iterative process in contrast to traditional sequential engineering approaches focusing first on solution issues. The use of models for verification and validation of measurable requirements to component solution integration is compliant with the last recommended best practices in industry (Fanmuy et al., 2012). The objective is to check the compliance of system requirements to stakeholder's requirements by execution of models to functionally design the required system as a whole before sending it to an architect by allocating commercial off-the-shelf (COTS) or specific components. PERCCOM students execute their system specification by models transformed with a SysML-based tool (Sysml-Harmony) and are trained for ongoing developments related to *tool independent exchange of simulation models* (Blochwitz et al., 2012).

ICT Expertise and Results

The ICT expert proposes methods and results for providing the required information on the three ICT architecture solutions.

Function: energy consumption estimation of ICT architecture: The per hour energy consumption of all devices implemented in the ICT architecture has been measured by ICT experts; the values obtained are 35 Wh for the switch and 100 Wh for the computer. The network works 24/7 and the ten computers 10 h per day. By using the Equation (3.2) (E_u), the total energy consumptions of ICT architectures during five years are:

- Solution A: 22,849 kWh
- Solution B: 24,382 kWh
- Solution C: 25,915 kWh

Function: energy cost estimation of ICT architecture: With an average price of French kWh (0.12€kWh), the total ICT costs of energy consumptions are:

- Solution A: 2741€
- Solution B: 2925€
- Solution C: 3109€

Function: carbon emission estimation of ICT architecture: By using the CO_2 factor per kWh in France (0.09), the ICT carbon emissions are [$\alpha_u E_u$ in Equation (3.5)].

- Solution A: 2056 kg $C0_2$
- Solution B: 2194 kg CO_2
- Solution C: 2332 kg CO_2

Function: energy consumption of ICT architecture life cycle: The analysis of energy consumption during the entire life cycle requires knowing the energy used for manufacturing and dismantling. According to EuPs (2007), the energy consumed to manufacture a computer is 900 kWh, and as approximation this value is used for computers and switches of architectures. The literature does not currently provide information about dismantling. Thus, 200 kWh is the empirical value selected.

By using the formula: $E_m + E_u + E_d$ (Equation (3.2)), the energy consumptions of ICT architectures are then:

- Solution A: 11,700+22,849+2600=37,149 kWh
- Solution B: 12,600+24,382+2800=39,782 kWh
- Solution C: 13,500+25,915+3000=42,415 kWh

Function: carbon emission of ICT architecture life cycle: Because the product is manufactured in China with a CO_2 factor per kWh at 0.97 and recycled in France, the carbon emissions are from the expression $\alpha_m E_m + \alpha_u E_u + \alpha_d E_d$ (Equation (3.5)):

- Solution A: (11,700 × 0.97)+(22,849 × 0.09)+(2600 × 0.09)=11,349+2056 +234=13,639 kg CO_2
- Solution B: (12,600 × 0.97)+(24,382 × 0.09)+(2800 × 0.09)=12,222+2194 +252=14,668 kg CO_2
- Solution C: (13,500 × 0.97)+(25,915 × 0.09)+(3000 × 0.09)=13,095+2332 +270=15,697 kg CO_2

The interest of this result is not the global quantity of CO_2 emission but the ratio between the manufacturing and use steps. The carbon emitted during the manufacturing corresponds around to three years of carbon emission. In the context of the planet pillar, it appears that the engineering effort should be mainly achieved at the manufacturing level, not the use level. Nevertheless, the strong increase of energy prices is a good motivation to develop sophisticated engineering in order to mitigate the energy costs (profit pillar).

Function: real-time measure of ICT energy consumption: The network architecture is based on Cisco products to monitor energy consumption of devices implementing EnergyWise MIB. A monitoring software was developed for collection by using SNMP all energy consumptions of devices compatible with EnergyWise. The interest in using SNMP is to gather other ICT information such the bandwidth use, the temperature of equipment, and so on, and to easily propose indicators of efficiency, such as the ratio between energy consumption of ICT architecture and its use, to create smart monitoring in analyzing the variation of energy consumption regarding the temperature, traffic, and network configuration (number of connected ports, port speed, etc.).

Raritan intelligent PDUs or the ICT devices that do not implement EnergyWise are used. The Raritan PDU especially enables collection of the energy consumption of computers in the platform.

Function: real-time measure of ICT carbon emission: Carbon emission is estimated from the measurement of energy consumed by ICT architecture and from the CO_2 factor per kWh produced in France provided in real time by RTE and from the Internet. Figure 3.6 shows the interface developed in this project by PERCCOM students.

Traceability Matrix

The design process of green ICT architectures must guarantee the sustainability of proposed solutions. The verification step is then crucial to check the process design. Table 3.3 shows that all functions were tested, confirming that the system requirements satisfy the stakeholder

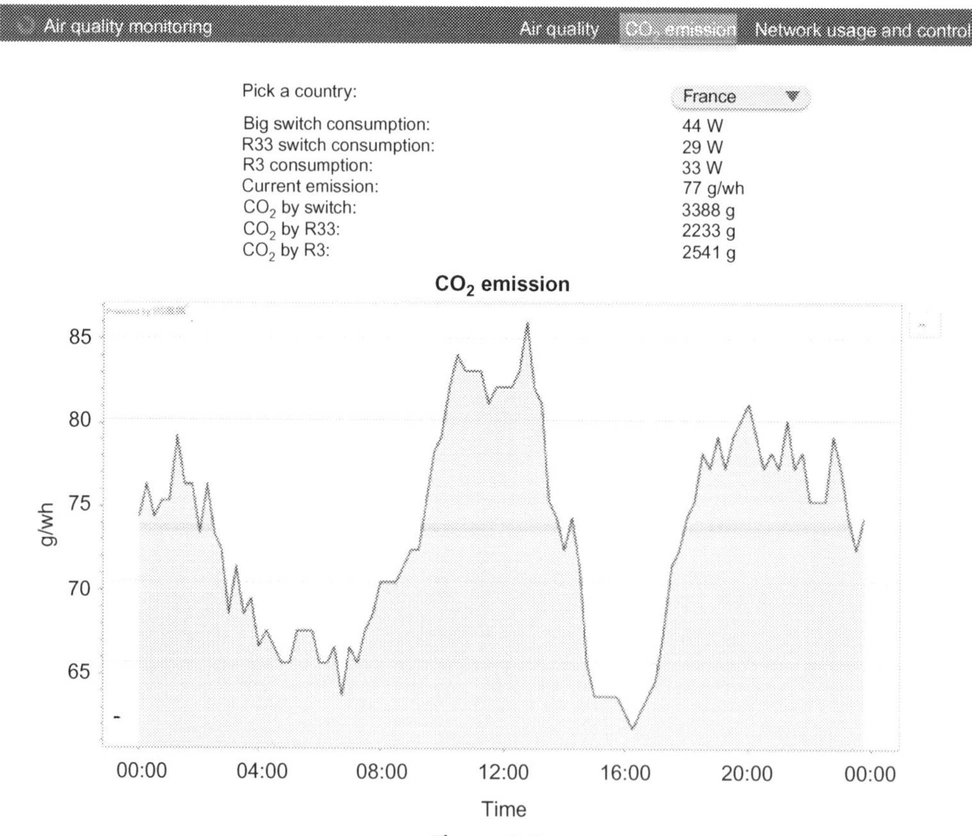

Figure 3.6
Monitoring ICT energy consumption and carbon emission.

Table 3.3 Traceability Matrix

Stakeholder Requirements	System Requirements	Functions	Verification Method	Test Case	Test Result
GICT0.1 Ecology_Pillar_ICT_Performance	GICT0.1.3.0 IT_Energy_Consumption	Energy consumption estimation of ICT architecture	Documentation	TC1	OK
		Energy consumption of ICT architecture life cycle	Documentation	TC2	OK
		Real-time measure of ICT energy consumption	Demonstration	TC3	OK
	GICT0.1.3.1 Ratio_CO2_Kwh	Carbon emission estimation of ICT architecture	Documentation	TC1	OK
		Carbon emission of ICT architecture life cycle	Documentation	TC2	OK
		Real-time measure of ICT Carbon emission	Demonstration	TC3	OK
GICT0.3 Economic_Pillar_ICT_Performance	GICT0.3.0 Ratio_Euro_Kwh	Energy cost estimation of ICT architecture	Documentation	TC1	OK

requirements. The study of traceability matrix avoids delivering ill-conceived projects by reconsidering the project. A redesign step should change the initial choices in selecting new ICT products to substitute nonadapted products, to develop software patches, to dismantle certain solutions, and so on. All these modifications generate premature and useless waste, consume additional energy, and generate wobbly, less sustainable solutions.

In conclusion, systems engineering provides good practices for the ecodesign of complex systems, especially to green the design of ICT projects.

Ecoefficiency Metrics

The results of the study described has obvious conclusions: Carbon emission and cost energy increase with the number of ICT products in the architecture. However, as Ascierto (2011) explained concerning PUE metric about data centers, the ICT system should be assessed in a global way when considering the intrinsic performances of ICT including the metrics presented in Section "ICT Technical Measures." For illustrating this comment, three stakeholder requirements are added in the study.

During the use of ICT architecture, the carbon emission must be less than 3000 kg CO_2, the energy cost must be less than 3000€, and the reliability must be SIL2 ($10^{-7} <$ failures per hour $< 10^{-6}$) (IEC Standard 61508, 2005).

Thus, ecoefficiency metrics coupling the green performances with other intrinsic ICT performances should be developed. These metrics bring together various kinds of information, making it difficult to define a global equation that includes all requirements. A radar diagram is an interesting representation to show both the network performance and the requirements. Each MoP then corresponds to one axis of the radar. To improve the radar readability, all axes are normalized and the interest points emphasized. The global view offered by the radar diagram should make the analysis of network solutions easier, both for the designer and the customer. The best solution is the one in which all MoP results are in the center of the radar, but the best one relative to the requirement (optimized solution) is where the MoP results are close and beneath the specification line.

For estimating reliable MoP, it is necessary to define n_i as being the number of items of equipment forming a path i. Item is either a link or a switch. Let λ be the failure probability per hour for any item and $\mu = 1 - \lambda$ as the nonfailure probability. It is considered that all switches and links have failures per hour equal to $\lambda = 10^{-5}$. The failure probability of a network (P) composed of p independent paths depends on the failure probability of each path i (π) such that:

$$P = \prod_{i=1}^{p} P_i = \prod_{i=1}^{p} (1 - \mu^{n_i})$$

For example, considering solution C, three paths are formed by three items (one switch plus two links). This gives:

$$P = \left(1 - \left(1 - 10^{-5}\right)^3\right)^3 = 2.7 \times 10^{-14}$$

The results are for solution A: 3×10^{-5} and for solution B: 9×10^{-10}.

All MoP and requirements are collected in a single radar diagram (Figure 3.7). The system engineer can then select the most appropriate solution. The radar diagram shows that only solution B satisfies all stakeholder requirements (green line) because the reliability of solution A is not acceptable and the energy cost of solution C is too high. If all solutions are unsatisfactory, the stakeholder requirements should be redefined, refined in an iterative way with the system and ICT engineers to converge to consensual and acceptable solutions.

Regarding the ethical consideration, this iterative process in the systems engineering approach provides professionalism and respects the ethical bases for IT decision making defined by Grupe et al. (2002).

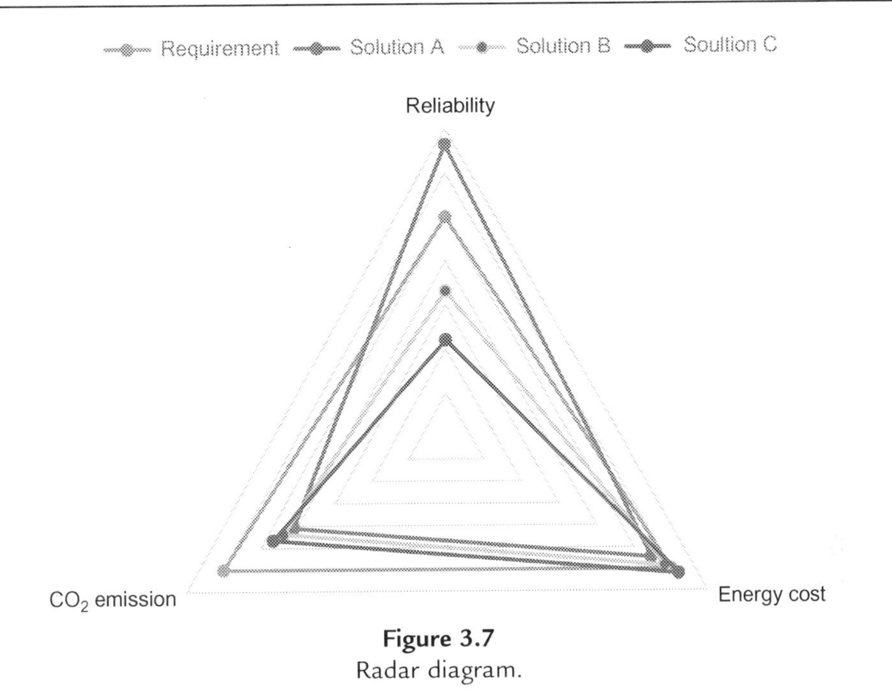

Figure 3.7
Radar diagram.

Conclusion

The design of green ICT-based solutions should consider the three pillars of sustainable development (Figure 3.1). The analysis is complex because it develops ICT solutions with the three objectives sometimes having conflicting criteria. Moreover, the analysis of the impact on the planet covers an unlimited scope that should include all pollution forms and the Earth's resource management, making a global assessment difficult during the entire ICT product life cycle. Pragmatically, green ICT mainly focuses on energy optimization. The major explanation is that it is a win-win activity because of its benefits on both planet and profit pillars. The recycling has also a positive result on both pillars because it enables the indefinite reuse of rare materials that are becoming more and more expensive. However, its implementation is very complex, and the return on investment is less immediate than energy. Regarding the people pillar, there are many contributions in the literature on computer ethics. However, from our best of knowledge, no significant contributions couple computer ethics with green ICT in a global sustainable development approach. People pillar should be analyzed at two levels: (1) how the ICT solution is designed and (2) how the developed ICT solution is used.

Nevertheless, the representation of sustainable development from a triangle is restrictive and especially hides the ICT performances corresponding to the fundamental activities of ICT engineers. Perhaps it is one factor explaining why the first green metrics in data centers do not include data center activities. The goal of the sustainable development triangle is to force ICT

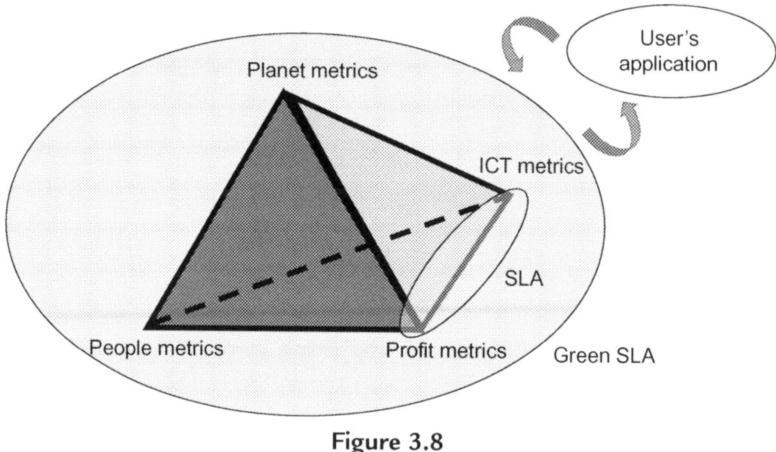

Figure 3.8
Green ICT metric pyramid.

engineers to analyze the interactions between the three pillars to find balanced solutions that satisfy all stakeholders. However, the base requirement for an ICT professional when creating new ICT solutions is to specify SLA with economy and ICT metrics. In Figure 3.8, a new pillar is added, enabling the integration of ICT metrics in green ICT systems engineering. From this green ICT metric pyramid, the ICT engineer can study the relationships between ICT metrics and the pillars of sustainable development triangle in order to propose ecoefficient green ICT metrics and to specify green SLA.

A systems engineering approach for green ICT projects is recommended for developing more sustainable ICT-based solutions gathering in the same study all stakeholder requirements and enabling verification of the conformity of solutions. Moreover, new constraints directly related to climate change make ICT engineering more complex. Indeed, the ICT sector contributes to climate change, so it must propose resilient ICT solutions to climate changes. ITU (2014) explains the risks posed by both acute weather events (e.g., short-term extreme events such as cyclones, intense rainfall, and flooding) and chronic trends (e.g., long-term changes in temperature, seasonality, and the rise of sea levels) to guarantee reliable telecommunication infrastructures. For example, the installation of antenna should be more robust to withstand extreme wind, lightning, and heavy snowfall. A checklist of new constraints is proposed to ensure that all these extreme climate conditions were envisaged in ICT engineering step (ITU, 2014).

Finally, the SMART 2020 (Smart 2020, 2008) report shows the ambivalence of the ICT sector because it was responsible in 2007 for 2% of global carbon emissions and participates in climate change, but the sector could contribute to mitigating the carbon footprint of other human activity sectors related to transport, buildings, power, and industry by improving their energy efficiency. In this context, the role of ICT engineers is to design and create new smart metering in order to analyze and control the energy consumption of applications in real time. These metrics defined by applications (transport, building, etc.) are not included in the green ICT metrics presented in this chapter.

References

Allenby, B.R., Cooper, W.E., 1994. Understanding industrial ecology from a biological systems perspective. Environ. Qual. Manage. 1520–6483 (3), 343–354.

Ascierto, R., 2011. From Money Pit to Profitability: The Business Case for Data center Efficiency. http://ovum.com/research/from-money-pit-to-profitability-the-business-case-for-data-center-efficiency-metrics/.

Benyus, J.M., 2002. Biomimicry: Innovation Inspired by Nature. Harper Perennial, New York. ISBN 0060533226.

Blochwitz, T., Otter, M., Akesson, J., Arnold, M., Claub, C., Elmqvist, H., Friedrich, M., Junghanns, A., Mauss, J., Neumerkel, D., Olsson, H., Viel, A., 2012. Functional mockup interface 2.0: the standard for tool independent exchange of simulation models. In: 9th International Modelica Conference, Munich, Germany, September 2012.

Bouffaron, F., Dupont, J.-M., Mayer, F., Morel, G., 2014. Integrative construct for model-based human-system integration: a case study. In: Paper Presented at the World IFAC Congress 2014, Cape Town, South Africa.

Cerri, G., De Leo, R., 2004. Base-station network planning including environmental impact control. IEE Proc. Commun. 22 (2004), 197–203.

Chapman, P.M., 2007. Determining when contamination is pollution—weight of evidence determinations for sediments and effluents. Environ. Int. 33 (4), 492–501.

Cohen, D., 2007. Earth's natural wealth: an audit. New Sci. Mag. (2605), 34–41.

Cui, S., Goldsmith, A.J., Bahai, A., 2005. Energy-efficiency of MIMO and cooperative MIMO techniques in sensor networks. IEEE J. Sel. Areas Commun. 22, 1089–1098.

Diederen, A.M., 2009. Metal Minerals Scarcity: A Call for Managed Austerity and the Elements of Hope. TN Defence, Security and Safety, Rijswijk, The Netherlands.

Donough, W., Braungart, M., 2002. Cradle to Cradle: Remaking the Way We Make Things, first ed. North Point Press, New York, ISBN 0865475873.

Drouant, N., Rondeau, E., Georges, J.P., Lepage, F., 2014. Designing green network architectures using the ten commandments for a mature ecosystem. Comput. Commun. 42, 38–46.

Emerson Network Power, 2009. Energy logic: reducing data center energy consumption by creating savings that cascade across systems. A White Paper from the Experts in Business-Critical Continuity.

Encyclopedia Britannica, 2012. http://www.britannica.com/.

EuPs, 2007. European Commission DG TREN Preparatory Studies for Eco-design Requirements of EuPs. ISSN: 1404-191X.

European Commission Report, 2010. Water Scarcity and Drought in the European Union, KH-30-09-180-EN-D.

Fanmuy, G., Fraga, A., Llorens, J., 2012. Requirements verification in the industry. In: Complex Systems Design & Management. Springer, pp. 145–160.

Foll, L.S., 2008. TIC et Énergétique: techniques d'estimation de consommation sur la hauteur, la structure et l'évolution de l'impact des TIC en France (Ph.D. thesis), Institut national des télécommunications.

Georges, J.P., Divoux, T., Rondeau, E., 2005. Confronting the performances of a switched ethernet network with industrial constraints by using the network calculus. Int. J. Commun. Syst. 18 (9), 877–903.

Graedel, T.E., Allwood, J., Birat, J.P., Buchert, M., Hagelüken, C., Reck, B.K., Sibley, S.F., Sonnemann, G., 2011. What do we know about metal recycling rates? J. Ind. Ecol. 15 (3), 355–366.

Green Grid, 2012. PUE™: a comprehensive examination of the metric. White Paper 49.

Green Peace International Report, 2012. How is Clean Your Cloud? JN 417.

Grupe, F.H., Garcia-Jay, T., Kuechler, W., 2002. Is it time for an IT ethics program? Information Management Strategy Systems and Technologies. Auerbach Publication, New York.

GS1, 2010. Implementation of GS1 EPC Global Standards in the Consumer Electronics Supply Chain. http://www.gs1.org/.

Herold, R., 2006. Introduction to computer ethics. Information Systems Security. Auerbach Publications, New York.

IEC Standard 61508, 2005. International Electrotechnical Commission, Functional Safety of Electrical/Electronic/Programmable Electronic Safety-related Systems.

IEEE, 2005. IEEE Standard for Safety Levels with Respect to Human Exposure to Radio Frequency Electromagnetic Fields, 3 kHz to 300 GHz, IEEE Std C95.1.

IEEE P1888, 2011. IEEE Standard for Ubiquitous Green Community Control Network Protocol, Ubiquitous Green Community Control Network Working Group, UGCCNet.

INCOSE, 2010. Systems Engineering Handbook. International Council on Systems Engineering, San Diego, CA.

ITU, 2011. Universal Power Adapter and Charger Solution for Mobile Terminals and Other Hand-held ICT Devices, Recommendation ITU-T L.1000.

ITU, 2014. Resilient Pathways: The Adaptation of the ICT Sector to Climate Change.

ITU-E800, 1994. Terms and Definitions Related to Quality of Service and Network Performance Including Dependability. http://www.itu.int/rec/T-REC-E.800.

ITU-T report, 2012. Toolkit on Environmental Sustainability for the ICT Sector, Chapter "General Specifications and Performances Indicators".

ITU-T L.1001, 2012. External Universal Power Adapter Solutions for Stationary Information and Communication Technology Devices.

Jin, Z., 2006. Revisiting the meaning of requirements. J. Comput. Sci. Technol. 21, 32–40.

Kesler, S.E., 2007. Mineral supply and demand into the 21st century. In: Briskey, J.A., Schulz, K.J. (Eds.), Workshop on Deposit Modeling, Mineral Resource Assessment, and Sustainable Development. U.S. Geological Survey Circular 1294, pp. 55–62 (paper 9).

Kubler, S., Voisin, A., Derigent, W., Thomas, A., Rondeau, E., Främling, K., 2014. Group fuzzy AHP approach to embed relevant data on "communicating material" Comput. Ind. 64 (4), 675–692.

Laszewski, G., Wang, L., 2010. Green IT Service Level Agreements, Grids and Service-Oriented Architectures for Service Level Agreements. Springer Science, New York, pp. 77-88.

Lawson, B., Wade, J., Hofkirchener, W., 2014. Systems Series Publishes Books Related to Systems Science, Systems Thinking, Systems Engineering and Software Engineering. http://www.collegepublications.co.uk/systems/syt/.

Lewis, A., 2013. The Four Key Environmental Factors of ICT: Energy, Carbon, E-Waste and Water. http://www.sustainability-perspectives.com/perspective/four-key-factors.

Makela, T., Luukkainen, S., 2013. Incentives to apply green cloud computing. J. Theor. Appl. Electron. Commer. Res. 8 (3), 74–86.

Michaut, F., Lepage, F., 2005. Application oriented network metrology: metrics and active measurement tools. IEEE Commun. Surv. Tutorials Second Quart. 7 (2), 2–24.

Perea, E., Oyarzabal, J.M., Rodríguez, R., 2008. Definition, evolution, applications and barriers for deployment of microgrids in the energy sector. Elektrotech. Informationstech. 125, 432–437, 0932-383X.

Pyster, A., Olwell, D., Anthony, J., Enck, S., Hutchison, N., Squires, A., 2011. A Guide to the Systems Engineering Body of Knowledge (SEBoK) version 0.75. Stevens Institute of Technology, Hoboken, NJ.

Reviriego, P., Sivaraman, V., Zhao, Z., Maestro, J., Vishwanath, A., Sanchez-macian, Russell, C., 2012. An energy consumption model for energy efficient ethernet switches. In: 10th Annual International Conference on High Performance Computing and Simulation, HPCS 2012, Madrid, 2-6 July 2012, pp. 98–104.

RICS QS & Construction Standards, 2012. Methodology to Calculate Embodied Carbon of Materials, P 32/2012, UK.

Ritz, T., Thalau, P., Phillips, J.B., Wiltschko, R., Wiltschko, W., 2004. Resonance effects indicate a radical-pair mechanism for avian magnetic compass. Nature 429 (6988), 177–180.

Seitz, H., Stinner, D., Eikmann, T., Herr, C., Roosli, M., 2005. Electromagnetic hypersensitivity (EHS) and subjective health complaints associated with electromagnetic fields of mobile phone communication—a literature review published between 2000 and 2004. Sci. Total Environ. 349, 45–55.

Smart 2020, 2008. http://www.smart2020.org.
Stevenson, C., Chouinard, G., Lei, Z., Hu, W., Shellhammer, S., Caldwell, W., 2009. IEEE 802.22: the first cognitive radio wireless regional area network standard. IEEE Commun. Mag. 47 (1), 130–138.
Tuppen, C., 2013. Circularity and the ICT Sector, Ellen MacArthur Foundation/Advancing Sustainability LLP. PERCCOM Seminar, Nancy.www.perccom.eu/files/2013/09/Nancy-ICT-CE-copy.pdf.
Uddin, M.B., Hassan, M.R., Tarique, K.M., 2008. Three dimensional aspects of corporate social responsibility. Daffodil Int. Univ. J. Bus. Econ. 3 (1), 199–212.
Val-Europe, 2012. http://www.valeurope-san.fr/info/UK/00.
Vatanski, N., Georges, J.P., Aubrun, C., Rondeau, E., Jämsä Jounela, S.L., 2009. Networked control with delay measurement and estimation. J. Control Eng. Pract. 17 (2), 231–244.
WEEE, 2003. Waste Electrical and Electronic Equipment, Directive 2002/96/EC of the European Parliament and of council of 27 January 2003. Official Journal of the European Union, Bruxelles.
World Health Organization, 2006. Base Stations and Wireless Technologies, Fact Sheet No. 304.
World Health Organization, 2011. Electromagnetic Fields and Public Health: Mobile Phones, Fact Sheet No. 193.
Williams, E.D., Sasaki, Y., 2003. Energy Analysis of End-of-life Options for Personal Computers: Resell, Upgrade, Recycle. In: Proceedings of the 2003 IEEE International Symposium on Electronics and the Environment. IEEE: Piscataway, New Jersey, pp. 187–192.

CHAPTER 4

The Law of Green IT

Eva Julia Lohse
Erlangen-Nürnberg University, Erlangen, Germany

General Remarks on Law and the Regulation of Environmental Behavior

Direct and Indirect Governance of "Green IT"

There exists at neither the national, European Union (EU) nor international level a single legislative act covering all aspects of "green IT" or with the aim to regulate green information technology (IT). One of the reasons is presumably that neither does there exist a defined area of regulation: The term *green IT* refers to various aspects of the protection of the environment, the most important principles being those of sustainability and conservation of natural resources. EU environmental law, however, has several legal instruments that directly or indirectly address different aspects of green IT:

(1) EU directives and regulations on energy efficiency that are relevant for servers, desktops, printers, and other devices and could lead to the development of more energy-efficient products and production processes.
(2) Law governing the production processes themselves (e.g., by prohibiting the use of certain materials, compounds or substances or regulating the sustainable use of natural resources in the production process).
(3) Legislation encouraging the development of technology for environmental management (e.g., IT equipment that detects energy loss) or the application of computers for home-office and telephone conferences, which reduces air pollution, traffic, and so on.
(4) Finally, different legal instruments, such as ecolabeling, certification and information in addition to public procurement and competition laws used to change the attitude of computer producers, retailers, consumers, and users.

Even though the EU follows a life cycle approach for its integrated product policy (IPP), most pertinent legislation for green IT deals with aspects of energy efficiency or energy saving. These aspects will be primarily addressed in this chapter.

Norm Addressees and Efficient Regulation

As in other areas of environmental law, a combination of legal instruments and of recurring nonlegal, mainly economic, instruments is used to govern the behavior of different stakeholders.

Producers and the Production Process

The most promising way for a sustainable product policy is to control producers, either directly by prohibiting certain forms of production, by demanding energy- or resource-efficient production (for the latter, see Reimer and Tölle, 2013, p. 596), and by setting product criteria, or indirectly by means of competition between private actors (Huppertz and Nusser, 2009, p. 629), by the use of ecocriteria in public procurement. Regulation 765/2008/EC (stating the requirements for accreditation and market surveillance relating to the marketing of products, OJ 2008 218/30) requires member states to guarantee an effective and efficient control of the entire market chain for products covered by Directive 2009/125/EC (establishing a framework for setting of ecodesign requirements for energy-related products, OJ 2009L 285/10; hereafter Ecodesign Directive), which means that public authorities may control not only producers and importers but also retailers (Dietrich and Akkerman, 2013, p. 276).

Different legal instruments affecting the behavior of producers regarding sustainability, energy efficiency, and conservation of natural resources use different definitions of *producer* (Huppertz and Nusser, 2009, p. 627) and might be implemented differently under the respective national laws of EU member states. Therefore, uniform requirements for green IT producers are hard to determine.

Consumers and Market-Related Mechanisms

A more market-related way to encourage the development of green products is to exert influence on consumers: They are free to choose rationally between different products. When provided sufficient information (e.g., on the energy efficiency of an IT device or on a label describing the impact of a product), consumers are believed to buy the product with the least impact on the environment, at least if this low impact is also reflected in an equal or lower price or a better price-value relationship.

Law does not govern these choices directly but enables and sometimes forces producers and retailers to supply sufficient information and to resort to certificates or labels. Duties of certification, information, or indication are on the one hand considered to be a main element within the regulation of green IT (Huppertz and Nusser, 2009, p. 627); on the other hand, labeling is claimed to be ineffective when modest savings for the consumer oppose high administrative burdens for producers and retailers (Sutherland, 2009, p. 180).

Nonstate Standard Setting and Voluntary Instruments (Codes of Conduct)

In recent times, the state increasingly has resorted to nonstate or cooperative legislation or to non-normative/voluntary instruments. Soft law instead of normative requirements draws on mechanisms of recommending best practices (e.g., for an improvement of energy-saving measures).[1] From there, common standards might evolve bottom-up instead of top-down. The efficiency of such instruments is doubtful when environmental law faces the problem that production without regard to environmental standards is less expensive. This means that consumers do not need to bear extra production costs that render those products cheaper.

Nonlegal standards and voluntary instruments, such as codes of conduct, may work in certain situations in environmental law (Friedrichs and Lohse, 2008, pp. 81-82), mostly if a dense normative order is lacking. Yet, they are often replaced by state legislation to enhance their functioning. In the field of green IT, the EU itself—not private actors—had adopted codes of conduct on external power supplies, uninterruptable power supplies, broadband equipment, and data center energy efficiency. Because such codes were willingly used by only a small number of producers (Sutherland, 2009, p. 180), they have been replaced or complemented by binding EU Commission regulations.

Another way to encourage interaction with nonstate actors is to leave a margin in state legislation for private actors to set standards or technical norms. The Ecodesign Directive in its Article 17, for example, refers to mechanisms of "self-regulation." In this situation, as with DI-Norms in Germany, manufacturers could establish standards for green IT. Those instruments mainly work on a voluntary basis. To become binding for all manufacturers, legal provisions that delegate standard setting to private institutions are needed because they might have a better understanding of suitable standards than legislative organizations. In this case, the standards will be applied uniformly for manufacturers to obtain mandatory certificates or permits or be admitted to a national market.

Additional instruments, also used on the EU level, are ecoaudit and ecomanagement methods. They are voluntary regulation methods that rely on marketing mechanisms and the guarantee of better procedures within industrial entities. Compliance with existing environmental legislation as well as certain ecofriendly procedures can be audited and certified. In the same way, ecolabels can be used voluntarily by manufacturers to increase consumer interest in their products and to guarantee respect for certain ecocriteria. In most cases, there are, however, no legal requirements for the sale or production of a product.

---

[1] Sources: http://www.green-it-wegweiser.de/Green-IT/Navigation/Der-Weg-zu-Green-IT/standardisierung,did=458114.html and http://iet.jrc.ec.europa.eu/energyefficiency/ict-codes-conduct/data-centres-energy-efficiency.

The Mechanisms of EU and National Law—Basics

EU law has developed into the most important legal instrument to regulate environmentally friendly behavior of individual consumers, producers, retailers, and state authorities of the member states. It can be divided into primary law (treaties), secondary law (directives and regulations under Article 288, Treaty on the Functioning of the European Union [TFEU]), and tertiary law (directives and regulations of the EU Commission under Article 290 (1) TFEU), each with a specific impact on the national legal systems. To understand the legal duties imposed on the respective countries, it is important to know about the basic functioning mechanisms and the relationship between national and union law.

Primary Law: Principles of Supremacy and Market Freedoms

Member states must agree to primary law in international treaties, which are mainly the Treaty on the European Union (TEU) and the TFEU. With relevance to our discussion, it is important to note that primary law contains principles on policies and goals regarding the environment (Article 191 TFEU) and regarding energy (Article 194 TFEU) and defines some general duties of the EU and its member states regarding the four market freedoms and the establishment of the single market.

Briefly, Articles 26-66 and 18-21 TFEU require the member states of the EU to guarantee the free movement of all production factors within the EU and to refrain from any measures that directly or indirectly, actually or potentially, hinder the free movement of those factors (so-called *Dassonville-Gebhart* formula, Case 8/74 *Procureur du Roi vs Benoît and Gustave Dassonville* [1974] ECR 837, para 5). Member states may maintain national measures to protect the environment, public health, or other elements for the public good; however, the measures must be proportionate and nondiscriminatory to products and services from other member states (so-called *Rewe* criteria, Case 120/78 *Rewe-Zentral AG vs. Bundesmonopolverwaltung für Branntwein* [1979] ECR 649, para 8). These provisions and their interpretation by the European Court of Justice (ECJ) result in a "negative harmonization" of national laws because national legal requirements not justifiable under the *Rewe* criteria become inapplicable in the EU context (Case 106/77 *Simmenthal* [1978] ECR 629), a process which contributes to a pool of harmonized national laws (Lohse, 2012, p. 287).

If a product has entered the market legally in one of the member states in conformity with national (or prevailing EU) requirements, it cannot be treated as illegal or be banned from sale in any other member state according to the *principle of mutual recognition*. Therefore, if a member state has stricter requirements for the production and trade of IT products, even in areas not covered by EU Law, it cannot prevent the sale of products that do not meet its national requirements within its jurisdiction as long as the have legally entered the market elsewhere in the EU.

Primary law is directly applicable for its addressees (mainly all organs of member states with legislative, administrative, or jurisdictional powers, even on the regional or local level, and the EU, hardly individuals) and prevails over the national laws of the member states (Case 6/64 *Costa vs E.N.E.L.* [1964] ECR 585, para 7). All state authorities must apply national law in accordance with EU law and set aside national law contrary to it on their own account (Case 103/88 *Fratelli Costanzo* [1989] ECR 1839; see Jans and Vedder, 2012, p. 212). If national law contradicts primary law, it becomes inapplicable within the scope of EU law but remains in force. It can thus regulate the legal relations with third-party countries (i.e., products or services for export outside the EU).

Secondary Law: Directives and Regulations and Direct Effect

Legal Bases in EU Environmental Law

Secondary law is issued by the EU Council and the European Parliament based on a transfer of legislative powers within the treaties. The most important legal bases in our context are Article 114 TFEU for the regulation of the functioning of the single market, Article 192 TFEU (environmental protection), and Article 194 (2) TFEU (energy).

Article 114 TFEU complements the mechanism of the market freedoms and the principle of mutual recognition: when negative harmonization is not sufficient because it can address only single obstacles to the common market, the EU can harmonize national laws by a secondary instrument, thereby setting at least a common minimum standard within the EU.

In environmental law, there have been claims that harmonization leads to a race to the bottom because the member state with the lowest environmental standard may determine the harmonized laws or—even worse—cannot merely maintain its low standards but encourage cheaper and less environmental protective production within its boarders under the use of the market freedoms. To aim for a high level of protection, the TFEU in Article 114 (5), (6) and Article 193 provides mechanisms for the member states to uphold higher national levels of protection despite harmonization. Those national standards may not be considered to contravene EU law if they are consistent with the market freedoms and do not hinder the functioning of the internal market. This means that in some circumstances, higher national environmental standards exist despite harmonization, which must be respected by their addressees.

Regulations

Regulations are directly applicable law for the respective addressees, who are often individuals, especially producers or retailers. They can impose direct duties (e.g., for the use of hazardous or rare materials in the production process, for the gain of certificates to be able to sell the product, or for the labeling of products). One relevant example is Regulation 1907/2006/EC concerning the Registration, Evaluation, Authorization, and Restriction of Chemicals (REACH) on the use of hazardous chemicals. However, regulations are used less frequently because they leave no or a quite small margin concerning the implementation by the member states and are therefore a strong instrument of harmonization.

Directives

Directives must be transformed into national law to become directly applicable within the member states. According to Article 288 TFEU, the EU sets only the goal, whereas the member states determine form and means of implementation. This results in a variety of national measures that are not identical but must realize the goals set in a directive and must not hinder those goals to be reached. If the meaning of a provision of a directive is unclear or a national court is unsure how to interpret and apply specific law, the CJEU may decide how to interpret the directive (not the national law!), Article 267 TFEU.

The directive states a duty of transposition only for the member states, but no direct duties for individuals (Case C-91/92 *Faccini Dori* [1994] ECR I-3325; C-148/78 *Ratti* [1979] ECR 1629). It can, however, grant directly applicable rights against the member state (not against individuals) when it fails to correctly implement the directive and if the provision of the directive is clear, defined, and does not need further implementation by national law.

A lack of correct implementation can also lead to state liability as when an individual suffers damage resulting from the non-implementation of the directive. Whereas direct effect is not very relevant for the case of green IT because most directives in this field require the establishment of duties in national law for individuals, not the state and there is no horizontal direct effect (between individuals) of directives (Case C-91/92 *Faccini Dori* [1994] ECR I-3325; Joined Cases C-397/01 to C-403/01 *Pfeiffer et al.* [2004] ECR I-8835; Joined Cases 372-374/85 *Traen* [1987] ECR 2141), there may be constellations when a producer or retailer suffers damages from an incorrectly transferred directive, especially if the placing on the market of a product is concerned.

Comitology and Implementing Measures: Delegated Legislation by the Commission

According to Article 290 TFEU, the EU Commission (but not regularly Council and Parliament) can pass legislative acts (directives and regulations) outside the regular procedure in Article 289 (1) TFEU if a directive or regulation delegates the power to the Commission. This procedure had been called *comitology* before the Treaty of Lisbon introduced Article 290, 291 TFEU but now is addressed in two different sources: Article 290 TFEU concerning delegated legislation by the Commission and Article 291 TFEU concerning specific Commission measures needed for uniform implementation of legally binding acts of the EU by the member states (Jans and Vedder, 2012, pp. 84-85).

Both forms enable the legislature to set only a framework, whereas the executive power may define more detailed requirements according to their more profound knowledge of the subject and more familiarity with the specific problems in the area of regulation. The process is also more flexible and faster, which seems especially important in those areas subject to technical changes and new findings in the related technologies.

This type of specific measure is foreseen in Article 15ff. Ecodesign Directive. In general, measures can be passed in the form of regulations and therefore contain direct duties for producers, retailers or even consumers or as directives, requiring further implementation by the member states. Art. 18, 19 Ecodesign Directive fixes a specific procedure for passing those regulations, for example by requiring the involvement of a consultation forum, which includes representatives from all interested parties concerned with the product—industry and environmental protection groups alike. This offers possibilities of public participation in EU decision-making procedures to enhance acceptance but also guarantees that available best practices are discussed and find their way into the harmonizing legal instruments, thus establishing a high standard of environmental protection throughout the EU (see also Article 191 (2) 1 TFEU: high level of protection).

Those mechanisms of public participation for creating legal instruments should be made use of by experts in the field to shape harmonized EU standards on green IT (Article 191 (3) TFEU). For member states and individuals, the implementation of measures by the Commission sets the same duties as regulations or directives. An implemented regulation therefore defines directly applicable product criteria for producers and retailers and should be more commonly used than a directive, which requires transformation by the member states and cannot therefore establish uniform criteria.

Sustainability in EU Law

EU environmental law is governed by several general principles, which not only direct the Commission, Council, and Parliament in their lawmaking but also through the application of EU law by the national authorities and must be taken into account by the member states in the process of implementation. For our purpose, the main focus is the principle of sustainability or sustainable development, respectively.

Article 3 TEU: Principle of Sustainable Development and Green IT

In general, Article 3 (3) 2 TEU states the principle of sustainable development and the aim to reach a high standard of environmental protection within the EU. According to Article 11 TFEU, sustainability is to be furthered by respecting environmental protection in all EU policies and measures. Therefore, the protection of the environment is the main instrument to guarantee sustainability. Article 191 TFEU defines several principles related to sustainable development, such as requiring a polluter to pay, and those for prevention and precaution (Epiney, 2013, pp. 156, 157; Appel, 2014, para 42). The EU Commission has established the Europe 2020 Strategy and the Sustainable Development Strategy to implement sustainable development in their policies.

The principle of sustainability reflects care for the planet and for future generations, an idea that becomes especially important in the context of green IT. Sustainable behavior can be summarized as efficiency, sufficiency, and consistency in the use of natural resources (Herrmann et al., 2012, p. 525), objectives that should also govern technical strategies for green IT. The prudent use of rare and nonrenewable materials, which can be produced or obtained only under polluting, nonsustainable circumstances, the recycling, or the reuse of parts of IT equipment, and, of course, the promotion of energy-efficient or energy-saving production and operation of IT equipment all contribute to a prudent and rational use of natural resources (Article 191 (1) TFEU) and to sustainable development of informational technologies. *Negajoules* (i.e., the amount of energy not used in production or operation) are considered the best opportunity for sustainability in energy policies (Schomerus, 2009, p. 418) and are also at the core of EU policies directed directly or indirectly at green IT in the context of IPPs. Also, on the national level, green IT and energy efficiency are closely related, and national programs are shaped to promote e-energy and green responsibility (Lejeune, 2008, p. R112).

Integrated Product Policy (IPP)

For the purpose of sustainable development, the EU Commission has defined five fundamental principles of IPP in a green paper (COM, 2001, 68 final) and a white paper (COM, 2003, 302 final). Those are (1) life cycle thinking, (2) working with the market, (3) stakeholder involvement, (4) continuous improvement, and (5) a variety of policy instruments.

Life cycle thinking should especially be enhanced by those promoting green IT. Life cycle thinking refers to a product policy that takes into account all steps from the exploitation of substances to the actual production process, the sale, the use, and the final disposal or recycling of this product. All steps should be as sustainable as possible, and all stakeholders should cooperate to make such a life cycle approach possible. The ecolabel regulation (Regulation 66/2010/EC of November 2009 on the ecolabel, OJ 2010L 27/1) in Consideration (5) and the Ecodesign Directive in Considerations (7) and (13) refer to the life cycle approach. In most cases, IPPs are understood as being related to the energy efficiency of products and to the strategies of the EU to reduce their CO_2 emissions.

Birth of a "Law for the Conservation of Natural Resources"

Even though it is not entirely clear what the term *natural resources* means (Herrmann et al., 2012, p. 524; Reimer and Tölle, 2013, p. 591), it is commonly accepted (in respect to Principle 2 of the Declaration of the United Nations Conference on the Human Environment [ILM 1972,

p. 1416]) to include not only water, land, flora, and fauna but also oil, gas, and chemical substances (Jans and Vedder, 2012, p. 37; Herrmann et al., 2012, p. 524); thus, it includes resources that can easily be exhausted if not prudently removed and used (Appel, 2014, para 42).

From the vague and general principle of sustainability, some conclude that a law for the conservation of natural resources combining the aim of IPP, energy efficiency, and prudent use of nonrenewable resources should be passed (see, inter alia, Herrmann et al., 2012; Reimer and Tölle, 2013, p. 590). At the same time, economic considerations such as the guarantee of a sufficient, long-term supply of commodities for the industrial sector may lead to an enhanced conservation of resources (Reimer and Tölle, 2013, p. 594), so one might question whether there is actually a predominant link to environmental law. Thus far, there is no single legal instrument—neither on national nor on international levels—for the conservation of resources in general, but Article 191 (1) TFEU demands an EU environmental policy governed by consideration of the prudent and rational use of natural resources. EU legislative acts should therefore reflect this approach.

The principle of sustainability understood as one for the conservation of natural resources is not specific to media or causes (passim Herrmann et al., 2012, p. 527), yet it relates to various fields of national and EU environmental law, such as waste and recycling legislation, mining law, water law, and energy law, and legislation regulating the production and marketing of specific products. Those areas are also relevant to the production and sale of IT products; therefore, the legislation summarized under "protection of resources" is considered green IT legislation. At the same time, it is doubtful that there is general legislation for the protection of resources because a common legal definition of *resources* already is lacking as are general principles governing this field of law. Rather, the protection of resources in a wide sense remains the main issue of the principle of sustainability and must be the guideline and objective for more specific legislation on the various natural resources to be used prudently and rationally.

Specific European Legal Instruments Relevant to Green Computing and Their Implementation

In law, green IT is often reduced to "technologies for energy efficiency" (Huppertz and Nusser, 2009, p. 625), mirroring the focus of IT sciences on saving energy in production and application process and using IT to help reduce energy consumption in other industrial or third sector processes. So far, the EU has passed little legislation on the energy efficiency of the production of IT products but offers several legal instruments addressing the design of green IT products and their sales. The EU follows a life cycle approach (i.e., comprehensive regulation of manufacture, use, and disposal [see also Jans and Vedder, 2012, p. 378]). The problem with the existing variety of legal instruments is that the instruments use different definitions (e.g., for *producer*) or result in one product falling under different provisions with partly ambiguous requirements (Nusser, 2010, p. 135). To solve this problem, a comprehensive legal instrument is needed, but the development seems to lead in the opposite direction.

Green IT legislation needs to fix at what point in the life cycle (manufacture, entering the market, sale, consummation/use) what type of provisions and standards and what type of regulation (voluntary instruments, no regulation in favor of market mechanisms, or top-down legislation) prove most efficient. It also needs to decide whether producers, retailers, consumers, or public authorities should be addressed and by which mechanisms compliance with the rules can be guaranteed. The idea should be to stimulate the demand for more sustainable goods and production technologies (EU Focus, 2008, p. 15). In the following section, some legal instruments relevant to green IT law are presented.

Public Procurement: Environmental Standards as Criteria for Tenders

Public procurement law allows public authorities to demand that IT providers use energy-efficient IT products and to recur to green IT services, in order to guarantee that certain environmental standards are met. This simultaneously offers a market mechanism for the implementation of green IT standards in the EU, and provides a way for public authorities to show their commitment to environmentally friendly behavior. In 2008, the EU Commission issued an action plan recommending among other things that member states provide fiscal incentives and adopt common green procurement practices (EU Focus, 2008, p. 16). In return, the EU is to provide, for example, model tender specifications with the aim to create harmonized procurement practices and to prevent a rundown on environmental standards relating to energy efficiency and product specifications in IT.

Risk of Market Distortion, Discrimination, and Intratransparency

This change in policies is so far surprising because for a long time it had been claimed that EU law prevents public purchasers to use environmental criteria such as for the assessment of tender responses (Kunzlik, 2013, p. 173; Jasper and Seidel, 2009, p. 56). Those criteria were seen not to be tender specific and therefore not allowed under public procurement law, which allowed only for technical and economic criteria. The reason for this was that other than tender- or tenderer-specific criteria bear the risk of distortion of competition, of (hidden) discrimination by setting nonobjective or nonmeasurable criteria, and of instrumentalizing public procurement for political purposes (Dreher, 2007, paras 97 and 182; Fehling, 2011, paras 97, 134, and 136).

On the other hand, it had always been possible to use environmental criteria as minimum for a tender, which had to be met by the tenderer to enter the procedure at all. In this case, the good or service purchased is determined to be per se environmentally friendly, excluding other products from the tender (Fehling, 2011, paras 97 and 142). One example in green IT could be to purchase only printers with high-energy efficiency, to demand a service provider of photocopies and printing to use only recycled paper, or to adhere to a certain ecomanagement method. National procurement legislation can also set specific criteria for the environmental requirements for the tendered good.

Strategy Europe 2020 and Renewed Sustainable Development Strategy

Yet both the Strategy Europe 2020 (COM [2010] 2020 final) and the Renewed Sustainable Development Strategy (European Council [2006] 10917/06) declare green public procurement (GPP) to be a major strategy for environmental protection in the EU. The 2008 Communication on Green Public Procurement aims at establishing a voluntary framework for all member states while pursuing economic goals (see Kunzlik, 2013, p. 176). There were also plans to adopt mandatory environmental standards for different sectors.

The computer and other office equipment sector was the first to be governed by these mandatory procurement standards. Regulation 106/2008 of the European Council and Parliament (OJ 2008L 39/1) concerns the ecolabeling of office equipment and connects the Energy Star requirements, which have been developed in the United State for certifying energy efficiency of office equipment, to mandatory public procurement standards for IT services (Jasper and Seidel, 2009, p. 59; Söbbing, 2013, p. 2; Kunzlik, 2013, p. 179). Likewise, Article 9 of the *Energy-Consumption-Labeling-Directive* (Directive 2010/30/EU, OJ 2010 L 153/1) allows tendering only for those products with the highest performance level and the best energy efficiency class.

New Approach in ECJ Case Law

The ECJ has rendered three seminal decisions on the use of environmental assessment criteria (Case C-513/99 *Concordia Bus Finland* [2002] ECR I-7213; Case C-448/01 *Wienstrom* [2002] ECR I-14527; Case C-368/10 *Commission/The Netherlands* (*EKO and MAX HAVELAAR*)). Differing from minimum standards, they are only admissible if further conditions are fulfilled: (1) the criteria must be related to the matter of the tender, (2) the criteria must be quantifiable by objective methods, (3) the criteria must be explicitly communicated before the tender to be transparent, and (4) the criteria must not contravene EU primary law, especially the market freedoms and the principle of nondiscrimination by using criteria that cannot, or can only hardly or with higher expenses be fulfilled by tenderers from other member states. In *Commission/The Netherlands* (Case C-368/10, paras 62 and 94), the CJEU clarified that the adherence to specified ecolabels as assessment criteria is not necessarily admissible. Article 2 of *Directive 2004/18 on the Coordination of Procedures for the Award of Public Works Contracts, Public Supply Contracts, and Public Service Contracts* (OJ 2004 L 134/114) demands that all tenderers be treated equally and nondiscriminatorily with regard to technical requirements and that the criteria are transparent and accessible for all in the same way. This seems problematic if the tender refers to ecolabels issued by nongovernmental organizations because it might be unclear for tenderers, whether their products meet those standards and the distribution criteria for those labels might be random. Also, products might meet those criteria without having applied for the label referred to and are therefore excluded.

Use of Certificates and Ecolabels as Assessment Criteria

The *Max Havelaar* case is important for public procurement in the field of green IT because methods such as Energy Star, EU ecolabel, or mandatory certification under the Ecodesign Directive provide valuable yardsticks for green IT products. As long as they are mandatory requirements for certification, there should not be a problem with transparency or discrimination because all products that fall under the scope of the pertinent directives need to conform to those criteria. The criteria are clearly determined in a legal instrument available to all manufacturers, service providers, and retailers. Therefore, a referral to those mandatory certificates is consistent with the requirement of transparency. Likewise, the Ecodesign Directive and Commission Regulation 1275/2008, repealed by Regulation 801/2013 and others, on energy efficiency of office equipment (Energy Star regulation) are nondiscriminatory because they apply in the same manner to all products marketed within the EU and constitute harmonized product standards throughout it. In practice, however, also nonmandatory standards, such as the German Blue Angel, the voluntary EU-ecolabel, and certificates by TCO (Tjänstemännens Centralorganisation, i.e., the Swedish Confederation of Professional Employees) and EPEAT (Electronic Product Environmental Assessment Tool, a project of the Green Electronics Council) form part of tenders (Söbbing, 2013, p. 2). It is to be seen to what extent this conforms with EU procurement law. In any case, authorities will have to watch to allow tenderers to prove conformity with the criteria "transported" by a certificate or label by other means than holding the certificate or label (see Case C-368/10, paras 94 and 96).

This way to promote green IT seems to be an especially good strategy for local authorities. By not only setting minimum standards for energy efficiency of new IT equipment (absolute protection by referring to certificates, EU-labels, or other EU environmental product standards) but also using energy efficiency or other sustainable product standards (such as the percentage of IT equipment being compatible with different versions of software and software updates) as assessment criteria, local or public authorities can encourage service providers to aim for the best (in relation to green IT) services and products (see also Jasper and Seidel, 2009, pp. 57-58).

Implementation in German Law

In Germany, the changes in EU case law and the economic approach to sustainable development have been reflected by a change in national public procurement law. Whereas formerly para 97 (4) GWB did not explicitly allow for environmental criteria to be included, it now explicitly lists environmental criteria to be taken into account. It therefore implements EU demands for GPP and might help to overcome shortcomings in this field that were listed in 2006; by then, only 7 of then 24 member states practiced GPP to a significant extent (EU Focus, 2008, p. 17), because of the uncertainty of whether the ecologic criteria are legal under EU law.

Ecodesign Directive: Regulation of Manufacturing and Encouraging Green Innovation

Based on the Green book by the Commission on IPP, the Ecodesign Directive has become the main instrument in the area of green IT.

Scope and Relevance for IT Equipment and Devices

The Directive is not limited to IT products. Because ecodesign as such aims at the "reduction in the environmental effects of products, in particular of energy consumption over the life cycle of a product" (Sutherland, 2009, p. 179), and because IT products consume energy, most tertiary legislative acts for the implementing the Ecodesign Directive directly or indirectly concern IT or IT-related products, though. The Directive provides a framework for compulsory minimum requirements as well as voluntary benchmarks (EU Focus, 2008, p. 15)—the first being set by the EU Commission and the second by manufacturers as self-regulatory instruments according to Article 17 of the Directive. So far, however, no self-regulatory instruments have been acknowledged by the Commission to equal its measures.

The 2009 Ecodesign Directive was amended in 2012 to include energy-related-products (i.e., those that do not themselves consume energy but have an impact on energy consumption during their use). The amendment seems not to be of great importance for the field of green IT because the 2009 directive had already covered most IT products.

Energy Efficiency or General Resource Efficiency and Sufficiency

From its outset, the Directive has not been limited to energy-efficient product design, but its focus lies in this area (Nusser, 2010, p. 130; Schomerus and Spengler, 2010, p. 54). It offers a margin of implementation for the conservation of all natural resources, which has so far seldom been used by member states and the Commission. This results in a lack of harmonized product requirements regarding the conservation of resources (Herrmann et al., 2012, p. 530). One of the reasons could be that EU environmental law already includes quite a few product requirements (e.g., for the indication of energy consumption [Directive 92/75/EEC, OJ 1992 L 297/16] or the use of hazardous substances in electronic devices [Directive 2002/95/EC, OJ 2003 L 37/19]). An inclusion of the same instruments into the implementing measures for the Eco-Design-Directive bears the risk of contradicting legislation. At the same time, however, a comprehensive regulatory instrument would enhance transparency and user friendliness for manufacturers. Several stakeholders had proposed the inclusion of requirements on disposal, recycling, conservation of natural resources, and hazardous substances, such as mercury, during the consultation for measures regarding computers (Schomerus and Spengler, 2010, p. 59). These proposals were not considered in the final draft by the Commission because it deemed the RoHS-Directive (2001/65/EC, OJ 2011 L 174/88) and the Commission regulations on external power supplies and on standby/off-mode use to be sufficient to improve the environmental performance of computers and servers (EU Commission Staff Working Document, Executive Summary, SWD [2013] 218).[2] Thus, a valuable chance for a comprehensive regulation on IT ecodesign regarding the life cycle approach was missed. Simultaneously, the Directive has a strong relation to the regulation

[2] The impact assessments and executive summaries are available at: http://www.ksb.com/linkableblob/fluidfuture-en/1997284-531501/data/Eco-design-legislation-data.pdf.

of the single market because it is the framework for harmonized product design requirements by which all energy-related products gain access to the single market.

Responsibilities for Compliance with Ecodesign Requirements and Supervision

The Directive demands that the EU Commission develops tertiary measures (Article 291 (2) TFEU) for the ecodesign of products in collaboration with a consultation forum and a committee (Articles 18 and 19 Ecodesign Directive) if the product is covered by the criteria in Article 15 (2) of the Directive. When a product falls under the scope of one of those measures, national law is to require manufactures to ensure that the product conforms to the mandated measure before placing it on the market. The responsibility lies therefore with the manufacturer, who is defined in Article 2, no. 6 Ecodesign Directive as the "person who manufactures the product"; he or she is obliged to monitor whether standards are met. The same applies to the importer that places the product on the market in a member state other than its state of manufacture. According to Article 14 manufacturers and importers also must make sure that the consumers are informed—in accordance with the respective measure—about their role in reducing the environmental impact of the product during its use.

Member state authorities are required to control manufacturers and must implement penalties in accordance with national law, which are effective, proportionate, and dissuasive (Article 20 Ecodesign Directive). A penalty might include the ban of the product from the market or its recall from the market (see, for example, the German para 7 EVPG[3]). In this case, the EU Commission is informed, meaning that the product cannot be sold anywhere in the EU. Products may be controlled by national authorities throughout the entire supply chain, not only during manufacture or import; this conforms to EU Regulation 765/2008 for efficient control (Dietrich, 2012, p. 602; Dietrich and Akkerman, 2013, p. 276).

However, the main responsibility for the adherence to ecodesign criteria lies with the manufacturer or other persons, such as importers, who place the product on the market for the first time. The German regulation in para 4 EVPG is somewhat stricter than its European model because the retailer also must ensure that all products offered conform with the ecodesign requirements, carry a CE mark, and are accompanied by a declaration of conformity whereas Articles 3-5 Ecodesign Directive only name the responsibilities of the manufacturers, importers, and national authorities.

Implementing Measures and Harmonized Product Design Requirements

The measures implemented by the Commission set the criteria for specific products with environmental impact following the procedure in Annex II for an analysis of the technical, ecological, and economic impact to establish how production procedures can be improved and

[3] Energieverbrauchsrelevante-Produkte-Gesetz, BGBl. I 2011, 2224.

products can be rendered "greener." Therefore, it addresses not only the manufacturing process as such but also planning and design (Lustermann, 2007, p. 895). This is meant to set legal incentives for technological innovations to prevent products not meeting the requirements from being placed on the market.

These measures can be directly applicable for manufacturers and importers in the member states if issued as a regulation; otherwise, they need to be transposed into national law by the member states. The aim of this delegation of legislative powers is to gain flexibility and to enable quick reactions to changes with product standards. Implementing measures relevant to green IT include *Commission Regulation (EU) 1275/2008, amended by Commission Regulation 801/2013 on standby, off-mode electric power consumption of electrical and electronic household and office equipment (OJ 2013L 225/1), Commission Regulation (EU) 278/2009 on no-load condition of electric power consumption and external power supplies (OJ 2009L 93/3), and Commission Regulation (EU) 617/2013 of (June 26, 2013) implementing Directive 2009/125/EC of the European Parliament and of the Council with regard to ecodesign requirements for computers and computer servers (OJ 2014 C 110/108).*

Guarantee of Harmonized Interpretation and Application

In order to guarantee a uniform application and interpretation of the ecodesign requirements and a harmonized control of compliance with requirements under respective national laws, Article 12 of the Directive demands that member states take appropriate measures to ensure cooperation between national authorities within in the EU. There should also be exchange of information between the authorities and the EU Commission. Most of this is implemented by the Ecodesign Administrative Cooperation for Market Surveillance group (EDD-ADCO) established in 2009 for this purpose. The interpretation of requirements of the Commission's implementing measures by the EDD-ADCO are not binding for national authorities or the Commission (a binding interpretation can be provided only by the CJEU in a preliminary reference proceeding initiated by a national court [Article 267 TFEU]), but lacking other common interpretations, the measures are usually referred to by national authorities. The measures are published and therefore provide a good starting point for their harmonized application (see also Dietrich and Akkerman, 2013, p. 278).[4] The European Ecodesign Compliance Project (Ecopliant) also provides best practices in ecodesign and can therefore serve as a reference for manufacturers and authorities.[5]

---

[4] See further http://ec.europa.eu/transparency/regexpert/index.cfm?do=groupDetail.groupDetail&groupID=2601&Lang=EN.

[5] http://www.ecopliant.eu/.

Ecopliant is especially useful for implementing measures whose interpretation regarding their scope (which product falls under the ecodesign requirements) and the specific criteria or requirements to be met by the product are questioned. One of the most problematic measures in this regard, which also is relevant to green IT, is the regulation on standby and off-mode use (Nusser, 2010, p. 134): It concerns all products named in Annex I. They apply primarily to IT products used in homes (defined by Article 2 (8)) in an area within 10 m of radio or television receivers. It remains unclear whether the provision applies to IT equipment used in offices within the reach of a receiver and what percentage of use is meant by "primarily." The products must also rely on energy from public networks; this poses the question as to whether notebooks using an accumulator or batteries, for example, are included. Nusser (2010, p. 134) finds that they are not. This problem might have been solved by the implementing measure on computers and computer servers under Commission Regulation (EU) 617/2013 that do not fall under the scope of the standby regulation; according to Consideration (12), computers have specific ecodesign requirements and those of Regulation 617/2013 are better suited to their specific demands. Notebooks are explicitly listed as computers, which should clarify the issue.

Energy Labeling Directive and Voluntary Ecolabeling

A completely different approach within the IPP and life cycle approaches is taken by using ecolabeling mechanisms.

Voluntary EU Ecolabel and Energy Star Label

Regarding green IT, there are two labeling methods, first *Regulation 106/2008/EC on a Community energy-efficiency labelling programme for office equipment* (OJ 2008L39/1), stemming from an agreement between the United States and the European Union on the use of the US Energy Star program, and second the EU-Ecolabel European Flower (Regulation 66/2010/EC, OJ 2010L 27/1), which is more broadly designed for products with lower/reduced impact on the environment in production, usage and/or disposal as compared to other products in the same (defined) product group ("eco-friendliness in relation to other products"; see Epiney, 2013, p. 327; Jans and Vedder, 2012, p. 378). Both methods are voluntary, meaning that products that meet the product-specific requirements can be entered into the program (also for marketing reasons), but that there might also be products on the market meeting the requirements, but not featuring a label.

National ecolabels can coexist as long as they set higher standards than the EU-labels, however the Commission encourages the use of the EU-label to avoid confusion of consumers and to reach harmonized product-related measures for the functioning of the internal market (see Considerations (6) and (15) of the Ecolabel regulation). One of the best-known national ecolabels is the German Blue Angel, after which the EU labeling method was modeled.

Commission Measures on the Requirements

The regulations on the ecolabel provide the framework for EU Commission decisions on specific requirements for product groups. The idea is that only products with the lowest impact on the environment during their life cycle can be awarded an ecolabel (EU Focus, 2008, p. 18). The following product specifications are of interest for the regulation of green IT:

(1) The Energy Star regulation among others concerns desktop-computers, notebooks, servers, displays, and monitors for the use in offices, whereby the specifications must be developed together with the US Environmental Protection Agency, which has initially set up the Energy Star Program (see *Council Decision 2013/107/EU on the signing and conclusion of the Agreement between the Government of the United States and the European Union on the coordination of energy efficiency labelling programmes for office equipments*, OJ 2013L 63/5). The Energy Star requirements are limited to energy efficiency and to office equipment, whereas the ecolabel establishes general ecological criteria for product groups.

(2) Under the ecolabel regulation, the EU Commission has so far set criteria for notebooks (2011/330/EU), desktop-computers (2011/337/EU), photocopy paper (2011/332/EU) and imaging devices (2013/806/EU), which also include printers and photocopy machines (Article 1 (1)). The criteria covered are the use of hazardous materials, recyclability, packaging, emission of noise and particles, and others as well as energy efficiency.

Critical Assessment of Functioning Mechanisms of Voluntary Labels

Ecolabeling is both trade and market related and is supposed to raise environmental awareness of consumers, retailers and manufacturers (Jans and Vedder, 2012, p. 378, whereas the ECJ emphasizes the single-market-aspect of product-related criteria in Case C-281/01 *Commission vs. Council* [2002] ECR I-2049, para 41; see also Gundel, 2013, para 84). Consumers are supposed to take a rational choice by buying products bearing a label, and manufacturers are supposed to aim for meeting the ecolabel criteria for better marketing and thus to develop ecofriendly products.

It has been doubted on various grounds that these mechanisms will prove effective: labeling was considered ineffective regarding energy efficiency of ICT equipment, as savings for consumers, who buy more energy-efficient products are modest and labeling is burdensome for manufacturers (Sutherland, 2009, p. 180). This was seen as one of the main reasons for the EU to adopt the compulsory Ecodesign Directive.

It is also said that setting "relative" criteria within a product group does not necessarily award the best possible ecofriendly products, but those which meet the set criteria at a certain point of time (Epiney, 2013, pp. 330-331). This is true insofar as revision of criteria might take a long time and thus an incentive for developing even more ecofriendly products is lacking.

However, the same is true about mandatory ecodesign requirements and is thus not a specific problem of voluntary labeling methods.

Finally, one can rightly criticize that there are several competing labeling methods (EU/national as well as within the EU). This can lead to confusion with the consumers and might weaken the effectiveness of these programs (Schomerus, 2009, p. 421). Therefore, some authors prefer the so-called top-runner approach, which makes the most energy-efficient product the benchmark for all products. This benchmark becomes mandatory after a certain time lapse (Schomerus, 2009, p. 420), therefore the development of green products does not rely on rational consumer choices and decisions by the manufacturers. This might prove to be an efficient way for some products, but it is—different from the labeling methods—a strong interference with the decisions taken by manufacturers and consumers.

The voluntary labeling methods can also work efficiently via public procurement (see section "Public Procurement: Environmental Standards as Criteria for Tenders" and Söbbing, 2013, p. 2), when those labels are used as minimum requirements or assessment criteria in tenders.

Danger of "Label Shopping" and Incoherent Requirements

Ecolabels are awarded by the respective national authorities through contracts with the applicants for those products manufactured in the member state, first marketed there, or imported from a third country into the member state. This avoids "label shopping" if the label has been refused by a national authority. Yet, at the same time, there is the risk of diversification in the awarding of labels as the national authorities might interpret the minimum EU requirements differently and it is not always transparent for the consumer, in which member state the label has been awarded.

A uniform interpretation has become even more important since the Ecodesign Directive has entered into force: according to Article 9 (3) and (4) Ecodesign Directive products bearing the EU ecolabel are considered to be in conformity with the requirements in the respective measure under the Ecodesign Directive, if the ecolabel requirements parallel these requirements. Therefore, a manufacturer only needs to prove conformity once, making it easier and more cost-efficient for manufacturers. However, the ecolabel must not become a mean to evade harmoniously and stricter interpreted requirements under the Ecodesign Directive. The EU aims at providing harmonized standards and test methods by setting the Ecodesign and the ecolabel requirements simultaneously and in accordance to each other (EU Focus, 2008, p. 18). If this will in fact lead to a harmonious interpretation, remains to be seen in the future.

Mandatory Indication of Energy Consumption by Retailers

There also exists a mandatory certification method for retailers to use following *Directive 2010/30/EU regarding the labeling and standard product information of the consumption of energy and other resources by energy-related products* (OJ 2010L 153/1). It is now linked to the

Ecodesign Directive (see Consideration (2)) and therefore is not restricted to household appliances and energy-consuming products as its predecessor Directive 92/75/EC was. According to Article 1, no. 2, all firsthand products with a direct or indirect impact on energy consumption are included—all computers, servers, notebooks, monitors, and other electronic devices using IT. The directive is thus highly relevant for retailers of IT equipment and products.

The directive is intended to function the same way as voluntary labeling methods by informing the consumer on the respective energy consumption of the devices and products on sale and by encouraging a reasonable choice for the most efficient product. As with the Energy Star label, it is doubtful that low cost efficiency alone will lead consumers to choose ecofriendlier IT devices, but raising awareness for saving energy might still affect greener consumer decisions regarding IT products (e.g., Nusser, 2010, p. 136).

The Directive considers it important that all products bear a label to avoid consumer confusion and to enable a comparison between the products on sale to be made; therefore, it introduces a mandatory method (Consideration (12)). The Directive specifies the content of the information given: it includes providing standard product information on the consumption of energy and other technical information in a microfiche and using a system of energy efficiency classes (A to G). The supplier placing the product on the market is responsible that the product conforms to these requirements; member states' authorities must be entitled to perform efficient controls on the matter.

Restriction of the Use of Hazardous Substances and Conservation of Natural Resources

Directive 2002/95/EC on the restricted use of certain hazardous substances in electric and electronic equipment (OJ 2003L 37/19) aims to prohibit certain substances dangerous to human or animal health or the environment. Its objective is not so much to conserve rare resources but—a rather classic aim of environmental law—to protect health and nature. Directive 2011/65/EU (OJ 2011L 174/88) revised it to reflect new scientific findings on hazardous substances; the list in Annex II containing the prohibited substances must in future be adapted to new scientific findings on a regular basis.

Prohibiting the use of certain substances already in the manufacturing process helps to solve problems with the disposal of such equipment containing hazardous substances and with the pollution of the environment by the production process itself. Prohibition might also be helpful for limiting the impact that such substances might have during the use of an electronic device because the user might be exposed to the substances. The prohibition might also be an incentive for manufacturers and scientists to develop greener alternatives to existing products by replacing the hazardous substances.

The RoHS-Directive is not yet considered to compete with the Ecodesign Directive because the Commission does not set substance-related criteria in its implementing measures. Because the RoHS-Directive is, however, not very product/group specific, it is doubtful whether it is the best way to handle an IPP (Schomerus and Spengler, 2010, p. 59).

Recycling and Disposal

Part of implementing the life cycle approach is to pass legislation on the recycling and disposal of IT devices and equipment. Because of IT's rapid development, products become outdated very quickly and are often disposed of even though they are fully functional, creating a large amount of waste, some of which is hazardous. Legislation can address this problem in three ways: try to set incentives to create less waste (inter alia in the packaging), or harzardous create an obligation to recycle the products and restrict the use of substances that are nonrecyclable (see section "Restriction of the Use of Hazardous Substances and Conservation of Natural Resources").

Although EU legislation on waste disposal and recycling is only specific to IT products, where it sets standards for eco-labeling or prohibits the use of dangerous substances in electronic equipment, general legislation on recycling and waste also applies to IT devices. National law implements the Framework Directive on Waste Disposal 2008/98/EC (OJ 2009L 312/3), which sets the avoidance of waste as top priority followed by different categories of recycling and the disposal of products (Epiney, 2013, p. 553). For green IT, Directive 2012/19/EU on waste electrical and electronic equipment, amending directive 2002/96 (OJ 2002L 37/24), is relevant because it seeks to encourage the avoidance of waste by requiring manufacturers to pay for separate collection and disposal of waste IT equipment and devices.

Article 191 (2) TFEU establishes the polluter-pays principle, placing the responsibility for the cost of disposing of a product on the manufacturer (Article 8 Framework Directive), ensuring that the general public or the end user does not pay the costs. In practice, however, the disposal costs will be added to the price of the product (Article 14 Framework Directive) and therefore become an incentive for the consumer to buy the more cost-efficient (and in the logic of EU environmental law, greener) product (Konzak, 2013, para 23 KrWG, para 1). This also serves as an incentive for manufacturers to avoid high costs for recycling or waste disposal so that the product can remain on the market (Jans and Vedder, 2012, p. 483). In this respect, even legislation on waste disposal can be seen as part of the conservation of resources (Herrmann et al., 2012, p. 530) because manufactures might try to find ways to reuse IT equipment or to design it in a way that new software does not necessarily require new hardware. It might also encourage ecodesign of IT products (Jans and Vedder, 2012, p. 488).

Conclusions

The Integrated Product Policy persued by the EU has established a number of different "green" regulative schemes different green regulations relevant to manufacturers and retailers as well as IT consumers. Some of these green legal instruments might be more efficient in the future than others, and there will definitely be revisions of specific criteria and—hopefully—consolidation of different legislative instruments to provide for a more coherent law on the conservation of IT resources. Energy efficiency is currently at the heart of most of regulations, but the EU tries increasingly to include "older" instruments by taking a life cycle approach. It is up to scientists, manufacturers, consumers, and environmental protection groups to enhance the awareness of green IT legislation and to partake in the mechanisms of public participation foreseen by some of the instruments in order to create even 'greener' product requirements in IT.

References

Anonymous, 2008. EU promotes sustainable products and technologies. EU Focus 239, 15–19.

Appel, I., 2014. Europäisches und nationales Umweltverfassungsrecht. In: Koch, H.-J. (Ed.), Umweltrecht, fourth ed. Verlag Franz Vahlen, München, pp. 41–113 (para 2).

Dietrich, S., 2012. Das Energieverbrauchsrelevante-Produkte-Gesetz. NVwZ, 598–604.

Dietrich, S., Akkerman, F., 2013. EU-Ökodesign-Richtlinie—Implementierung—Umsetzung—Überwachung. ZUR, 274–278.

Dreher, M., 2007. GWB Allgemeine Grundsätze. In: Immenga, F., Mestmäcker, E.-J. (Eds.), Wettbewerbsrecht: GWB, fourth ed. C.H. Beck, München (para 97).

Epiney, A., 2013. Umweltrecht der Europäischen Union. Nomos, Baden-Baden.

Fehling, M., 2011. GWB Allgemeine Grundsätze. In: Pünder, H., Schellenberg, M. (Eds.), Vergaberecht. C.H. Beck, München (para 97).

Friedrichs, J., Lohse, E.J., 2008. Revisiting the junctures of international and domestic administration in times of new forms of governance: modes of implementing standards for sustainable development and their legitimacy challenges. Eur. J. Legal Stud. 2 (1), 49–86.

Gundel, J., 2013. Europäisches Energierecht. In: Danner, W., Theobald, Ch. (Eds.), Energierecht, 79. Ergänzungslieferung. C.H. Beck, München.

Herrmann, F., Sanden, J., Schomerus, T., Schulze, F., 2012. Ressourcenschutzrecht—Ziele, Herausforderungen, Regelungsvorschläge. ZUR, 523–531.

Huppertz, P., Nusser, J., 2009. Green IT—Ökodesignanforderungen an die ITK-Produktgestaltung und Möglichkeiten der wettbewerbsrechtlichen Durchsetzung. CR, 625–632.

Jans, J.H., Vedder, H.H.B., 2012. European Environmental Law. European Law Publishing, Groningen.

Jasper, U., Seidel, J., 2009. Umweltkriterien in der kommunalen Vergabe. KommJur 15 (2), 56–59.

Konzak, O., 2013. KrWG. In: Giesberts, L., Reinhardt, M. (Eds.), BeckOK Umweltrecht, 31st ed. C.H. Beck, München (para 23).

Kunzlik, P., 2013. Green public procurement—European law, environmental standards and 'what to buy' decisions. J. Environ. Law 25 (2), 173–202.

Lejeune, M., 2008. Neue Herausforderungen für IT und IT-Recht. CR, R112–R113.

Lohse, E.J., 2012. The meaning of harmonisation in the context of European Union law—a process in need of definition. In: Andenas, M., Baasch Andersen, C. (Eds.), Theory and Practice of Harmonisation. Edward Elgar Publishing, Cheltenham (UK) and Northampton (MA, USA), pp. 282–313.

Lustermann, H., 2007. Klimaschutz durch integrierte Produktpolitik—die neue EuP-Richtlinie. NVwZ, 895–900.

Nusser, J., 2010. Zwei Jahre EBPG—Erste Erfahrungen mit der Umsetzung der Ökodesign-Richtlinie. ZUR, 130–136.

Reimer, F., Tölle, S., 2013. Ressourceneffizienz als Problembegriff. ZUR, 589–598.

Schomerus, T., 2009. Rechtliche Instrumente zur Verbesserung der Energienutzung. NVwZ 418–423.

Schomerus, T., Spengler, L., 2010. Die Erweiterung der Ökodesign-Richtlinie - auf dem Weg zur "Super-Umweltrichtlinie"? EurUP, 54–61.

Söbbing, T., 2013. Auswirkungen der EU-Ökodesign-Richtlinie auf die IT-Branche. ITRB, 1–2.

Sutherland, E., 2009. Regulating the energy efficiency of ICT equipment. Comput. Telecommun. Law Rev. 15, 179–180.

CHAPTER 5

Quantitative and Systemic Methods for Modeling Sustainability

Amin Hosseinian-Far[1], Hamid Jahankhani[2]
[1] *Leeds Beckett University, Leeds, UK*
[2] *GSM London, UK*

Introduction

First UN conference on the Human Environment was held in Stockholm, Sweden in 1972 (United Nations, 1972). This conference which was also known as the Stockholm Conference, paved the way for future discussions on international treaties relevant to environmental sustainability. The Brundtland Report (WCED, 1987), titled "Our Common Future," established the grounds for primary generic sustainable models. This United Nations World Commission on Environment and Development (WCED) report introduced the term sustainable development, and the elements of this new concept were subsequently used in various sustainability representations. Sustainable development (SD) often denotes a maintainable environmental system whereas sustainability connotes a broader context (Hosseinian-Far et al., 2012). Sustainability is considered in a range of applications, including but not limited to environment, banking, businesses and organizations, manufacturing, computing, and so on (Khare, 2005). In certain disciplines (e.g., ecology), resilience is the word used instead of sustainability (Ludwig et al., 1997).

Complexity

Todorov and Marinova (2009) state that all existing models of sustainability—regardless of whether they lie in the environmental sector or the economic domain—seek to address its complexity. However, they also propose that the use of coevolution should be considered "including the role of humans as sustainability guardians." Shastri et al. (2008) consider the role of humans in ecosystem sustainability and use nonlinear and uncertainty models to analyze the success of environmental sustainability.

Systemic models are intended to demonstrate the systems' endurance regardless of their application. This knowledge representation not only implies the use of artificial intelligence techniques as in knowledge management systems but also can be implemented by means of other methods such as integrated assessment, system dynamics, econometrics, and so on (Seuring, 2013). Regardless of the approach used, the systems' attributes remain unchanged. In most cases, a major attribute of the considered system is its complexity (Myung, 2000). *Complexity* refers to the state of a system being complicated and twisted (Ladyman and Lambert, 2012). This complexity arises from several sources. A complex system that might be selected for modeling can be adaptive, chaotic, and nonlinear (Lansing, 2003). Because of this complexity and perhaps nonlinearity, some elements of the system cannot be reflected fully in the model. Dismissing systems' components could result in a biased output and lead to ineffective decision making. The complexities arise not only from individual domains of a system under consideration but also from the interactions between these social, environmental, and financial aspects (Hosseinian-Far et al., 2012).

Complex systems can be divided into these types: chaotic systems (affected by the butterfly effect); nonlinear systems, which do not reflect the behaviors and emergence of the subsystems and are modeled nonlinearly; and the complex adaptive systems (CAS), which learn from the environment. The adaptive characteristics of CAS would be beneficial in reflection of policy making and previous market trends (Rocha, 2004; Lansing, 2003).

Modeling Approaches

According to O'Riardan (2001), the traditional and universal models for sustainability are the Venn diagram explanation and the Russian doll model. The Venn diagram visualizes the economic, environmental, and social domains that have overlapping circles and shows sustainability development being achieved where the common ground exists in the center. On the other hand, the Russian doll model, which has a Boolean lattice structure, does not consider the economic domain as a separate area and believes that the economic and social domains are the subsets of the environment. According to the criteria of Chapman and Eames (2007), the Russian doll model is a stronger sustainability model. There are many other different ways to illustrate sustainability in a model. It can be elucidated simply by qualitative analysis of all three domains.

There are some other modeling tools and methodologies for the representation and analysis of complex systems. There are also some agreed upon frameworks for modeling sustainability. Integrated assessment (IA) or integrated sustainability assessment (ISA) "is a cyclical, participatory process of scoping, envisioning, experimenting and learning through which a shared interpretation of sustainability for a specific context is developed and applied in an integrated manner in order to explore solutions to persistent problems of unsustainable development" (Hinterberger and Jäger, 2008).

Among the quantitative techniques, agents and multiagent systems are quite useful because they can bring together the components of the system using agents' characteristics. The artificial intelligence techniques used in multiagent systems would enable the system to adapt itself to the changes faced from outside the boundary (Helbing and Balietti, 2012). However, there is a major drawback in using multiagent systems: They are too expensive for building large decision support system (DSS) applications. Although they can theoretically fit fairly well in a CAS methodology, in practice, they are not responsive to all of the components. Moreover, evaluation of such systems is challenging (Gleizes, 2011).

The use of neural networks as a tool for modeling the inputs and outputs of a network of linked components sounds very functional. But the major drawbacks with neural networks are found in training because the training of the network in a sustainability scenario is almost impossible; this is more so in backpropagational neural networks (Burger, 2010). Taking a longitudinal study might provide help for the training, but long intervals and a prolonged wait time are required for testing. The training can be done through time intervals by inputting new data to the system. Hence, they are not practical in large complex systems where model training is nearly impossible.

System Dynamics and Control

The system dynamics models provide further quantitative techniques for sustainability modeling. The difficult nature of developing a system using system dynamics also has made it impossible for developing the model. A system dynamics might be useful for modeling some small trade-offs in an environment in which an increase in one component would lead to a drop in another. Stella might be a good Integrated Development Environment (IDE) for these small projects; but again, if the system becomes large with various elements in the knowledge base and analysis, further design and development of the model would be unattainable. The bottom-up approach used in systems dynamics studies is often valuable in overcoming some degree of complexity in the system under consideration; however, for a system very complex in nature, the degree of error and also the difficulty in building a system dynamics model will be inevitable. Stella (mentioned earlier), ithink, Vensim, and Insight Maker are some well-known integrated modeling environments. Most of the development environments enable the modeler to create both casual loop diagrams (CLD) and stock-flow diagrams. The nodes in a CLD are connected using arrowed arcs that carry either a positive or a negative sign. A positive sign on an arc means that an increase in the first node would lead to an increase to the second node; a negative sign symbolizes a decrease in the second node as result of an increase in the first node (Kirkwood, 2012). Other symbols might be used for defining feedback loops in a CLD (e.g., B [balancing] for a negative sign and R [reinforcing]) for a positive sign (Kirkwood, 2012). A CLD may later be transformed into a stock-flow model in which some variables can be converted to rates. System dynamics follow basic rules of applied systems sciences and systems thinking, which was introduced initially by Peter Checkland (1981).

Influence Diagrams and Knowledge Representation

The rapid expansion of electronic computation persists in confronting our ability to conceptualize and explain the world around us. Mathematical tools and formal descriptions produce ineffective results when used as a communication device because most people not trained in or accustomed to a mathematical means of expression. Yet virtually everyone has information that could be considered useful in the solution of their own or others' problems if only it could be tapped. Influence diagrams (ID) can be both a formal description of the problem to be solved by computers and a representation easily understood by people in all walks of life and with varying degrees of technical proficiency. IDs can be considered as bridges between qualitative description and quantitative specification (Howard and Matheson, 2005). The reason for the power of this representation is that it can serve at the three levels of specification (relation, function, and number) and in both deterministic and probabilistic cases (Mathys et al., 2011).

Despite the presence of various modeling techniques in the field of sustainability, the use of probabilistic inference and Bayesian network models should be given high priority. Because sustainability scenarios are often complex, an understanding of the systems' resilience and of the boundary paradox is essential when analyzing and designing case studies (Jahankhani et al., 2012).

There are many similarities between system dynamics modeling and probabilistic modeling as in Bayesian networks and IDs (Daly et al., 2011). The adaptive nature of the model is an issue with probabilistic networks as the base, but that can be easily fixed by incorporating an agent-based engine. Further study into this could investigate the use of goal-oriented knowledge management systems theories (Xiao and Greer, 2006). The model is a formal one in which replication, testing, and validation are possible using the available algorithms. The predictive nature of the IDs, the excellent graphical interface for nonexpert users, and the accurate mathematical and probability layer would create an ideal approach to modeling sustainability systems with complex natures. It would still fit into the CAS structure if the agents were incorporated to give the system the adaptively feature (Sherif, 2006). The use of knowledge management theories would help to build the model's knowledge base. The final model can be implemented as a DSS for the stakeholders (Cai et al., 2009). It also can be disseminated as a web-based DSS tool by web services techniques (O'Donnell and Arnott, 1994).

There are number of algorithms for evaluating IDs and resolving the models, such as the reversal and removal algorithm introduced by Shachter (1986). This algorithm solves the algorithm in two stages:

- First reversing the direction of the arcs
- Second removing the nodes one by one to assess the effect and deterministic value of that node

After the introduction of the Shachter's set of rules, other scholars developed algorithms with other rules. Most of them still had the reversal and removal and only tried to make the algorithm more efficient in more complex systems. Pearl's algorithm (Pearl, 1998) and Zhang's algorithm (Zhang, 1998) are some instances of later sets of rules.

Influence Diagram vs. Decision Tree

Influence diagrams can construct a decision problem, but the fact is that only the surface of the decision problem is covered. However, decision trees (DT) reveal more details about the decision problem. Each route in a DT represents probability branches (Burton, 2014).

According to DECIDE-IT user manual (2011), the order of nodes in an ID is not as important as it is in a DT. This flexibility gives preference to using IDs rather than DTs. According to Decision IT, this suppleness gives the decision maker the opportunity to adjust decision variables accordingly at different stages of modeling.

IDs and DTs can be converted into each other. There is slight difference in constructions because the DTs cannot be constructed with the chance node in the beginning of the diagram. For instance, chance node alpha should not precede the decision node beta because the decision variable in a DT cannot depend on a chance node. Similarly, adding nodes would become problematic because the addition of decision nodes, which depend on an uncertain chance, are not valid (DecideIT, 2011).

The majority of DTs and IDs can be interchangeable; therefore, the consideration and selection of a specific tool might not be an issue for all problem domains; both DT and ID can capture the logic of the decision scenario and develop the DSS.

IDs are perfect for showing a decision's organization but encapsulate some details. A DT reveals more details of the decision scenario compared with an ID. The expansion of the decisions can lead to a probability tree (Howard and Matheson, 2005).

Unlike the nodes in a DT, the nodes in an ID do not require being totally ordered, nor do they require depending directly on all predecessors. This freedom from total ordering allows for convenient probabilistic assessment and computation. This ability gives the possibility of rescheduling the observation of some variables to the decision maker. It also provides the possibility of adding extra nodes without interrupting some of the orders.

Lastly, IDs have power in both deterministic and probabilistic cases. In the deterministic cases, the relationships between nodes mean that one variable can depend in a general way on several others. For example, profit is a function of revenue and cost. At the level of function, we specify the relationship; namely, that profit equals revenue minus cost. At number level, we specify the numerical values of revenue and cost, and hence determine the numerical value of profit. In probabilistic cases at the level of relation, we mean that given the information available, one variable is probabilistically dependent on certain other variables and probabilistically independent of even more variables. At the level of function, the probability distribution of each variable is assigned conditionally on the values of the variables on which it depends. Finally, at the number level, unconditional distributions are assigned on all variables that do not depend on any other variable and hence determine all joint and marginal probability distributions. As an example of the probabilistic case, we might assert at the level of relation that income depends on age and education and that education depends on age. Next, at the level of function we would assign the conditional distribution of income given age and education as well as the distribution of education given age. Different individuals can make the successive degrees of specification. Thus, an executive may know that sales depend in some way on price but may leave to others the probabilistic description of the relationship. Because of its generality, the ID is an important tool not only for decision analysis but also for any formal description of relationship and thus for all modeling work.

Modeling Approaches and Decision Support Systems

A generalized delineation for DSSs would be an interactive information system that assists the decision makers and planners in their business and decision-making activities.

Some analysis has already been done for some domains using experiments and computer simulations. The economy and the environmental analysis using the RetScreen tool (Excel-based software using its macros and many imported libraries) is one example (NRCAN.gc.ca, 2014), which was implemented by Canadian Natural Resources. This seems to be one of the more advanced tools in this field. But it does not consider the management domain or the new policies, patents, and technologies that affect the overall model.

IDs can be used to design DSS. IDs are visual representations of probability and uncertainty. Decisions, related objectives, uncertainties, and elements can be visualized and inferred using ID. The main advantage of an ID is its provision of visual conceptual representations of the problem domain. ID can be employed in a DSS in various contexts, such as in an environmental sustainability framework. In such a scenario, DSS would be designed with probabilistic networks, and ID would assist decision makers in precisely predicting the effects of climate change policy plans. Hence, the integration and use of ID would provide a more accurate platform for planning. Additionally, fewer planning failures and less investment fiascos together with thorough sustainable growth would assist the current economy recovery. IDs are great tools for modeling. Sustainability modeling using a DSS designed by means of ID and probabilistic networks can easily overcome the boundary paradox of the system under study.

Adding nodes to the system considering the time ordering of the nodes will overcome the boundary issue. Also, a visual and graphical presentation of the knowledge with the quantitative probability functions and utilities would precisely predict the system function. The inference in the ID can be easily done using existing algorithms. The complexity (which is the main drawback) can also be analyzed. In sustainability modeling, these facts are not considered.

Criticisms

The positivist approaches to modeling sustainability often use a quantitative method; however, this depends on the subject area. These methods are typically systems methodologies although microeconomic or financial techniques can also be used. Regardless of the approach taken, there are limitations in each modeling techniques mentioned earlier. The complexity of the systems under consideration is so immense that many assumptions should be made prior to analysis and design. The degree of error in some cases is also fairly high. Moreover, the boundary predicament is seen as a dilemma for analysts and designers. The predictive approaches in sustainability modeling have proved this in many scenarios in which long-term predictions were mistaken and deviated with the actual events. The typical argument is that the main aim of sustainability modeling is for risk minimization and maximization of success, profit, energy, resources, and so on There are also judgmental frameworks that encourage the incorporation of sustainability values into individual, corporate, and governmental strategies (Department for Communities and Local Government, 2012).

Many experts believe that technology led to concerns about environmental sustainability and sustainability development (Filho, 2009). Using technology to solve technology-developed problems seems to be an impossible predicament. On the other hand, there is a minority group that cogitate global warming and climate change as a myth (Robinson and Robinson, 1997). Some consider the problem broadly as a natural outcome of the earth's geologic time scale (Holmes et al., 2011). However, there is solid evidence demonstrating the correlation between the industrial era, rise of CO_2 emissions, and global warming (Donohoe, 2007). But already developed predictions based on simulations or quantitative analyses have not always demonstrated a correct outcome. Instances of these missed calculations are apparent in many research studies and even in some key research works. Hubert oil peak's (1967) conceptual prediction exhibited the trend of oil peak, but inaccuracies regarding the year of oil reserves' depletion were noticeable. Many scholars with recent simulations believe that predicting the oil reserves in plateaus is impossible because of the lack of data for certain variables of simulations.

Predictive information may also result in losses or gains for certain groups in a society. Conversely, some prototypes or models intend to simplify the complex nature of a system or an environment. There is a widely accepted code of ethics for such multidisciplinary study. Many other disciplines have an ethical code of conduct that provides a clearer guideline for their researchers. Although there are ethical procedures introduced by some societies (ACM ethical code of practice [1992], IEEE code of practice [ACM Council, 1992], and [IEEE, 2007]), these do not always cover multidisciplinary modeling such as sustainable modeling subject areas. Knowing that sustainability modelers need to incorporate variables from the society domain of sustainable development, analyses of behaviors therefore may be involved in such research works. Behavioral analyses can be seen as the key to social and psychological factors affecting the final framework and many ethical dilemmas that may emerge from this sphere of study.

Conclusions

Despite having various modeling techniques in the field of sustainability, probabilistic inference and Bayesian network models should be given high priority. The complexity of sustainability scenarios and understanding the systems' resilience and boundary paradox are essential when analyzing and designing case studies. Among the quantitative techniques, agents and multiagent systems are quite useful because they can bring together the components of the system using agents' characteristics. Furthermore, the artificial intelligence techniques used in multiagent systems would enable the system to adapt itself with the changes from outside the boundary. But there is a major drawback with multiagent systems: They are too expensive for building large DSS applications. Although in theory they can fit fairly well in a CAS methodology, in practice they are not responsive to all components. Use of neural networks as the tool for modeling the inputs and the outputs based on a network of linked components sounds very functional. But the major drawback with the neural networks involves its training. Training of the network in a sustainability scenario is almost impossible. A longitudinal study may help address the training issue, but long intervals and wait time are required for the testing. Hence, they are not practically usable in large complex systems because training is nearly impossible. The system's dynamic models are also other quantitative techniques for sustainability modeling. The difficult nature of developing a system using systems dynamic also has made it impossible to develop the model. System dynamics might be useful for modeling some small trade-offs in the environment in which an increase in one component would lead to a drop in another. Stella might be a good IDE for use in these small projects, but if the system becomes large with various elements in the knowledge base, the analysis, design, and development of the model would be unattainable.

References

ACM Council, 1992. ACM Code of Ethics and Professional Conduct. Association of Computing Machinery.

Burger, J., 2010. A Basic Introduction to Neural Networks. [Online] Available at: http://pages.cs.wisc.edu/~bolo/shipyard/neural/local.html [Accessed 2014].

Burton, D., 2014. Influence Diagrams & Basic Decision Trees. [Online] Available at: http://faculty.weber.edu/lburton/MHA%206350/Influence%20Diagrams%20&%20Dec%20Trees.ppt [Accessed 2014].

Cai, Z., Sun, S., Yannou, B., Si, S., 2009. An Influence Diagram for Maintenance Strategy Decision Making. IEEE Intl. Workshop in Intelligent Systems and Applications, Wuhan.

Chapman, D., Eames, C., 2007. Position Paper: Backgrounding New Guidelines for EE/EfS. [Online] Available at: http://nzaee.org.nz/ee-forum/efs-position-paper/ [Accessed 2014].

Checkland, P., 1981. Systems Thinking, Systems Practice. Wiley.

Daly, R., Shen, Q., Aitken, S., 2011. Learning Bayesian networks: approaches and issues. Knowl. Eng. Rev. 26 (2), 9–157.

DecideIT, 2011. DecideIT Decision Tool User manual. DecideIT.

Department for Communities and Local Government, 2012. National Planning Policy Framework. Department for Communities and Local Government, London.

Donohoe, M., 2007. Global warming: a public health crisis demanding immediate action. World Affairs Summer 11 (2), 44–58.

Filho, W.L., 2009. Communicating climate change: challenges ahead and action needed. Int. J. Clim. Change Strat. Manage. 1 (1), 6–18.

Gleizes, M.-P., 2011. Self-Adaptive Complex Systems. Institut de Recherche en Informatique de Toulouse, Toulouse.

Helbing, D., Balietti, S., 2012. Agent-based modeling. In: Understanding Complex Systems. Springer, Berlin, pp. 25–70.

Hinterberger, F., Jäger, J., 2008. Methods and Tools for Integrated Sustainability Assessment; Insights from the MATISSE project. Sustainable Europe Research Institute, Amsterdam.

Holmes, J., Lowe, J., Wolff, E., Srokosz, M., 2011. Global and Planetary Change. Global Planet. Change, 2–8.

Hosseinian-Far, A., Pimenidis, E., Jahankhani, H., 2012. Influence diagrams: predictive approach in decision support systems. Int. J. Strat. Manage. Decis. Supp. Syst. Strat. Manage. 17 (4), 16–22.

Howard, R.A., Matheson, J.E., 2005. Influence diagrams. Decis. Anal. 2 (3), 127–143.

Hubert, M.K., 1967. Degree of advancement of petroleum exploration in United States. AAPG Bull. 51 (11), 2207–2227.

IEEE, 2007. IEEE Code of Ethics. Institution of Electrical and Electronic Engineers Computer Society.

Jahankhani, H., Pimenidis, E., Hosseinian-Far, A., 2012. A probabilistic knowledge-based information system for environmental policy modeling and decision making. IFIP 1, 136–145.

Khare, A., 2005. Emerging Dimensions of Environmental Sustainability. Fachbuck Verlag Winkler, Edingen.

Kirkwood, C.W., 2012. New Product Dynamics. C. W. Kirkwood.

Ladyman, J., Lambert, J., 2012. What is a Complex System? University of Bristol.

Lansing, S.J., 2003. Complex adaptive systems. Annu. Rev. Anthropol. 23, 183–204.

Ludwig, D., Walker, B., Holling, C.S., 1997. Sustainability, stability, and resilience. Conserv. Ecol. 1(1).

Mathys, C., Daunizeau, J., Friston, K.J., Stephan, K.E., 2011. A Bayesian foundation for individual learning under uncertainty. Front. Human Neurosci.. 5(39)http://dx.doi.org/10.3389/fnhum.2011.00039.

Myung, J., 2000. The importance of complexity in model selection. J. Math. Psychol. 44 (1), 190–204.

NRCAN.gc.ca, 2014. RetScreen Suite. [Online] Available at: http://www.retscreen.net/ang/home.php [Accessed 2014].

O'Donnell, P.A., Arnott, D.R., 1994. Influence diagrams as a tool for decision support system design. In: Arnott, D., O'Donnell, D. (Eds.), Readings in Decision Support Systems. Monash University, Melbourne.

O'Riardan, T., 2001. Globalization, Localism and Identity. Earthscan, London.

Pearl, J., 1998. Probabilistic Reasoning in Intelligent Systems: Networks of Plausible Inference. Morgan Kaufmann Publishers Inc., San Mateo.

Robinson, A.B., Robinson, Z.W., 1997. Science has spoken: Global warming is a myth. Wall Street J. http://stephenschneider.stanford.edu/Publications/PDF_Papers/RobinsonAndRobinson.pdf.

Rocha, L., 2004. Complex System Modelings. [Online] Available at: http://informatics.indiana.edu/rocha/complex/ [Accessed 2014].

Seuring, S., 2013. A review of modeling approaches for sustainable supply chain management. Decision Support Syst. 54 (4), 1513–1520.

Shachter, R.D., 1986. Evaluating Influence Diagrams. Oper. Res. 34 (6), 871–882.
Shastri, Y., Diwekar, U., Heriberto, C., Williamson, J., 2008. Is sustainability achievable? Exploring the limits of sustainability with model systems. Environ. Sci. Technol, 6710–6716.
Sherif, K., 2006. An adaptive strategy for managing knowledge in organizations. J. Knowl. Manage. 10 (4), 72–80.
Todorov, V.I., Marinova, D., 2009. Models of Sustainability. 18th World IMACS/MODSIM Congress, Cairns, Australia.
United Nations, 1972. Report of the U.N. Conference on the Human Environment. United Nations Publications, Stockholm.
World Commission on Environment and Development (WCED), 1987. Our Common Future. Oxford University Press, Oxford. ISBN 019282080X.
Xiao, L., Greer, D., 2006. A hierarchical agent-oriented knowledge model for multi-agent systems. Int. J. Comput. Appl. Technol.
Zhang, N.L., 1998. Probabilistic inference in influence diagrams. Comput. Intell. 14, 475–497.

SECTION III

Sustainable Computing, Cloud and Big Data

CHAPTER 6

Sustainable Cloud Computing

Konstantinos Domdouzis
Sheffield Hallam University, Sheffield, UK

Introduction

Globalization has resulted in the development of new opportunities in the process of economic growth and has led to the resolution of many social problems. New businesses are created and social issues are reported in a matter of seconds. However, this progress has negative impacts. The most important consequences of globalization can be shown to be on the natural environment. Some decades ago, environmental issues were considered only as possible future problems, but today they are recognized as current issues as governments try to take measures or explore environmentally friendly technologies to stop the constant environmental disaster. The creation of the World Wide Web and the Internet was the driving force behind globalization. These two technologies managed to transform the entire planet into a digital village and created new routes for scientific research and knowledge. However, the consequences of digital technologies were also negative. The maintenance of a huge number of computer servers and the daily use of millions of personal computers have contributed to the increase in the Earth's carbon footprint. New technologies, such as cloud computing, have emerged and provide hope that computing power can be harnessed so that it does not affect the environment.

Clouds include more than the Internet. According to Mell and Grance (2011), cloud computing is a model for enabling a shared pool of computing resources on demand that can be rapidly released with minimal management effort. Dikaiakos et al. (2009) define *cloud computing* as the transfer of information technology (IT) services from the desktop and into large data centers. Qian et al. (2009) have defined the architecture of cloud computing used in organizations as one with two main layers, the core stack and the management layer. The core stack layer includes three sublayers: applications, platform, and resource (Figure 6.1).

There are different categories of cloud computing, such as hardware as a service (HaaS), software as a service (SaaS), platform as a service (PaaS), and infrastructure as a service (IaaS). HaaS allows users to rent an entire data center on a pay-as-you-go subscription. SaaS hosts a software application that customers can use without installing the software on their local

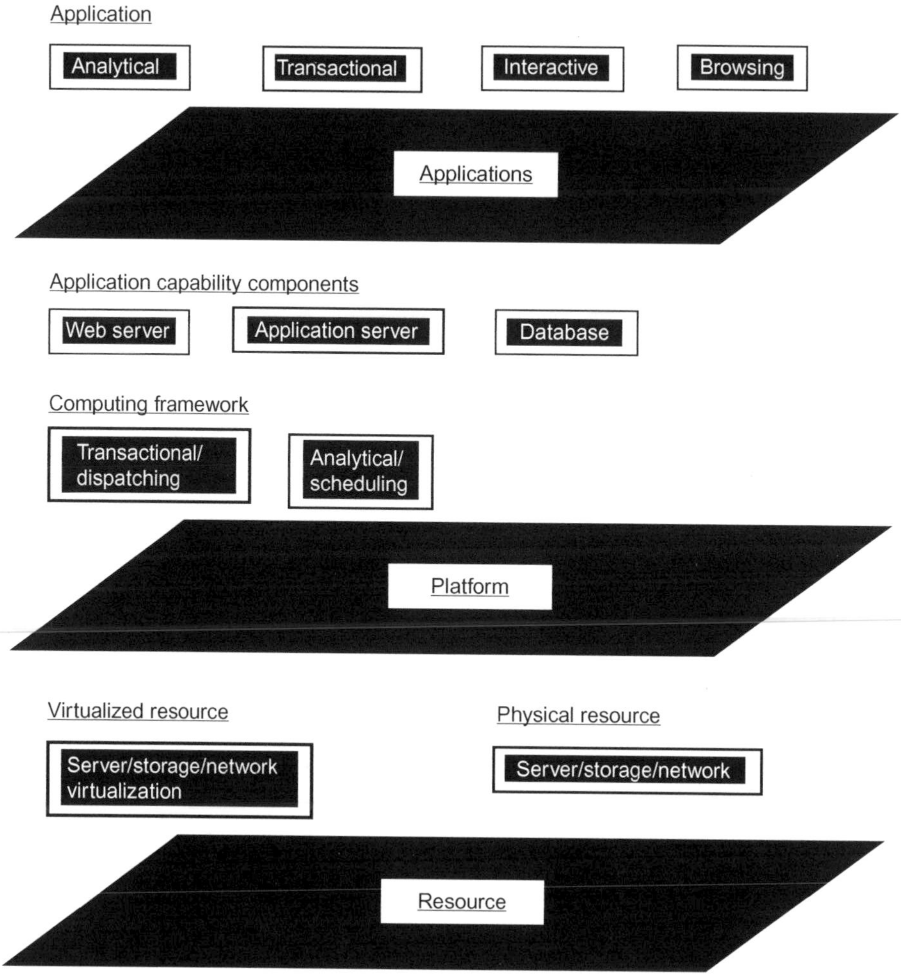

Figure 6.1
Cloud computing architecture. *Source: Adapted from Qian et al. (2009).*

computers (Wang and von Laszewski, 2008). Cloud systems can provide the software platform on which to run a system (PaaS) (Vaquero et al., 2009). With infrastructure as a service (IaaS), the user can deploy and run his or her own operating system in addition to the virtualization software offered by the provider (Prodan and Ostermann, 2009). Table 6.1 lists some technologies for the different cloud computing platforms.

The advantages of using cloud technologies are numerous. They provide flexibility by providing IT services, such as software updates and troubleshooting security issues, and allow the development of shared applications, thus promoting online collaboration. Cloud computing does not require the use of high-quality equipment and is easy to use. In addition, it enables the sharing of data between different platforms (Zhang et al., 2010). Cloud computing

Table 6.1 Cloud Computing Platforms

Cloud Computing Framework	Example Technology
SaaS	Google Apps, Facebook, YouTube
PaaS	MS Azure, Google AppEngine, Amazon SimpleDB/S3
IaaS	(Infrastructure) Amazon EC2, GoGrid, Flexiscale
	(Hardware) Data Centers
HaaS	Servers, backup

provides new horizons to business competitiveness because it enables small businesses to use new technology in order to compete more efficiently with larger competitors by using IT services more rapidly and in a more secure way. Through cloud computing, companies use the server space they need, therefore reducing the consumption of energy in comparison to that used by on-site servers.

Cloud computing can be considered an example of green IT. Jenkin et al. (2011) define *green ITs* as the technologies that directly or indirectly promote environmental sustainability in organizations. The term *sustainability* refers to the fair resource distribution between present and future generations and between agents in the current generation to maintain a scale of the current economy in relation to the ecological system that supports it (Costanza, 1994). The term *environmental sustainability* refers to the preservation of the natural environment's ability to support human life. This can be achieved only by making appropriate decisions. Environmental sustainability has a long history. In 1842, Jean Baptiste Joseph Fourier provided the first reference to greenhouse gases. In 1896, Svante Arrhenius suggested that levels of carbon dioxide in the atmosphere impact global temperatures via the greenhouse effect. The first major conference on environmental sustainability, the Stockholm Conference for the Human Environment, was held in 1972. In 1992, the United Nations Framework Convention on Climate Change (UNFCC) was held in Rio de Janeiro. The Kyoto protocol signed in 1997 legally binds 37 industrialized countries to specific measures (Boone and Ganeshan, 2012).

In order to achieve environmental sustainability, fundamental changes in production must be realized. These changes are parallel to the continuous evolution of technological systems, especially the creation of innovative technological systems. Hekkert et al. (2007) specify that the rate of technological change is specified by the competition among existing innovation systems, not only different technologies. In order to understand innovative systems, a definition of technological systems must be provided. Technological systems involve a network of agents that interact in the economic/industrial area under a particular institutional infrastructure and support the generation and production of technology (Carlsson and Stankiewicz, 1991). Innovative technological systems therefore must be used in order to achieve sustainability. They can be required to be sustainable by using technologies such as those that manage resources in such a way that performance and power are optimized. Additionally, they enable the realization of research with more efficient use of global energy sources. Cécile et al. (2002)

specify that innovative policies for sustainable development should promote technological diversity and long-term innovation capacity with parallel use of clean technologies. Watson et al. (2010) mention that it is important to follow a scientific approach in the development of IT for environmental sustainability and to stop using approaches that involve management and policy formation (Watson et al., 2010).

Challenges in the Use of Cloud Computing As Green Technology

There are a number of challenges associated with the use of green cloud computing. The main challenge is to minimize energy use and satisfy the requirements related to quality of service. However there are a number of issues when cloud computing is considered for environmental applications:

- *Energy-aware dynamic resource allocation*: In cloud computing, the excessive power cycling of servers could negatively impact their reliability. Furthermore, in the dynamic cloud environment, any interruption of energy can affect the quality of the provided service. Also, a virtual machine (VM) cannot record the timing behavior of a physical machine exactly. This can lead to timekeeping problems and inaccurate time measurements within the VM, which can result in incorrect enforcement of a service-level agreement (SLA) (Kalange Pooja, 2013).
- *Quality of service (QoS)-based resource selection and provisioning*: QoS-aware resource selection plays a significant role in cloud computing. Better resource selection and provision can result in energy efficiency (Kalange Pooja, 2013).
- *Optimization of virtual network topologies*: Because of VM migrations or the machines' nonoptimized allocation, communicating VMs may ultimately be hosted on distant physical nodes, and as a result, the cost of data transfer between them may be high (Kalange Pooja, 2013).
- *Enhancing awareness of environmental issues*: The users of green cloud computing technologies should become aware of the use of the technology in the resolution of specific environmental problems, such as the reduction of carbon emissions.
- *SLAs*: These provide for the replication of one application to multiple servers. Cloud customers should evaluate the range of parameters of SLAs, such as data protection, outage, and price structure, offered by different cloud vendors (Padhy et al., 2011).
- *Cloud data management*: Cloud operation is characterized by the accumulation of large amounts of data. Cloud service providers rely on cloud infrastructure providers to achieve full data security. In addition, VMs can migrate from one location to another; therefore, any remote configuration of the cloud by service providers could be insufficient (Padhy et al., 2011).
- *Interoperability*: Many public cloud systems are closed and are not designed to interact with each other. Industry standards must be created in order to allow cloud service providers to

design interoperable cloud platforms. The Open Grid Forum is an industry group that is working on open cloud computing interface to provide an application program interface (API) for managing different cloud platforms (Padhy et al., 2011).
- *Security*: Identity management and authentication are very significant, especially for government data. Governments, specifically the US government, have incorporated cloud computing infrastructures into the work of various departments and agencies. Because the government is a very complex entity, the implementation of cloud computing involves making policy changes, implementing dynamic applications, and securing the dynamic environment (Paquette et al., 2010).

Cloud Computing and Sustainability

When combined with specific characteristics of other technologies, such as the distributed resource provision of grid computing, the distributed control of digital ecosystems, and the sustainability from green computing, cloud computing can provide a sociotechnical conceptualization for sustainable distributed computing. Grid computing is a form of distributed technology in which a virtual supercomputer includes a cluster of networked computers performing very large tasks (Foster and Kesselman, 2004). Digital ecosystems are sociotechnical systems characterized by self-organization, scalability, and sustainability (Briscoe and De Wilde, 2006; Briscoe, 2009). Their purpose is to extend service-oriented architecture (SOA), thus supporting network-based economies (Newcomer and Lomow, 2005). Green computing is the efficient use of computing resources, which respects specific values for societal and organizational success. These values are people, planet, and profit (Marinos and Briscoe, 2009).

Social, economic, and environmental sustainability can also be achieved through the development of a green infrastructure (Benedict and McMahon, 2002). Weber et al. (2006) define *green infrastructure* as the abundance of landscape features that in combination with ecological processes contributes to human health. Lafortezza et al. (2013) identify a green infrastructure framework (GIF) that includes five functions: ecosystem services, biodiversity, social and territorial cohesion, sustainable development, and human well-being. These elements interact with each other

A number of cloud computing business models have been developed to ensure sustainability. Examples are the cloud cube model that enables collaboration in cloud formations used for specific business needs and the hexagon model, which provides six main criteria (consumers, investors, popularity, valuation, innovation, and get the job done [GTJD]) for business sustainability and shows how cloud computing performs according to these criteria (Chang et al., 2010). The cloud can lead to business sustainability through business improvement and the transformation and creation of new business value chains (Berman et al., 2012).

Focal companies are responsible for the supply chain; they provide direct contact to the customer. Additionally, they design the product or service offered. Focal companies are also responsible for the environmental and social performance of their suppliers; therefore, there is increased need for sustainable supply chain management (Seuring and Müller, 2008; Handfield and Nichols, 1999; Schary and Skjøtt-Larsen, 2001). There is a direct linkage between suppliers, focal companies, and customers. The integration of environmental thinking into supply chain management results in green supply chain management. In this case, sustainable information systems can be used to provide sustainable information services in the supply chain. Research on green information systems can be classified to different categories. The first category involves the examination of how the software development life cycle can be modified to reduce the potential negative environmental impacts of systems (Haigh and Griffiths, 2008). The second involves research on environmental reporting, measurement and accounting systems, and the use of knowledge management for environmental sustainability. Some research studies are also related to the consideration of environmental parameters when designing new products. A cloud computing platform can be characterized as a green information system that provides green information services.

The sustainability offered by cloud computing can be shown by the fact that the specific technology allows small business organizations to access large amounts of computing power in a very short time; as a result, they become more competitive with larger organizations. Third-world countries can also be significantly benefit by cloud computing technologies because they can use IT services that they previously could not access because they lacked the resources. Cloud computing accelerates the time in the businesses market because it allows quicker access to hardware resources without any up-front investment. In this case, there is no capital expenditure (capex), only operational expenditure (opex). Cloud computing makes possible the realization of new, innovative applications such as real-time, location-, environment-, and context-aware mobile interactive applications; parallel batch processing used for the processing of large amounts of data during very short periods of time; and business analytics for customer behavioral analysis (Marston et al., 2011).

Sustainable Applications of Cloud Computing

The architecture, engineering, and construction (AEC) sector is a highly fragmented, project-based industry with very strong data sharing. Beach et al. (2013) describe how cloud computing can be used in the AEC sector for better data management and collaboration. In order to create an efficient architecture for a cloud computing prototype, the authors used a building information model (BIM) data representation that is a complete 4-D virtual repository of all the data related to a construction project, such as 3-D models of building structure, construction data management information such as plans and schedules, information about all items within the building, and data about the progress of the construction project. The specific cloud computing platform is based on CometCloud, which is an autonomic computing engine for cloud and grid environments. Specifically, it is useful in the development of clouds with resizable computing capability that both integrates local computational environments and public cloud services and enables the capability to develop a range of programming applications (Kim and Parashar, 2011). The CloudBIM prototype was built using CometCloud's master/worker programming model and includes three main elements: client, masters, and workers. The function of CloudBIM is based on the interaction between masters and workers. The workers store data and are responsible for the validation of each query they receive. The client is responsible for the provision of the interface between the users and the local master node. This interface converts users' actions into queries (Beach et al., 2013).

Social sustainability can also be achieved through the use of social networks. This type of network can be based on cloud platforms or cloud applications and can be realized in social networks. The creation of social networking sites provides easier and less expensive ways for sustainable development communities to develop a wide variety of new communities. Examples of social networking sites for social sustainability are TakingITGlobal, which supports youth-led action using blogs, online groups and event calendars, UnLtd that supports social entrepreneurship, and People for Earth, a social network that provides advice on how people should live more environmentally friendly lives.

Busan is South Korea's second largest city and the fifth largest container globally. A cloud infrastructure has been provided to Busan based on Cisco Unified Computing System. The developed solution is the Busan Smart+Connected Communities, which aims to deliver social, economic, and environmental sustainability. It connects the Busan metropolitan government, the Busan Mobile Application Centre, and five local universities. During 2014, it is expected that the Busan cloud platform will create 3500 new job opportunities and 300 start-up companies focused on the Mobile Application Center development (Cisco, 2012).

Cloud computing has been also applied to government. The US government has made efforts to introduce cloud computing to the General Services Administration (GSA), the National Aeronautics and Space Administration (NASA), the Department of the Interior, the Department of Health and Human Services (HHS), the Census Bureau, and the White House. Specifically, the GSA can provide cloud-based hosting of the federal government's primary e-government portals—USA.gov—and its Spanish-language companion site, GobiernoUSA.gov. NASA uses the Nebula cloud computing platform, which can provide transparency in the involvement with space efforts. The US Department of the Interior's National Business Center (NBC) has introduced several cloud-based human resources management applications, including Web-based training, staffing, and recruitment programs. The NBC also offers cloud-based financial and procurement software (U.S. Department of the Interior, NBC, 2009). The Program Support Center (PSC) of the Department of HHS has selected Salesforce.com for the SaaS pilot, and within weeks, it had a working pilot online (Gross, 2009). The US Census Bureau uses Salesforce.com's SaaS to manage the activities of about 100,000 partner organizations. The White House uses also cloud computing to allow US citizens to express their opinions about President Obama (Hart, 2009). The UK government has also introduced the use of cloud computing through the development of G-cloud, which is a governmentwide cloud computing network (Glick, 2009). Petrov (2009) identified efforts from specific European countries, such as Sweden, France, and Spain, focused on the implementation of cloud computing for economic development, health and education services, and transportation networks.

Cloud computing is currently used for the advancement of scientific research. An example is the Cumulus project, which is an ongoing cloud computing project established at the Steinbuch Centre for Computing of the Karlsruhe Institute of Technology (KIT). The aim of Cumulus is the provision of virtual infrastructures for scientific research (Juve and Deelman, 2010). Cloud computing is extremely significant for the European Organization for Nuclear Research (CERN), which realizes a number of experiments (e.g., the ALPHA experiment, the Isotope mass separator on-line facility [ISOLDE] experiment, the large Hadron collider (LHC), the total elastic and diffractive cross-section measurement) and generates amazingly extreme amounts of data (approximately 15 petabytes) useful for simulation and analysis. The CERN Data Centre is responsible for the collection of these data. In 2002, the CERN Data Centre required the use of the worldwide LHC computing grid (WLCG), which is a distributed computing infrastructure in tiers that provides to 8000 physicists real-time access to LHC data. However, CERN is currently focused on the use of cloud technology for better handling these colossal data sets (CERN, 2014). It is very probable that CERN will adopt the use of the OpenStack cloud computing software in order to update its data management operations (Purcell, 2014). Other examples of the use of cloud computing for scientific research are Red Cloud from Cornell University, Wispy from Purdue University, and the San Diego Supercomputer Center (SDSC) cloud (University of Chicago and Argonne National Laboratory, 2014).

Global Forest Watch is an online mapping tool created by Google, the World Resources Institute, and 40 other partners. It is based on the analysis of Landsat7 satellite images that are processed by Google Earth Engine, the company's cloud platform for geodata analysis. The amount of image data that were processed was about 20 tera-pixels, which required 1 million central processing unit (CPU) hours on 10,000 computers operating in parallel for a period of several days. Google estimates that a single computer would need 15 years to perform the tasks realized by Global Forest Watch. The aim of the mapping system is also the enhancement of satellite images with social data, which will document possible forest abuse (Claburn, 2014).

Cloud computing is used to process an inversion process for magnetotellurics, a geophysic technique used for the characterization of geothermal reservoirs and mineral exploration. The entire inversion process is implemented on the Amazon Elastic Compute Cloud (EC2) cloud using available EC2 instances The determination of subsurface electrical conductivity is significant in a range of applications covering tectonic evolution and mineral and geothermal exploration. Mudge et al. (2011) describe how a Fortran program was developed in order to abstract logic and process modeling calculations. The program is embedded in a web application. The whole inversion process is implemented on the Amazon EC2 cloud using available EC2 instances. The whole process is further packaged as a MT 3D inversion software product, which is accessible anywhere through a secure login. The magnetotellurics method is used for understanding the processes involved during the Enhanced Geothermal Systems (EGS) fluid system.

Lawrence Berkeley National Laboratory and Northwestern University in the United States created a modeling tool that shows the energy savings from moving local network software and computing to serve cloud farms. The specific tool is called Cloud Energy and Emissions Research Model (CLEER). The aim of the model is the provision of a framework for the assessment of the net energy consumption and greenhouse gas emissions of cloud computing in comparison to traditional systems (Irfan and ClimateWire, 2013). Cloud computing saves energy through virtualization, which allows virtual machine consolidation and migration, heat management, and temperature-aware allocation; these are techniques that result in the reduction of power consumption (Garg and Buyya, 2012). The large-scale deployment of virtualized server infrastructure can result in the balance of computer and storage loads across physical servers. The optimized design of cloud data centers allows servers to run at optimal temperature and use. Furthermore, cloud computing provides dynamic provisioning of infrastructure capacity and sharing of application instances between client organizations that results to the reduction of peak loads (accenture and WSP, 2010).

Eye on Earth is an environmental information sharing system developed by the European Environment Agency (EEA), which is based on Microsoft Windows Azure cloud platform. Information sharing is important in efficiently and effectively addressing environmental issues. The users of the environment can be policy makers, communities or individuals, environmental organizations, and emergency responders. Different environmental data, such as air and water quality are examined and can be represented in a visual format. For example, an intelligent map allows users to discover environmental information (European Environment Agency, 2014).

The US government has launched the Climate Change Initiative to help organizations and citizens to use public data more efficiently to better prepare for the effects of climate change. Google has donated 50 million hours of cloud computing time on the Google Earth Engine geospatial analysis platform to the project. Google Earth Engine can detect changes on Earth's surface using satellite imagery. Examples of its application are the development of the first global maps of deforestation and a nearly real-time system that identifies deforestation resulting from climate change (Google, 2014). The engine will be used to manage the agricultural water supply and model the impacts of sea level rises (Weiss, 2014).

Information and communication technologies (ICTs) play an important role in emergency response systems during natural disasters. Alazawi et al. (2011) suggest the use of an intelligent disaster management system based on the use of vehicular ad-hoc networks (VANETs) and mobile and cloud technologies. VANETs are the most advanced technology of intelligent transportation systems (ITS). They are vehicles that are equipped with sensors and wireless communication capability. The main focus of the application of VANETs is safety and transportation efficiency. The cloud-based vehicular emergency response system includes three main layers: the cloud infrastructure, the Intelligence, and the system interface layers. The cloud infrastructure layer is the foundation for the base platform for the emergency system. The intelligence layer provides the necessary computational models to develop optimum emergency response strategies. The system interface layer collects data from the Internet, social media, and even roadside masts. The system was implemented in Ramadi, Iraq.

Suakanto et al. (2012) suggest the use of cloud computing infrastructure for detecting environmental disasters early and monitoring environmental conditions. Depending on the type of environmental application in which the system is used, different sensors are used. For example, for air monitoring applications, carbon dioxide or air quality control sensors are used; for water monitoring applications, water pH measurement sensors can be used. A remote terminal unit (RTU) is used in order to collect data from the different types of sensors in analogue and digital form. The units are placed in areas that are prone to disasters. A cloud platform is used for central data storage and processing (Figure 6.2).

Data collected by the sensors are sent to the listener service of the virtual machines. The main function of this service is to read and process sensor data and store them in the cloud storage repository. When users are online, they can see the stored data by using application services installed on virtual machines.

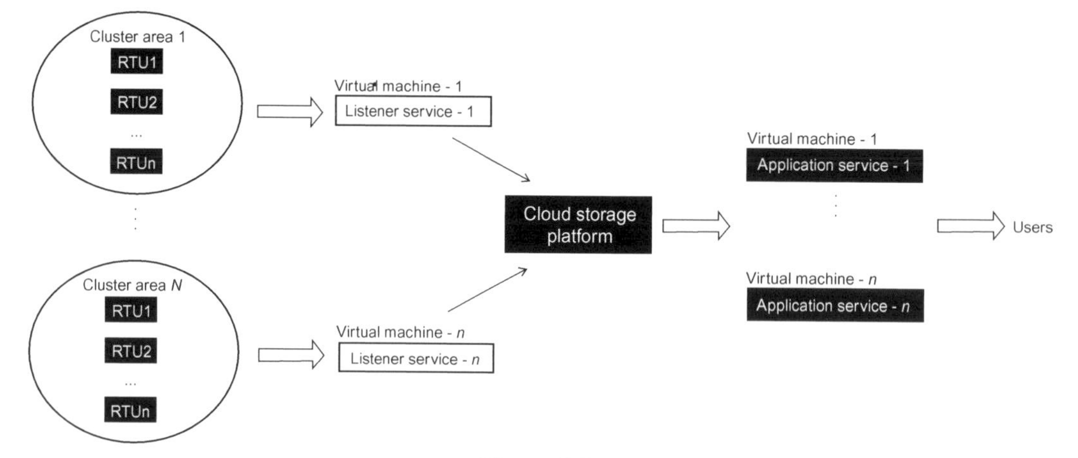

Figure 6.2
Cloud platform for environmental situations. *Source: Adapted from Suakanto et al. (2012).*

Agricultural modernization has great significance for China's growing economy. Even though the country has realized improvements in crop cultivation and animal and plant breeding, agricultural production is a decentralized operation characterized by the use of low-level information technologies. Agricultural modernization involves the use of modern agricultural equipment, modern agricultural planting and breeding technology, and modern forms of production organization and management. Cloud computing can lead to the establishment of an information network services platform and the integration of isolated production facilities, technical equipment, and information services. Applications of cloud computing are the integration and sharing of agricultural information, the real-time monitoring of agricultural production, the provision of agricultural technology services, construction and improvement of agricultural supply chain, and the tracking of the quality of agricultural products (Zhu et al., 2013).

Technologies Associated With Sustainable Cloud Computing

Some technologies have the same characteristics as cloud computing or are related to cloud computing are:

- *Grid computing*: This involves a large network of computers used to create large supercomputing resources. In these types of networks, large and complex computing operations can be performed.
- *Virtualization*: Virtualization allows the consolidation of multiple applications into virtual containers located on a single or multiple servers (Padala et al., 2007). Virtualization is the foundation of cloud computing. It enables users to access servers without knowing details of those servers.
- *Utility computing*: This term defines a "pay-per-use" model for using computing services.
- *Autonomic computing*: The term *autonomic* refers to self-management, in this case, of computers. In autonomic computing, computers can correct themselves without human intervention.
- *Web services*: Web services simplify the application-to-application interaction using a systematic framework based on existing Web protocols and open XML standards. This framework includes three mail elements: communication protocols, service descriptions, and service discovery (Curbera et al., 2002).

Future Prospects of Sustainable Cloud Computing

The future of cloud computing in relation to its impact to sustainability is excellent. There is huge potential for additional applications of the technology to different industries, such as manufacturing, health care, and education. Examples of this potential are the provision of access to global data resources through cloud computing, the realization of low-cost simulation experiments, provision of massive and flexible computing power for drug discovery, and real-time health monitoring. With its rapid prototyping and collaborative design and the improvement of manufacturing processes, cloud computing can also contribute significantly to supply chain coordination. Furthermore, the technology can be the basis for highly interactive, collaborative learning (World Economic Forum, 2010).

The European Union (EU) envisions the realization of the global cloud ecosystem. It can offer new features to support cloud employment and to improve adoption of cloud computing. It could provide supporting tools that would cover issues related to building and supporting new platforms easily, new programming models, and tools that deal with distribution and control, improved security and data protection, efficient data management, energy efficiency, and easy mash-ups of clouds exposing a single-user interface. The EU can also exploit the capabilities offered by existing cloud systems to enhance the capabilities of products and services offered by European industry (European Commission, 2014). Developing countries, such as India, can seek ways to enhance sustainability through the use of green IT. In fact, India spent 18.2 billion for green IT in 2013. A study by Global e-Sustainability Initiative and Microsoft showed that running services over the cloud can be 95% more efficient than those run in other ways (Thomond, 2013).

Reflections on Sustainable Cloud Computing Applications

Sustainability is a topic of great importance regarding continuous environmental, social, and economic problems. Cloud computing can be applied to all aspects (social, business, environmental) of sustainability. Specific characteristics of the technology, such as sharing data, allow small businesses to have access to huge computing power and as a result, these businesses can become more competitive. Cloud technology can be adjusted to meet requirements of sustainability. It offers dynamic provisioning of resources, multitenancy (the serving of multiple businesses using the same infrastructure), server use, and the power efficiency of data centers (accenture and WSP, 2010). The technology has huge potential in shaping sustainability in different fields. Especially related to environment, it promotes environmental research, development of emergency response systems, and implementation measures to prevent climate change.

However, there are some concerns in relation to the energy efficiency of cloud computing. When files to be transmitted over a network are quite large, the network will be a major contributor to energy consumption. Furthermore, the VM consolidation could reduce the number of active servers but will put an excessive load on few servers whose heat distribution can become a major issue. A green cloud framework that considers these problems can be developed. In such a framework, users submit their cloud service requests through a middleware green broker that manages the selection of the greenest cloud provider to serve the user's request.

A user request can be for software, platform, or infrastructure. The cloud providers register their services in the form of offers to a public directory and the green broker assesses them. These offers include green services, pricing, and time when the service should be assessed for the least carbon emission. The green broker calculates the carbon emission of all cloud providers. It then selects the services that will result in the least carbon emission and buys them on behalf of users. Some additional steps need to be taken: the design of software at various levels (OS, compiler, algorithm, application) that facilitates energy efficiency, the measurement of data center power, and the deployment of the data centers near renewable energy sources (Garg and Buyya, 2012).

Conclusions

Cloud computing is an advanced technology with great impact on different fields. Michael Dertouzos, late director of Massachusetts Institute of Technology Laboratory of Computer Science, had the vision to make computing an indispensable part of human life, just like oxygen. Therefore, he initiated Project Oxygen for the development of human-centered, pervasive computing. Cloud computing is the best example of such technologies. It has revolutionized the way services are offered through the World Wide Web by reducing the dependence to hardware. Most importantly, applications can run independently from specific computer configurations.

Cloud computing is a significant factor for achieving sustainability. It promotes academic research, the development of businesses, and the strengthening of business competition. Governmental agencies have started integrating the technology in their existing infrastructure, and social media have gained widespread acceptance. The impact that cloud computing can have in the future for the environment is significant. Applications such as the realization of emergency response systems during natural disasters and the early detection of environmental disasters can be efficiently realized through the use of cloud computing. Social networking can also be used for the development of teams that promote environmental consciousness or provide information that can be useful during environmental crises. However, more research needs to be conducted future to fully comprehend the energy efficiency provided by the use of this specific technology.

References

accenture and WSP, 2010. Cloud computing and sustainability: the environmental benefits of moving to the cloud. Technical report [online]. Available at: http://www.accenture.com/SiteCollectionDocuments/PDF/Accenture_Sustainability_Cloud_Computing_TheEnvironmentalBenefitsofMovingtotheCloud.pdf (accessed 01.03.2014.).

Alazawi, Z., Altowaijri, S., Mehmood, R., Abdljabar, M.B., 2011. Intelligent disaster management system based on cloud-enabled vehicular networks. In: ITS Telecommunications (ITST) 11th International Conference, pp. 361–368. http://dx.doi.org/10.1109/itst.2011.6060083.

Beach, T.H., Rana, O.F., Rezgui, Y., Parashar, M., 2013. Cloud computing for the architecture, engineering & construction sector: requirements, prototype & experience. J. Cloud Comput. Adv. Syst. Appl. 2, 8.

Benedict, M.A., McMahon, E.T., 2002. Green infrastructure: smart conservation for the 21st century. Renewable Resour. J. 20, 12–17.

Berman, S.J., Kesterson-Townes, L., Marshall, A., Srivathsa, R., 2012. How cloud computing enables process and business model innovation. Strategy Leadersh. 40 (4), 27–35.

Boone, T., Ganeshan, R., 2012. By the numbers: a visual chronicle of carbon dioxide emissions. In: Boone, T., Jayaraman, V., Ganeshan, R. (Eds.), Sustainable Supply Chains - Models Methods, and Public Policy Implications. Springer, New York.

Briscoe, G., 2009. Digital Ecosystems. Ph.D. dissertation, Imperial College London, London.

Briscoe, G., De Wilde, P., 2006. Digital ecosystems: evolving service-oriented architectures. In: Conference on Bio Inspired Models of Network, Information and Computing Systems. IEEE Press, Cavalese, Italy. [online]. Available at: http://arxiv.org/abs/0712.4102.

Carlsson, B., Stankiewicz, R., 1991. On the nature, function and composition of technological systems. J. Evol. Econ. 1 (2), 93–118.

Cécile, P., Valenduc, G., Warrant, F., 2002. Technological innovation fostering sustainable development. Levers for a Sustainable Development, Policy.

CERN, 2014. Computing-experiments at CERN generate colossal amounts of data. The Data Centre stores it, and sends it around the world for analysis [online]. Available at: http://home.web.cern.ch/about/computing (accessed 10.04.2014.).

Chang, V., Wills, G., De Roure, D., 2010. Cloud business models and sustainability: impacts for businesses and e-research. In: UK e-Science All Hands Meeting 2010, Software Sustainability Workshop, Cardiff, GB, 13-16 Sep 2010, p. 3.

Cisco, 2012. City transforms economic sustainability with public cloud, customer case study. Available at: http://smartcitiescouncil.com/system/tdf/public_resources/cisco_busan%20ec%20dev.pdf?file=1&type=node&id=247 (accessed 18.04.2014.).

Claburn, 2014. Google Powers Forest Protection Effort. [online]. Available at: http://www.informationweek.com/government/cloud-computing/google-powers-forest-protection-effort/d/d-id/1113950 (accessed 18.04.2014.).

Costanza, R., 1994. Chapter 24: three general policies to achieve sustainability. In: Jansson, A., Hammer, M., Folke, C., Costanza, R. (Eds.), Investing in Natural Capital. Island Press, Washington, DC.

Curbera, F., Duftler, M., Khalaf, R., Nagy, W., Mukhi, N., Weerawarana, S., 2002. Unraveling the web services web—an introduction to SOAP, WSDL, and UDDI. IEEE Internet Comput. 6 (2), 86–93.

Dikaiakos, M.D., Katsaros, D., Mehra, P., Pallis, G., Vakali, A., 2009. Cloud Computing: Distributed Internet Computing for IT and Scientific Research. IEEE Internet Comput. 13 (5), 10–13. http://dx.doi.org/10.1109/MIC.2009.103.

European Commission, 2014. The Future of Cloud Computing—Opportunities for European Cloud Computing Beyond 2010. In: [online]. Available at: http://cordis.europa.eu/fp7/ict/ssai/docs/cloud-report-final.pdf, (accessed 19.04.2014.).

European Environment Agency, 2014. Eye on Earth Briefing. [online]. Available at: http://www.eea.europa.eu/about-us/what/seis-initiatives/eye-on-earth-briefing/view, (accessed 20.03.2014.).

Foster, I., Kesselman, C., 2004. The Grid: Blueprint for A New Computing Infrastructure. Morgan Kaufmann, San Francisco, CA.

Garg, S.K., Buyya, R., 2012. Green cloud computing and environmental sustainability. In: Murugesan, S., Gangadharan, G.R. (Eds.), Harnessing Green IT Principles and Practices. Wiley, Chichester, West Sussex, pp. 315–337.

Glick, B., Glick, B., 2009. Digital Britain commits government to cloud computing. Computing. [online]. Available: http://www.computing.co.uk/computing/news/2244229/digital-britain-commits (accessed 15.06.2013.).

Google, 2014. Helping Our Communities to Adapt to Climate change. [online]. Available at: http://google-latlong.blogspot.co.uk/2014/03/helping-our-communities-adapt-to.html (accessed 14.04.2014.).

Gross, G., 2009. Gov't agencies embrace cloud computing: government agencies say they're moving toward an embrace of cloud computing and software-as-a-service. PC World, February 25, 2009. [online]. Available: http://www.pcworld.com/article/160233/article.html, (accessed 22.02.2014.).

Haigh, N., Griffiths, A., 2008. The environmental sustainability of information systems: considering the impact of operational strategies and practices. Int. J. Technol. Manage. 43 (1/2/3), 48–63.

Handfield, R.B., Nichols, E.L., 1999. Introduction to Supply Chain Management. Prentice-Hall, New Jersey.

Hart, K., 2009. Tech firms seek to get agencies on board with cloud computing. *The Washington Post*, March 31, 2009. [online]. Available at: http://www.washingtonpost.com/wp-dyn/content/article/2009/03/30/AR2009033002848.html (accessed 15.04.2014.).

Hekkert, M.P., Suurs, R.A.A., Negro, S.O., Kulhmann, S., Smits, R.E.H.M., 2007. Functions of innovation systems: a new approach for analysing technological change. Technol. Forecasting Social Change 74, 413–432.

Irfan, U., ClimateWire, 2013. Cloud computing saves energy. Sci. Am., June 12, 2013. [online]. Available at: http://www.scientificamerican.com/article/cloud-computing-saves-energy/ (accessed 22.04.2014).

Jenkin, T.A., Webster, J., McShane, L., 2011. An agenda for 'Green' information technology and systems research. Inf. Organ. 21 (1), 17–40.

Juve, G., Deelman, E., 2010. Scientific workflows and clouds. ACM Crossroads 16 (3), 14–18.

Kalange Pooja, R., 2013. Applications of green cloud computing in energy efficiency and environmental sustainability. J. Comput. Eng. 1, 25–33.

Kim, H., Parashar, M., 2011. CometCloud: An Autonomic Cloud Engine, Cloud Computing: Principles and Paradigms. Wiley (chapter 10), pp. 275-297.

Lafortezza, R., Davies, C., Sanesi, G., Konijnendijk, C., 2013. Green Infrastructure as a tool to support spatial planning in European urban regions. iForest 6, 102–108.

Marinos, A., Briscoe, G., 2009. Community cloud computing. In: Proceedings of the 1st International Conference on Cloud, Computing, pp. 472–484.

Marston, S., Li, Z., Bandyopadhyay, S., 2011. Cloud computing—the business perspective. Decis. Support Syst. 51 (1), 176–189.

Mell, P., Grance, T., 2011. The NIST definition of cloud computing—recommendations of the national institute of standards and technology. National Institute of Standards and Technology, Special Publication 800-145.

Mudge, J.G., Chandrasekhar, P., Heinson, G.S., Thiel, S., Mudge, J.G., Chandrasekhar, P., Heinson, G.S., Thiel, S., 2011. Evolving inversion methods in geophysics with cloud computing. In: Proceedings of IEEE eScience 2011, Stockholm.

Newcomer, E., Lomow, G., 2005. Understanding SOA with Web Services. Addison-Wesley, Upper Saddle River, NJ.

Padala, P., Zhu, X., Wang, Z., Singhal, S., Shin, K.G., 2007. Performance Evaluation of Virtualization Technologies for Server Consolidation. [online]. Available at: http://137.204.107.78/tirocinio/site/tirocini/Tirocinio-Zuluaga/Documents/virtualizzazione/Technologies%20for%20Server.pdf, (accessed 03.05.2014.).

Padhy, R.P., Patra, M.R., Satapathy, S.C., 2011. Cloud computing: security issues and research challenges. Int. J. Comput. Sci. Inf. Technol. Secur. 1 (2), 136–146.

Paquette, S., Jaeger, P.T., Wilson, S.C., 2010. Identifying the security risks associated with governmental use of cloud computing. Gov. Inf. Q. 27, 245–253.

Petrov, O., 2009. Backgrounder: financial crisis and cloud computing—delivering more for less. Demystifying cloud computing as enabler of government transformation. World Bank, Government Transformation Initiative, June 16, 2009. [Online]. Available at: http://www.siteresources.worldbank.org/.../BackgrounderFinancialCrisisCloudComputing.doc (accessed 11.04.2014.).

Prodan, R., Ostermann, S., 2009. A survey and taxonomy of infrastructure as a service and web hosting cloud providers. In: 10th IEEE/ACM International Conference on Grid Computing, October 13-15, pp. 17–25.

Purcell, A., 2014. Tim Bell on the Importance of OpenStack for CERN. [online]. Available at: http://www.isgtw.org/feature/tim-bell-importance-openstack-cern, (accessed 10.04.2014.).

Qian, L., Luo, Z., Du, Z., Guo, L., 2009. Cloud computing: an overview. Cloud Comput. Lect. Notes Comput. Sci. 5931, 626–631.

Schary, P., Skjøtt-Larsen, T., 2001. Managing the Global Supply Chain, second ed. Copenhagen Business School Press, Copenhagen.

Seuring, S., Müller, M., 2008. From a literature review to a conceptual framework for sustainable supply chain management. J. Cleaner Prod. 16 (5), 1699–1710.

Suakanto, S., Supangkat, S.H., Roberd Saragih, S., Arief Nugroho, T., Gusti Bagus, I., Nugraha, B., Suakanto, S., Supangkat, S.H., Roberd Saragih, S., Arief Nugroho, T., Gusti Bagus, I., Nugraha, B., 2012. Environmental and disaster sensing using cloud computing infrastructure. In: International Conference on Computer and Communications Security—ICCCShttp://dx.doi.org/10.1109/ICCCSN.2012.6215712.

Thomond, P., 2013. The Enabling Technologies of a Low-Carbon Economy: A Focus on Cloud Computing. In: [online]. Available at: http://gesi.org/assets/js/lib/tinymce/jscripts/tiny_mce/plugins/ajaxfilemanager/uploaded/Cloud%20Study%20-%20FINAL%20report_2.pdf, (accessed 20.04.2014.).

U.S. Department of the Interior, National Business Center (NBC), NBC's Federal Cloud Playbook, August 2009. [Online]. Available at: http://api.ning.com/files/Rn9mXJiYBdsR1iwPJ68dWWEwO6TEkRCO18Xu-Hj7QfIbqZ-*Oc4fswll48ShViNARu3RaoZ8vfUrRxe5SUKnMr7jcLSbQ6I2/NBCCloudWhitePaperFinalWebRes.pdf (accessed 01.04.2014.).

University of Chicago and Argonne National Laboratory, 2014. Clouds. [online]. Available at: http://scienceclouds.org/infrastructure-clouds/, (accessed 10.04.2014.).

Vaquero, L.M., Rodero-Merino, L., Caceres, J., Lindner, M., 2009. A break in the clouds: towards a cloud definition. ACM SIGCOMM Comput. Commun. Rev. 39 (1), 50–55.

Wang, L., von Laszewski, G., 2008. Scientific cloud computing: early definition and experience. In: 10th IEEE International Conference on High Performance Computing and Communications (HPCC '08), September 25-27, Dalian, China.

Watson, R.T., Boudreau, M.-C., Chen, A.J., 2010. Information systems and environmentally sustainable development: energy informatics and new directions for the IS community. MIS Quart. 34 (1), 23–38.

Weber, T., Sloan, A., Wolf, J., 2006. Maryland's green infrastructure assessment: development of a comprehensive approach to land conservation. Landscape Urban Plan. 77 (1–2), 94–110.

Weiss, T.R., 2014. Google Provides Cloud Computing Resources for Climate Change Research. [online]. Available at: http://www.eweek.com/cloud/google-provides-cloud-computing-resources-for-climate-change-research.html, (accessed 15.04.2014.).

World Economic Forum, 2010. Exploring the Future of Cloud Computing: Riding the Next Wave of Technology-Driven Transformation. [online]. Available at: http://www.weforum.org/pdf/ip/ittc/Exploring-the-future-of-cloud-computing.pdf (accessed 15.04.2014.).

Zhang, S., Zhang, S., Chen, X., Huo, X., Zhang, S., Zhang, S., Chen, X., Huo, X., 2010. Cloud computing research and development trend. In: Second International Conference on Future Networks, (ICFN '10), January 22-24, 2010, Sanya, China, pp. 93–97.

Zhu, Y., Wu, D., Li, S., 2013. Cloud computing and agricultural development of china: theory and practice. IJCSI Int. J. Comput. Sci. 10 (1), 7–12.

CHAPTER 7

Sustainable Software Design

Michael Engel
Leeds Beckett University, Leeds, UK

Overview and Scope

Sustainability is a broad term encompassing a large number of concepts. In the area of software engineering, a number of different definitions exist. For instance, Tate (2005) defines sustainable software development as follows:

> Sustainable software development is a mindset (principles) and an accompanying set of practices that enable a team to achieve and maintain an optimal development pace indefinitely.

In the context of this chapter, however, the software development process plays only a minor part. Here, *sustainability* is used in an ecological context, as defined by Dick et al. (2010):

> Sustainable Software is software, whose direct and indirect negative impacts on economy, society, human beings, and environment that result from development, deployment, and usage of the software are minimal and/or which have a positive effect on sustainable development.

Adhering to this definition, in the following text we analyze the direct as well as indirect effects that software has on the sustainability of combined hardware/software systems throughout the product life cycle.

Evaluating Sustainability Effects

In order to evaluate and compare the environmental impact of software, a common base must be established. In general, the overall environmental impact of a product is difficult to assess. Especially during production, a large number of factors influence this impact. During the production of LCD screens, for example, an inorganic compound, the gas NF_3 (nitrogen trifluoride), is used, which is estimated to be about 17,000 times more effective as a greenhouse gas than carbon dioxide (CO_2), according to Weiss et al.(2008). In order to achieve comparability between the impact that different chemical elements and processes have on the environment, most studies break this impact down to an equivalent of CO_2 impact, measured in kg of CO_2 emitted into the atmosphere.

However, there is no universally accepted method to assess the environmental friendliness of software. The related metrics found in current literature differ in terms of measures of resources, the kind of result generated, the analyzed environment and application domain, and their usage. Bozzelli et al. (2013) performed an extensive analysis of metrics used to assess the "greenness" of software. Their study concludes that most researchers exclusively concentrate on metrics related to energy consumption at runtime but neglect the production phase of the overall product. Nevertheless, this study is useful to obtain an overview of the different approaches and to assess the suitability of a given metric for a specific context.

Sustainability and the Product Life Cycle

An assessment of the sustainability properties of software must take place in the context of the product the software is intended for. As shown in Figure 7.1, the life cycle of all IT products consists of a number of stages, starting from the sourcing of raw materials and ending with the disposal or recycling of the product. For the impact analysis of software, this life cycle can roughly be categorized into three sources: the production—including the sourcing and processing of raw materials and its distribution to the end customer—use, and disposing or recycling of a product.

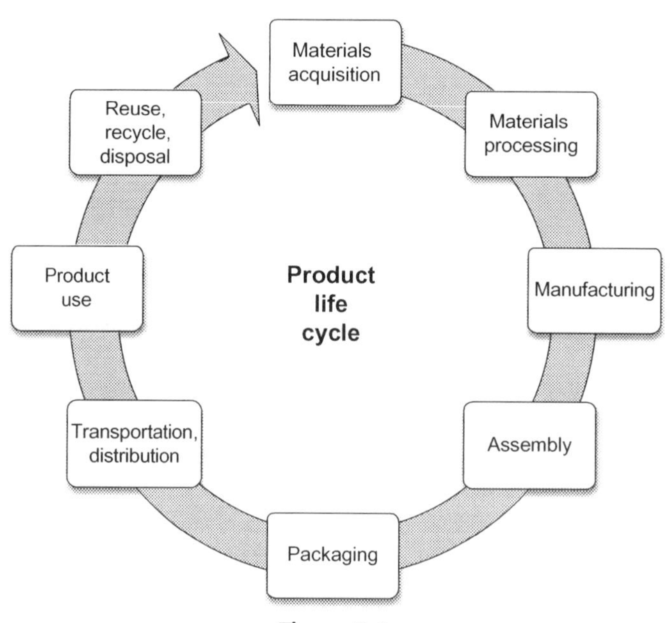

Figure 7.1
The life-cycle of IT products.

According to recent analyses by Marwedel and Engel (2010) and Deng et al. (2011), the manufacturing of current IT devices, such as laptop computers and flat-panel displays, plays an important role when considering a product's overall sustainability. Overall, manufacturing is responsible for up to 50% of the overall environmental impact of a product (when converting the impact to CO_2 impact), whereas the use of a product contributes largely to the remaining impact. Other influences, such as transporting a product from the factory to the customer or recycling a device, have only negligible impact. Figure 7.2 shows an example of the distribution of environmental impact over the life cycle of a typical desktop PC, assuming a typical lifetime of five years. Transportation and assembly make up only 6% of the overall environmental impact, whereas production and use generate the largest impacts. However, the environmental impact depends on additional factors, such as the energy mix used to power the operation of the devices. Figure 7.3 shows the different CO_2 impacts of operating a PC in a number of European countries with a different mix of energy sources.

This distribution of environmental impact over a product's life cycle implies that two different means of impact of software products must be considered. On the one hand, the *direct* impact results from the use of the software during the use of the product it is integrated into. On the other hand, a software can have a number of *indirect* effects when it requires the exchange of a physically working product because of increased performance requirements of updated versions or manufacturers cease to provide updated versions of software for older hardware generations, leading to effects such as increased vulnerability to security problems.

Next we discuss the different types of impact and give an overview of methods to reduce the environmental effect and discuss the potential related benefits and drawbacks of each method.

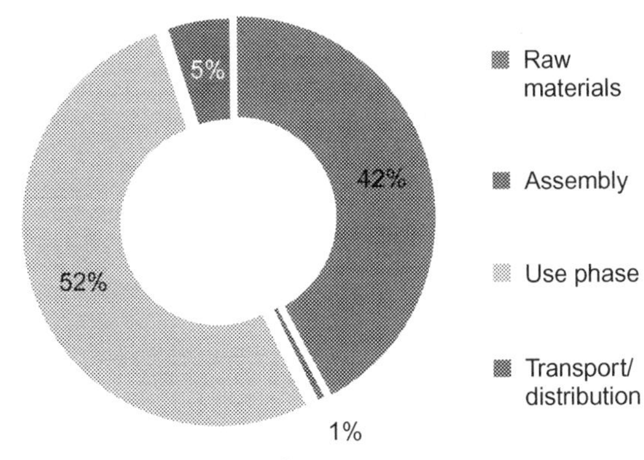

Figure 7.2
Distribution of CO_2 impact to PC life-cycle phases according to Fujitsu (2010).

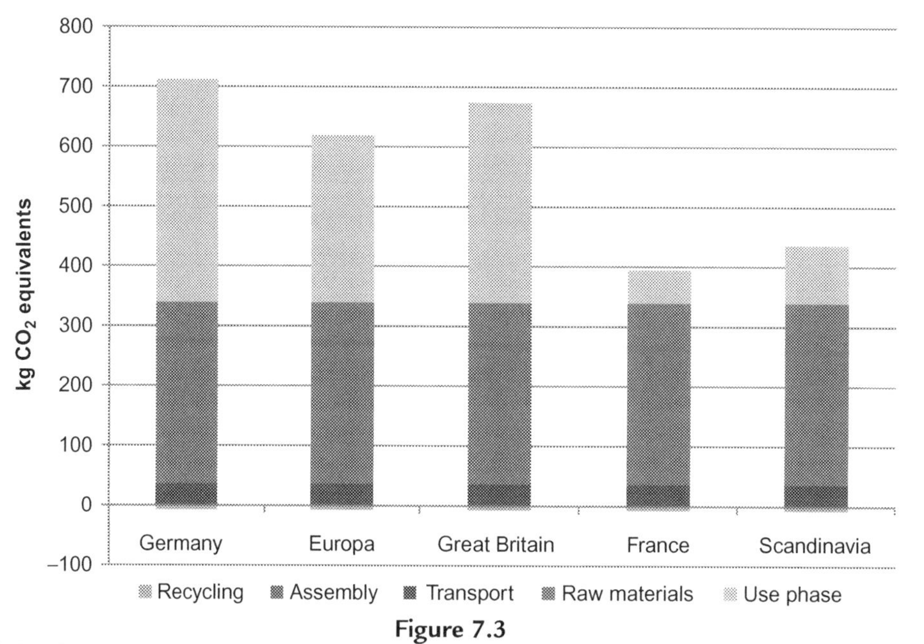

Figure 7.3
CO_2 impact over the PC life-cycle in different European countries according to Fujitsu (2010).

Direct Effects: Sustainability During Use

Improving sustainability during the use of a device by reducing its energy has effects directly noticeable by its users. Especially for battery-operated portable devices, such as laptop computers and mobile phones, a reduced energy consumption during runtime will result in an extended runtime on battery power.

Vargas (2005) and Carroll and Heiser (2010) analyzed the distribution of energy consumption in mobile phones by categorizing the energy use into wireless network, central processing unit (CPU), RAM, graphics, display, and peripheral use. Both analyses show that the fraction of energy used by the CPU and RAM (i.e., the direct effects of executing software on the device) amounts to 25-40% of the device's overall energy consumption.

Another application area that is extremely critical in terms of energy consumption is the sensor network. These networks consist of a large number of small sensor nodes, embedded computing devices that communicate over a wireless network. The sensor nodes are commonly distributed over a large area (e.g., a field in case of monitoring agricultural parameters) and are either battery powered or rely on energy-harvesting methods (Alvarado et al., 2012). Exchanging depleted batteries or performing maintenance on malfunctioning energy-harvesting devices In these applications is infeasible because of economical as well as physical constraints. As a consequence, sensor network researchers have invented a large number of methods to improve the runtime of sensor nodes on battery or on limited harvested energy using a number of scheduling approaches. Although we do not concentrate specifically on sensor networks in this chapter, some of the methods discussed are also applicable to these environments, and we expect the topic of efficient sensor devices to gain increased relevance because of the increased interconnection of devices in the internet of things (IoT).

Runtime Energy Consumption Basics

The overall energy consumption of a device is $E = \int P dt$ (i.e., low-power design is one way to reduce its energy consumption). On the hardware level, process technology and circuit design improvements during the last decades have enabled the reduction of power consumption while allowing higher processing speeds. However, this development hits a roadblock because of physical limitations that result in the lack of feasibility for scaling semiconductors according to Moore's law (Moore, 1965) in the future. Although using large-scale process technologies, modern CMOS circuits consume an amount of power approximately proportional to their operating frequency, $E \sim f$, Kim et al. (2003) predicted the increasing relevance of static power consumption because of leak currents when migrating semiconductors to smaller structure sizes. A more detailed analysis and prediction of these effects of nonideal scaling of semiconductors can be found in the International Technology Roadmap for Semiconductors (ITRS) roadmap (Kahng, 2013). For future complex circuits, the effect of leakage power will be paramount, in semiconductors with a large number of transistors it is no longer feasible to supply power to all transistors at the same time. As a consequence, only a part of such a chip will be active at any given time. This effect, termed "dark silicon" (Esmaeilzadeh et al., 2011), is expected to gain increasing significance for future highly complex circuits.

Because of these technological limitations on hardware, software methods to reduce the power consumption of devices have increasing relevance. Most these software approaches exploit a physical property or a specific hardware functionality in order to reduce the power consumption.

Analyzing the Energy Consumption of an Application

An obvious method for assessing the energy consumption of a given application is to measure the power consumption of the device executing the code (Tiwari et al., 1994; Russell and Jacome, 1998; Šimunić et al., 1999). Although the measuring approach results in a rather precise modeling of energy consumption, it is not practically applicable in a large number of cases. Especially in the area of embedded systems, *design space exploration* methods are used to find the optimal configuration (i.e., processor types and operating frequencies and details of the memory hierarchy) of a multicore platform. Building all required platforms in order to evaluate the energy consumption is obviously not feasible.

Energy models allowing for a flexible architecture have been developed for memories and processor architectures. The CACTI framework (Thoziyoor et al., 2008) predicts energy consumption for caches whereas DRAMsim2 (Rosenfeld et al., 2011) provides a cycle-accurate behavior and energy model for dynamic memories. The Wattch framework (Brooks et al., 2000) performs power estimation for processors on the architectural level without considering the underlying circuit or its layout. All these memory models are derived from specifications instead of measurements using a real physical device. As a consequence, they tend to be less precise.

In order to overcome this problem, *combined energy models* were developed; they predict the energy consumption of processors, communication channels, and memories. The first model was developed by Steinke et al. (2001) as a mixed model using measurements and predictions.

Steinke et al. (2001) described the overall energy consumption of a system as consisting of energy for CPU instruction execution and data processing as well as energy required for reading instructions and reading or writing data from and to memory:

$$E_{\text{total}} = E_{\text{cpu\_instr}} + E_{\text{cpu\_data}} + E_{\text{mem\_instr}} + E_{\text{mem\_data}}$$

Each class of energy consumption was subsequently refined as to its dependency on operand. For example, the cost for a sequence of m instructions is given as:

$$\begin{aligned} E_{\text{cpuInstr}} = & \sum \text{MinCostCPU}(\text{Opcode}) + \text{FUCost}(\text{Instr}_{i-1}, \text{Instr}_i) \\ & + \alpha_1 \times \sum w(\text{Imm}_{i,j}) + \beta_1 \times \sum h(\text{Imm}_{i-1,j}, \text{Imm}_{i,j}) \\ & + \alpha_2 \times \sum w(\text{Reg}_{i,k}) + \beta_2 \times \sum h(\text{Reg}_{i-1,k}, \text{Reg}_{i,k}) \\ & + \alpha_3 \times \sum w(\text{RegVal}_{i,k}) + \beta_3 \times \sum h(\text{RegVal}_{i-1,k}, \text{RegVal}_{i,k}) \\ & + \alpha_4 \times \sum w(\text{IAddr}_i) + \beta_4 \times \sum h(\text{IAddr}_{i-1}, \text{IAddr}_i) \end{aligned}$$

with w representing the number of "1" bits, h describing the Hamming distance between two values (i.e., the number of bits that flip when transitioning from one value to the next), FUCost the cost of switching functional units, and α and β parameters determined through experiments.

The work by Steinke et al. (2001) provides some interesting insights into CPU energy consumption that enabled the simplification of the resulting CPU energy model. One important result was that the Hamming distance between adjacently accessed addresses plays a major role, as shown in Figure 7.4. A number of factors could be ignored in the model. For example, it is not important which bits of an address are set to "1" or how many "1" bits an address on the memory address bus contains. Also, the cost of flipping a bit on the address bus is independent of the position of that bit. It is also irrelevant which data bit is set to "1"; however, the number of "1" bits on the data bus has a small effect (in the range of 3%) on the energy consumption. Finally, the cost of flipping a bit on the data bus is independent of the bit's position.

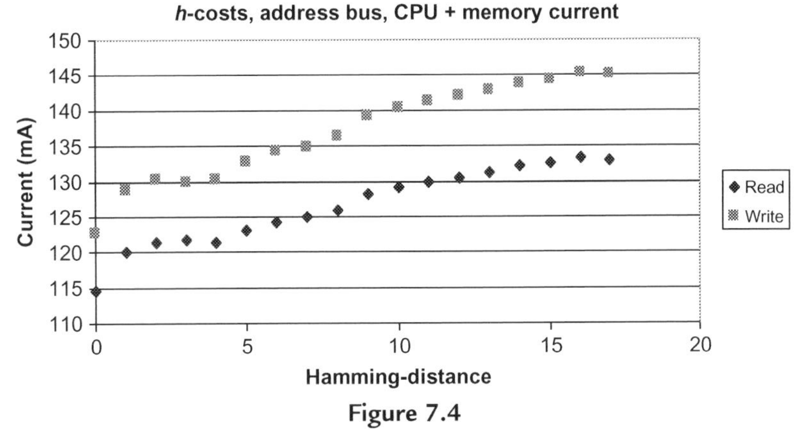

Figure 7.4

Relation of the Hamming distance between data words and the required current.

The CPU energy model was extended by model-based values for memories supplied by CACTI, which were validated against existing measurements of real static memory components. Using the model by Steinke et al. (2001) enabled the analysis of the energy consumption of a given sequence of machine language instructions for the ARM7TDMI processor. This model was subsequently used as the basis of research on energy-optimizing compilers.

Energy Consumption Reduction Using Physical Properties of Semiconductors

One of the commonly used approaches to conserve energy is *dynamic voltage-frequency scaling* (DVFS) (Le Sueur and Heiser, 2010). Lowering the operation frequency of a processor reduces the power consumption but will not lead to a direct reduction of energy consumption because a given program requires proportionally more time to execute. Energy savings can be achieved only because semiconductors commonly require a lower voltage supply when operating at lower frequencies. The relation of active power (ignoring leakage power) to the supply voltage V_{DD} is

$$P \sim V_{\mathrm{DD}}^2 \times f,$$

resulting in a superlinear reduction of dynamic power consumption when reducing the operating frequency and supply voltage simultaneously.

A number of additional hardware-based methods are available to reduce the power consumption. Two methods commonly used in modern semiconductors are *clock gating* (Wu et al., 2000) and *power gating* (Shi et al., 2007). Clock gating disconnects a circuit or a part of it from its driving clock(s). Because the dynamic power consumption of CMOS circuits is dependent on the operating frequency, this can be reduced to zero using clock gating. However, because of effects such as leakage currents, static power consumption is still relevant.

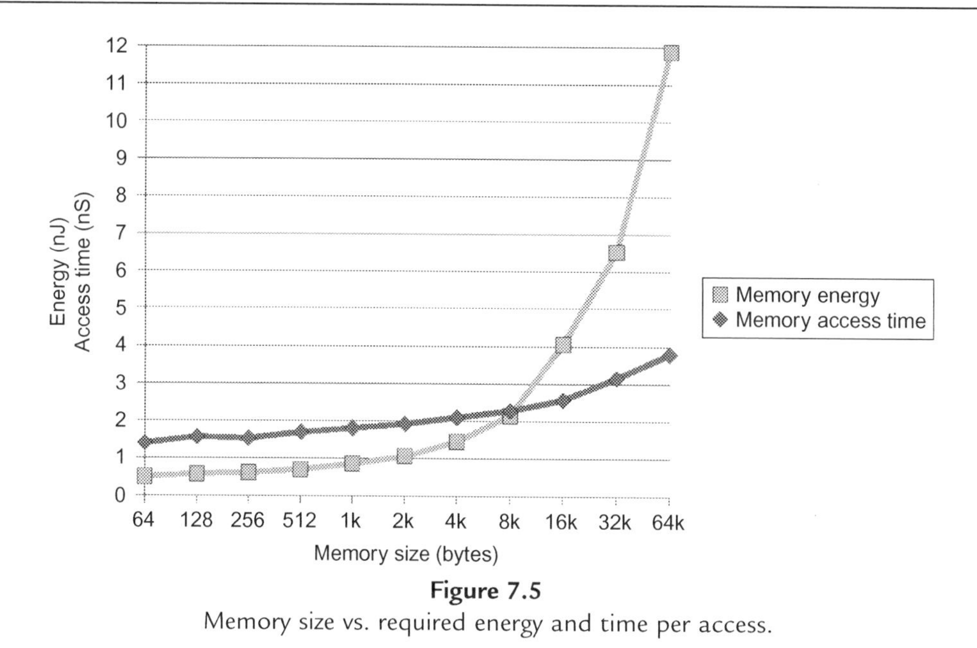

Figure 7.5
Memory size vs. required energy and time per access.

The dynamic and static power consumption of a circuit can be reduced to zero using power gating, which switches of the supply voltage to a circuit or a part of it. Although power gating has the obvious advantage of significant power reduction, it requires all data stored in the power gated circuit (e.g., in flip-flops or embedded memories) to be restored after the power is reapplied. Consequentially, this takes significantly longer compared to reapplying the frequency to a clock gated circuit.

Another method to reduce the energy consumption is to exploit differences in power consumption in the memory hierarchy of a system. Small static memories, called *scratchpad memories* (SPM) (Banakar et al., 2002) or *tightly coupled memories* (TCM), are significantly more energy efficient than large dynamic memories. An example for the relation of memory size to the energy required per access is given in Figure 7.5 (Marwedel et al., 2004). An approach to reduce the memory energy can rely on the locality of reference principle (Denning, 2005). This principle describes the phenomenon that a given value or related storage locations are frequently accessed by programs. *Temporal locality* refers to the reuse of specific data within a small time frame whereas *spatial locality* refers to the use of data objects within storage locations that are close to each other. Both locality patterns can be exploited to conserve energy by placing the related data objects into scratchpad memory.

Optimizing the Energy Consumption of an Application: Compiler Techniques

Based on precise energy models, approaches to reduce the energy consumption can be performed either using compiler-based static analysis techniques or dynamically, using system software to adapt operational parameters of the hardware at runtime.

Early approaches to reduce the energy consumption of programs exploited standard compiler optimization techniques. Optimizations such as instruction scheduling (Parikh et al., 2000), the strength reduction of operations (e.g., replacing multiplications by a sequence of shifts and additions), and the exploitation of small bit widths of load and store operations tend to improve the performance as well as the energy consumption of code. Based on the Steinke et al.'s (2001) energy models, the first energy-optimizing compiler, *encc*, utilized an energy profiler to analyze the generated code and varied the set of applied optimizations based on feedback from the profiler (Marwedel et al., 2001). Subsequently, the topic of compiler-based energy optimization on various levels of abstraction was discussed in a number of publications. Kadayif et al. (2005) devised a method for compiler-directed high-level energy estimation and optimization. Kandemir et al. (2002) developed low-level compiler back-end optimizations for energy reduction and concluded that compiling for power/energy is different from compiling for execution cycles.

Recently, approaches that apply postcompiler optimization using genetic algorithms have been introduced for x86 systems (Schulte et al., 2014). This approach combines methods from profile-guided optimization, superoptimization, evolutionary computation, and mutational robustness. It searches for program variants that improve the nonfunctional behavior while retaining the original code semantics using characteristic workloads and predictive modeling to guide the search.

However, energy savings as a result of applying compiler back-end transformations remained rather limited in scope. One promising approach was to exploit compiler transformation and static analysis techniques to explore the locality principle and assign code and data elements to SPM.

A first approach to exploit SPMs using a nonoverlaying static allocation of code and data was presented in Steinke et al. (2002). For each memory object k, the allocation method determines its size s_k and potential gain g_k. The optimization target is then to maximize the total gain $G = \sum g_k$ under the constraint that the sum of all objects allocated to the SPM is no larger than the SPM size SSP: $\mathrm{SSP} \geq \sum s_k$. This Steinke et al. approach used a Knapsack problem solver algorithm to find an optimal partitioning of the SPM.

Optimizing the Energy Consumption of an Application: Runtime Approaches

Although static approaches to optimize energy consumption work well in environments that use only one or a small number of well-defined applications, static approaches lose their optimization potential when applied to more complex systems. In these systems, the behavior and interaction of the programs on a system may change dynamically during their execution. Thus, dynamic approaches that analyze and optimize at runtime are required.

The use of DVFS has been exploited in a large number of publications. For predictably scheduled systems, such as real-time systems, recent work was published, for example, by Lee et al. (2009), Li and Broekaert (2014), and Kim et al. (2013). Using DVFS in general-purpose systems is more complex because the requirements of tasks and their order is not known in advance. Because switching between different voltage and frequency levels requires a significant amount of energy (Muhammad et al., 2008), an incorrect prediction of future computing power requirements may result in unnecessary switching, thus reducing the possible energy savings. DVFS for general-purpose systems has been investigated, for example, in Ayoub et al. (2011), Carroll and Heiser (2014), and Curtis-Maury et al. (2008).

For memories, dynamic allocation of SPMs is more complex than its static variant. Depending on changing circumstances, code and data objects must be frequently copied into and out of the SPM, depending on the current working set of the applications. The optimization must ensure that the cost of copying objects from and to the SPM does not offset the optimization potential achieved by accessing these objects in the SPM. A first approach to implement scratchpad sharing was proposed by Verma et al. (2005) and later integrated into the scheduler of a real-time operating system by Pyka et al. (2007). Egger et al. (2008) presented a scratchpad memory manager for a preemptive multitasking system and evaluated the energy saving potential using a large number of benchmark applications. Current research focuses on optimizing systems using SPMs-introduced dynamic scratchpad memory virtualization to improve power consumption and performance of multiprocessor systems (Bathen et al., 2011).

Energy consumption can also be reduced by exploiting the heterogeneous nature of current multiprocessor systems on chip (MPSoCs). Such an MPSoC as provided by the ARM big. LITTLE architecture (Jeff, 2012) consists of a number of processor cores with different internal architectures and clock frequencies. Code can be executed using low power on a simple, slow core or, if required, accelerated on a fast, complex core. Cordes et al. (2012) presented approaches to extract task and pipeline parallelism inherent in sequential C code and perform a multicriterial optimization to find a compromise between execution time and energy consumption. This parallelization approach is unique because it is able to generate code for a heterogeneous parallel platform that is as fast as required—in contrast to as fast as possible—under given energy constraints.

Optimizing the Energy Consumption of an Application: Probabilistic Approaches

A different approach to optimize the energy consumption relies on the fact that a large number of algorithms from such diverse areas as multimedia, machine learning, and signal processing are resilient against small variations or noise in the data they operate on. Several approaches reduce the number of significant digits in floating point calculations using clock or power gating in order to save energy.

Chakrapani et al. (2006) introduced probabilistic CMOS (PCMOS), an architecture that supplies the gates of an ALU that calculate on less significant bits of a word with increasingly lower supply voltages, resulting in power savings but introducing errors into the less significant bits with a high probability because of propagation delays. The efficiency of PCMOS was evaluated on small benchmark applications as well as a large-scale scientific application for weather forecasts (Düben et al., 2014). However, Heinig et al. (2012) has shown that it is difficult to achieve using PCMOS in low-power processors using integer data types.

Sampson et al. (2011) presented EnerJ, an approximate computing framework for Java. EnerJ extends Java with type qualifiers that indicate that data in the annotated variable can be operated on using approximate computation hardware and stored in low-power memories. Like PCMOS, EnerJ also requires hardware extensions to support energy management.

Taken together, a large number of different approaches are available to dynamically optimize the energy consumption of systems at compile time as well as runtime. However, combining these approaches is not straightforward. An optimization that, considered alone provides an improvement in energy consumption, can be canceled out by another optimization operating on different hardware components. An overall framework for systemwide energy optimization that integrates a large number of the existing approaches and can perform multicriterial optimizations for a complete system is still missing and an interesting target for future research.

Indirect Effects: Sustainability vs. Production

In contrast to the direct effects of executing software, which can be measured or modeled as described previously, the indirect effects of software are harder to assess because they are densely intertwined with the life cycle of a product.

Based on their physical properties, current computing devices, such as PCs, mobile phones, and printers, can have a lifetime of ten years or more. However, most devices are used for significantly shorter periods of time. A study by Entner (2011) analyzed the frequency of handset replacement cycles. Mobile phones were on average replaced every 18.7 months in the United States, the country in the study with the highest per capita income. In contrast, in India, a handset was used for 322.1 months on average. European countries averaged between 24.5 (United Kingdom) and 53.3 (Italy) months in this study.

Obviously, except for corner cases such as accidents or stolen phones, there is no physical requirement to buy a new handset every 18 months in the United States. However, in a mostly saturated market with an estimated smartphone market penetration of more than 55% in 2014

(Deloitte, 2013), manufacturers actively search for ways to sell the next generation of handsets. The situation is similar for many electronic devices, such as computers and printers. Originally, this approach was associated with the policy of US auto makers to change their models every year to create an incentive to buy new cars. Bulow (1986) analyzed this so-called planned obsolescence, the production of goods with uneconomically short useful lives so that customers must make repeat purchases.

Thus, the question is what role software plays in the context of planned obsolescence. First, software can be used to actively *enforce* (rather than just encourage) the obsolescence of a piece of hardware. The most prominent examples are printers. Many low-cost inkjet printers employ a chip in the ink cartridges that not only ensures that only original cartridges can be used but also implements a counter that prevents them from being used after a certain threshold (number of pages, time, etc.), even though the cartridge may still contain usable ink or could be refilled. Although this is an example of software used to actually *decrease* the sustainability of printers, there is also software providing countermeasures by, for example, resetting the chip inside the cartridge. Although this software can definitely help to improve the ecological footprint—as well as the economy of operating the printer—for obvious reasons, we will not link to related resources here.

The ever-increasing demands of upcoming generations of system and application software can be interpreted as a more subtle way of planned obsolescence. Previous generation hardware often does not work or does not work well with newer software releases because they are written with respect to the capabilities of the currently available hardware. This is often used as an excuse not to provide system software upgrades after a short amount of time, especially in the Android mobile phone market. Current data on this is hard to find unless one counts the large number of complaints from users on diverse online forums. A study from 2011 (theunderstatement.com, 2011) analyzed the currentness of the Android system version for 18 different handsets on the US market at that time:

- 7 of the 18 Android phones never ran a current version of the OS.
- 12 of 18 ran a current version of the OS for only a matter of weeks or less.
- 10 of 18 were at least two major versions behind well within their two year contract period.
- 11 of 18 stopped getting any support updates less than a year after release.
- 13 of 18 stopped getting any support updates before they even stopped selling the device or very shortly thereafter.
- 15 of 18 do not run Gingerbread, which shipped in December 2010.
- When Ice Cream Sandwich comes out, every device on it will be another major version behind.
- At least 16 of 18 handsets will almost certainly never get Ice Cream Sandwich.

However, the largest problem forcing upgrades in this situation is not the fact that the devices are not able to run newer versions of apps—in fact, many owners of a phone use only a small number of built-in apps[1] (Pew Research, 2011). A much larger problem is that no security fixes are provided for older phones. With the increasing amount of malware for the Android platform (Castillo, 2011), this leaves the user in a state of fear and uncertainty and possibly more willing to upgrade a perfectly working device early.

There are a number of similar cases, for example, in the PC industry. Although newer versions of operating systems, such as Windows 7 and 8, could in theory run on older hardware, in practice a lack of drivers for older components, such as graphics cards or network controllers, prevents the installation of these newer system versions.[2]

How software can be of help here depends on the expectations and requirements of the user and often also on the user's technical abilities. However, in general, the use of an open-source operating system helps to extend the lifetime of an older piece of equipment because support for a hardware platform is driven by developers' interest rather than economical requirements of the manufacturer. Although in the PC world, an old Windows installation could be replaced by, for example, an installation of Linux or FreeBSD, this solution will work for only a small number of users because of compatibility and usability concerns. For Android-based systems, however, compatible upgrades of the original mobile phone software are possible because of the open-source nature of the underlying system.[3] Systems such as CyanogenMod (Powers, 2011) are developed by a large number of volunteers and are made freely available for a number of older as well as current handsets, providing current software versions and bug fixed for devices without manufacturer support. However, these approaches also have their limits because some hardware components of phones rely on closed-source drivers, which are not compatible with more recent versions of the Linux kernel used in newer Android versions. Nevertheless, even in this area, some researchers provide software tools to enable the reverse engineering of binary drivers in order to enable compatibility (Chipounov and Candea, 2010).

As a consequence, the sustainability of devices could be improved if users were able to use them for a longer period of time, as also stated by Marwedel and Engel (2010). On the software side, the indirect effects of sustainability with respect to production could be mitigated when devices were provided with updates for a reasonable time after their release. Because there are no obvious economic advantages to this, support for this from manufacturers cannot be expected. Nevertheless, open-source projects such as CyanogenMod could help to improve the situation, at least for technically proficient users.

[1] The study from November 2011 found out that roughly 80% of Android phone owners do not use more than ten apps on a regular basis whereas one-third of the users use only three to five apps regularly.

[2] In the case of Windows, this effect interestingly also works in the other direction—older versions of Windows (2000, XP) cannot be installed on recent hardware because of a lack of drivers.

[3] Android is based on a Linux kernel and a number of additional open-source components.

Conclusions and Outlook

This overview has shown that software has a large impact on the sustainability of electronic devices from tiny embedded sensor nodes up to PCs and servers. The major contributors to CO_2 emissions are the manufacturing of a device and its operation whereas factors such as transport and recycling are comparatively negligible.

On the one hand, the reduction of energy consumption is possible for direct causes by applying optimization methods at design and runtime of a system. These optimization approaches are based on energy measurements and models and are rather closely related to the hardware of a system and its operation modes. Although optimization approaches for single criteria have been intensively researched, an overall systemwide optimization framework that considers all parts of a system and the respective trade-offs between its components with regard to energy use is still not available.

On the other hand, a significant amount of energy can also be saved when viewing the life cycle of not only one product but also a series of products replacing each other. Because the production of a device can cause up to 50% of the overall CO_2 emissions, a device generated during its life cycle enabling its longer useful life, reduces or delays the requirement to produce a new one. This is often blocked by the manufacturers' desire to sell new devices to saturated markets by introducing planned or even forced obsolescency. Software plays an important role because limiting the availability of software upgrades to currently sold devices creates incentives to upgrade.

Some of the problems with planned obsolescency could be cured or mitigated by the use of open-source software. However, this is in no way a panacea because closed specifications of hardware devices, locked or encrypted bootloaders, or hardware fuses can provide manufacturers another way to enforce obsolescence.

The question of where we are heading remains. The limited availability of materials essential for modern electronic devices, such as raw earth elements and the ever-increasing cost of energy, will enforce sustainable production and operation of devices. Software will probably play an ever-increasing role in this because semiconductor fabrication technology heads toward the end of profitable scaling according to Moore's law.

One effect that can have interesting effects on the sustainability of software is the trend to connect mobile and stationary systems—as well as embedded systems in the IoT—to the cloud. This would enable the outsourcing of computing capacity to virtualized server farms, which can be operated more efficiently, even using a large share of renewable energy. The end user device itself would then be primarily relegated to being a sort of intelligent terminal. Some developments, such as simple laptops running Google's ChromeOS, which is essentially an execution platform for Web browser-based applications, may be pioneers of this new era.

However, also this must be taken with a grain of salt. In addition to the obvious privacy concerns, the pervasive use of such devices will significantly increase the energy consumption of the mobile and wired Internet infrastructure, in turn requiring research into network energy optimization as discussed in a different chapter of this book.

References

Alvarado, U., Juanicorena, A., Adin, I., Sedano, B., Gutiérrez, I., de No, J., 2012. Energy harvesting technologies for low-power electronics. Trans. Emerg. Telecommun. Technol. 23 (8), 728–741.

Ayoub, R.Z., Ogras, U., Gorbatov, E., Jin, Y., Kam, T., Diefenbaugh, P., Rosing, T., 2011. OS-level power minimization under tight performance constraints in general purpose systems. In: Proceedings of the 17th IEEE/ACM International Symposium on Low-Power Electronics and Design (ISLPED'11). IEEE Press, Piscataway, NJ, USA, pp. 321–326.

Banakar, R., Steinke, S., Lee, B., Balakrishnan, M., Marwedel, P., 2002. Scratchpad memory: a design alternative for Cache on-chip memory in embedded systems. In: Proceedings of the Tenth International Symposium on Hardware/Software Codesign. ACM, New York, NY, USA, pp. 73–78.

Bathen, L.A.D., Dutt, N.D., Shin, D., Lim, S., 2011. SPMVisor: dynamic scratchpad memory virtualization for secure, low power, and high performance distributed on-chip memories. In: Proceedings of the Seventh IEEE/ACM/IFIP International Conference on Hardware/software Codesign and System Synthesis (CODES+ISSS '11). ACM, New York, NY, USA, pp. 79–88. http://dx.doi.org/10.1145/2039370.2039386.

Bozzelli, P., Gu, Q., Lago, P., 2013. A systematic literature review on green software metrics. Technical report, VU Amsterdam.

Brooks, D., Tiwari, V., Martonosi, M., 2000. Wattch: a framework for architectural-level power analysis and optimizations. SIGARCH Comput. Archit. News 28 (2), 83–94. http://dx.doi.org/10.1145/342001.339657.

Bulow, J., 1986. An economic theory of planned obsolescence. Q. J. Econ. 101 (4), 729–749. http://dx.doi.org/10.2307/1884176.

Carroll, A., Heiser, G., 2010. An analysis of power consumption in a smartphone. In: Proceedings of the 2010 USENIX Conference on USENIX Annual Technical Conference (USENIXATC'10). USENIX Association, Berkeley, CA, USA, p. 21ff.

Carroll, A., Heiser, G., 2014. Mobile multicores: use them or waste them. SIGOPS Oper. Syst. Rev. 48 (1), 44–48. http://dx.doi.org/10.1145/2626401.2626411.

Castillo, C.A., 2011. Android Malware: Past, Present, and Future, McAfee White Paper. http://www.mcafee.com/us/resources/white-papers/wp-android-malware-past-present-future.pdf.

Chakrapani, L.N., Akgul, B., Cheemalavagu, S., Korkmaz, P., Palem, K.V., Seshasayee, B., 2006. Ultra efficient embedded SOC architectures based on probabilistic CMOS (PCMOS) technology. In: Design Automation and Test in Europe Conference (DATE).

Chipounov, V., Candea, G., 2010. Reverse engineering of binary device drivers with RevNIC. In: Proceeding of the 5th ACM European Conference on Computer Systems, Paris, France, April.

Cordes, D., Engel, M., Marwedel, P., Neugebauer, O., 2012. Automatic extraction of multi-objective aware pipeline parallelism using genetic algorithms. In: Proceedings of CODES+ISSS, pp. 73–82.

Curtis-Maury, M., Shah, A., Blagojevic, F., Nikolopoulos, D.S., de Supinski, B.R., Schulz, M., 2008. Prediction models for multi-dimensional power-performance optimization on many cores. In: Proceedings of the 17th International Conference on Parallel Architectures and Compilation Techniques (PACT '08). ACM, New York, NY, USA, pp. 250–259. http://dx.doi.org/10.1145/1454115.1454151.

Deloitte, 2013. The state of the global mobile consumer, 2013—Divergence deepens.

Deng, L., Babbitt, C.W., Williams, E.D., 2011. Economic-balance hybrid LCA extended with uncertainty analysis: case study of a laptop computer. J. Clean. Prod. 19 (11), 1198–1206.

Denning, P.J., 2005. The locality principle. Commun. ACM 48 (7), 19–24.

Dick, M., Naumann, S., Kuhn, N., 2010. A model and selected instances of green and sustainable software. In: HCC'10, pp. 248–259.

Düben, P.D., Joven, J., Lingamneni, A., McNamara, H., De Micheli, G., Palem, K.V., Palmer, T.N., 2014. On the use of inexact, pruned hardware in atmospheric modelling. Philos. Trans. R. Soc. A Math. Phys. Eng. Sci. 372 (2018), 20130276.

Egger, B., Lee, J., Shin, H., 2008. Scratchpad memory management in a multitasking environment. In: Proceedings of the 7th ACM International Conference on Embedded Software (EMSOFT), Atlanta, USA, December, pp. 265–274.

Entner, R., 2011. International Comparisons: The Handset Replacement Cycle. Recon Analytics, Massachusetts.

Esmaeilzadeh, H., Blem, E., St. Amant, R., Sankaralingam, K., Burger, D., 2011. Dark silicon and the end of multicore scaling. In: Proceedings of the 38th Annual International Symposium on Computer Architecture (ISCA '11). ACM, New York, NY, USA, pp. 365–376. http://dx.doi.org/10.1145/2000064.2000108.

Fujitsu, 2010. Life Cycle Assessment and Product Carbon Footprint—Fujitsu ESPRIMO E9900 Desktop PC. http://fujitsu.fleishmaneurope.de/?attachment_id=2148.

Heinig, A., Mooney, V.J., Schmoll, F., Marwedel, P., Palem, K.V., Engel, M., 2012. Classification-based improvement of application robustness and quality of service in probabilistic computer systems. In: Proceedings of ARCS 2012, Munich, Germany.

Jeff, B., 2012. Big.LITTLE system architecture from ARM: saving power through heterogeneous multiprocessing and task context migration. In: Proceedings of DAC. ACM, New York, NY, USA, pp. 1143–1146.

Kadayif, I., Kandemir, M., Chen, G., Vijaykrishnan, N., Irwin, M.J., Sivasubramaniam, A., 2005. Compiler-directed high-level energy estimation and optimization. ACM Trans. Embed. Comput. Syst. 4 (4), 819–850. http://dx.doi.org/10.1145/1113830.1113835.

Kahng, A.B., 2013. The ITRS design technology and system drivers roadmap: process and status. In: Proceedings of the 50th Annual Design Automation Conference (DAC '13). ACM, New York, NY, USAhttp://dx.doi.org/10.1145/2463209.2488776 Article 34, 6 pages.

Kandemir, M., Vijaykrishnan, N., Irwin, M.J., 2002. Compiler optimizations for low power systems. In: Graybill, R., Melhem, R. (Eds.), Power Aware Computing. Kluwer Academic Publishers, Norwell, MA, USA, pp. 191–210.

Kim, N.S., Austin, T., Baauw, D., Mudge, T., Flautner, K., Hu, J.S., Irwin, M.J., Kandemir, M., Narayanan, V., 2003. Leakage current: Moore's law meets static power. IEEE Comput. 36 (12), 68–75. http://dx.doi.org/10.1109/MC.2003.1250885.

Kim, S.I., Kim, H.T., Kang, G.S., Kim, J.-K., 2013. Using DVFS and task scheduling algorithms for a hard real-time heterogeneous multicore processor environment. In: Proceedings of the 2013 Workshop on Energy Efficient High Performance Parallel and Distributed Computing (EEHPDC '13). ACM, New York, NY, USA, pp. 23–30. http://dx.doi.org/10.1145/2480347.2480350.

Le Sueur, E., Heiser, G., 2010. Dynamic voltage and frequency scaling: the laws of diminishing returns. In: Proceedings of the 2010 International Conference on Power Aware Computing and Systems (HotPower'10). USENIX Association, Berkeley, CA, USA, pp. 1–8.

Lee, W.Y., Ko, Y.W., Lee, H., Kim, H., 2009. Energy-efficient scheduling of a real-time task on DVFS-enabled multi-cores. In: Proceedings of the 2009 International Conference on Hybrid Information Technology (ICHIT '09). ACM, New York, NY, USA, pp. 273–277. http://dx.doi.org/10.1145/1644993.1645046.

Li, S., Broekaert, F., 2014. Low-power scheduling with DVFS for common RTOS on multicore platforms. SIGBED Rev. 11 (1), 32–37. http://dx.doi.org/10.1145/2597457.2597461.

Marwedel, P., Engel, M., 2010. Plea for a holistic analysis of the relationship between information technology and carbon-dioxide emissions. In: Workshop on Energy-aware Systems and Methods (GI-ITG), Hanover/Germany.

Marwedel, P., Steinke, S., Wehmeyer, L., 2001. Compilation techniques for energy-, code-size-, and run-time-efficient embedded software. In: Int. Workshop on Advanced Compiler Techniques for High Performance and Embedded Processors, Bucharest, Hungary.

Marwedel, P., Wehmeyer, L., Verma, M., Steinke, S., Helmig, U., 2004. Fast, predictable and low energy memory references through architecture-aware compilation. In: ASP-DAC 2004, pp. 4–11.

Moore, G.E., 1965. Cramming more components onto integrated circuits. Electron. Mag. 38 (8), 114–117. http://dx.doi.org/10.1109/jproc.1998.658762.

Muhammad, F., Muller, F., Auguin, M., 2008. Precognitive DVFS: minimizing switching points to further reduce the energy consumption. In: Proc. 14th IEEE Real-Time and Embedded Technology and Applications Symposium, WIP Session, St. Louis, MO, USA.

Parikh, A., Kandemir, M., Vijaykrishnan, N., Irwin, M.J., 2000. Instruction scheduling based on energy and performance constraints. In: 2000 Proceedings. IEEE Computer Society Workshop on VLSI. IEEE Comput. Soc, pp. 37–42. http://dx.doi.org/10.1109/IWV.2000.844527.

Powers, S., 2011. CyanogenMod 7.0—gingerbread in the house. Linux J.. 2011(206).

Purcell, K., 2011. Half of adult cell phone owners have apps on their phones. Pew research report. http://pewinternet.org/Reports/2011/Apps-update.aspx.

Pyka, R., Faßbach, Ch., Verma, M., Falk, H., Marwedel, P., 2007. Operating system integrated energy aware scratchpad allocation strategies for multiprocess applications. In: Proceedings of 10th International Workshop on Software & Compilers for Embedded Systems (SCOPES), Nice, France, pp. 41–50.

Rosenfeld, P., Cooper-Balis, E., Jacob, B., 2011. DRAMSim2: a cycle accurate memory system simulator. IEEE Comput. Archit. Lett. 10 (1), 16–19. http://dx.doi.org/10.1109/L-CA.2011.4.

Russell, J.T., Jacome, M.F., 1998. Software power estimation and optimization for high performance, 32-bit embedded processors. In: ICCD 1998, pp. 328–333.

Sampson, A., Dietl, W., Fortuna, E., Gnanapragasam, D., Ceze, L., Grossman, D., 2011. EnerJ: approximate data types for safe and general low-power computation. In: PLDI'11.

Schulte, E., Dorn, J., Harding, S., Forrest, S., Weimer, W., 2014. Post-compiler software optimization for reducing energy. SIGPLAN Not. 49 (4), 639–652. http://dx.doi.org/10.1145/2644865.2541980.

Shi, K., Lin, Z., Jiang, Y.-M., 2007. Practical Power Network Synthesis for Power-Gating Designs. EETimes Design How-To, May 2007.

Šimunić, T., Benini, L., De Micheli, G., 1999. Cycle-accurate simulation of energy consumption in embedded systems. In: Proceedings of the 36th Annual ACM/IEEE Design Automation Conference, pp. 867–872.

Steinke, S., Knauer, M., Wehmeyer, L., Marwedel, P., 2001. An accurate and fine grain instruction-level energy model supporting software optimizations. In: PATMOS, Yverdon, Switzerland.

Steinke, S., Wehmeyer, L., Lee, B., Marwedel, P., 2002. Assigning program and data objects to scratchpad for energy reduction. In: Proceedings of the Conference on Design, Automation and Test in Europe (DATE), Paris, France, March, pp. 409–415.

Tate, K., 2005. Sustainable Software Development: An Agile Perspective. Addison-Wesley Professional, Boston, MA, USA.

theunderstatement.com, 2011. Android Orphans: Visualizing a Sad History of Support. http://theunderstatement.com/post/11982112928/android-orphans-visualizing-a-sad-history-of-support.

Thoziyoor, S., Ahn, J.H., Monchiero, M., Brockman, J.B., Jouppi, N.P., 2008. A comprehensive memory modeling tool and its application to the design and analysis of future memory hierarchies. SIGARCH Comput. Archit. News 36 (3), 51–62.

Tiwari, V., Malik, S., Wolfe, A., 1994. Power analysis of embedded software: a first step towards software power minimization. IEEE Trans. Very Large Scale Integr. VLSI Syst. 2 (4), 437–445.

Vargas, O., 2005. Minimum power consumption in mobile-phone memory subsystems. Portable Des. 11 (9), 24–25, Infineon Technologies.

Verma, M., Petzold, K., Wehmeyer, L., Falk, H., Marwedel, P., 2005. Scratchpad sharing strategies for multiprocess embedded systems: a first approach. In: Proceedings of IEEE 3rd Workshop on Embedded System for Real-Time Multimedia (ESTIMedia), Jersey City, USA, September, pp. 115–200.

Weiss, R.F., Mühle, J., Salameh, P.K., Harth, C.M., 2008. Nitrogen trifluoride in the global atmosphere. Geophys. Res. Lett. 35, L20821. http://dx.doi.org/10.1029/2008GL035913.

Wu, Q., Pedram, M., Wu, X., 2000. Clock-gating and its application to low power design of sequential circuits. IEEE Trans. Circuits Syst. I: Fundam. Theory Appl. 47 (3), 415–420. http://dx.doi.org/10.1109/81.841927.

CHAPTER 8

Achieving the Green Theme Through the Use of Traffic Characteristics in Data Centers

Kiran Voderhobli
Leeds Beckett University, Leeds, UK

Introduction

Green IT and the Cloud

Computing has evolved from a client-server approach to distributed computing. This evolution has further given rise to many innovations and business models with cloud computing being one of them. Cloud computing has opened new avenues of possibilities and limitless powerful resources to end users. Today, it is feasible for businesses and end users to demand storage, processing resources, services, and hardware and software on an ad hoc as-needed basis from the cloud. Therefore, it is easy to understand why computing in general has become cheaper, which in turn has helped many business embrace the advantages offered by cloud-based infrastructures. The concept of being able to instantiate software and hardware resources to various users is the fundamental basis for how a cloud infrastructure is able to cater to the computing needs of various users. Cloud infrastructures exist as public and private, depending on accessibility profiles. Anyone could subscribe to the services offered by a cloud and use its resources on a "pay-as-you-go" basis. Users are also isolated from the technical intricacies that build the cloud.

In addition to the benefits of accessibility to resources, another major element of advantage to the cloud concept is better sustainability. The advent of cloud computing has been monumental in realizing the ambitions of implementing green computing. It is now possible to virtually provide a network without actually having to build one physically. Cloud infrastructures not only save money on hardware and software but also allow various levels of virtualization that help to create multiple instances of resources that can be shared. The fact that the cost of computing hardware is becoming less expensive while also becoming more powerful is a reason for adopting cloud infrastructure for sustainable information technology (IT). Because of the need for less physical computing equipment, there is also less demand for localized hardware cooling systems and other initiatives to conserve energy. Hence, compared to traditional models of networked computing, cloud-based virtualization helps conserve energy. Widespread adoption of clouds for everyday computing fits with current policies related to sustainability being pushed by many governments and companies. The aim is to control carbon footprints and greenhouse gases contributed by the everyday use of information and communication technologies (ICTs).

Although the use of virtualized resources saves energy, a cloud data center is still an epicenter of high-energy consumptions because of the parallel working of arrays of servers, processes, and networking equipment. Because of the fact that a cloud comprises many hardware and software elements placed in a distributed fashion, it is very difficult to precisely identify one area of energy optimization. Many contributing factors add to overall energy use. The figures presented by Sun and Lee (2006) and Hussein and Burak (2013) reiterate that a data center could use anywhere between 2000 and 3000 kW h (m^2 a year). With the global rise in data centers, the problem of high-energy consumption clearly is compounded. Although the presence of some of the contributors to this energy consumption is inevitable, there are still opportunities to minimize it. The immediate ideas that spring to mind when discussing solutions for sustainable computing are external cooling systems, changes to chip-level design, and lower power infrastructure. But sustainability must also be looked at from the viewpoint of local processes and network operations. After all, the idea is to optimize algorithms and data center activities, which when combined with other hardware-based solutions, will yield substantial energy savings. Although the complexity of computing has moved away from the user to the data center, significant overhead is still present at the servers that comprise the data center. Research into these processes for energy optimization has gained much importance in recent years.

Most network-related processes are the result of operations between an end user and a virtualized entity. Transmission of packets from point A to point B consumes energy. It is easy to imagine how energy intensive a data center can be because it is the centerpiece of network activities. Packet transmission resulting from virtualization should not be overlooked. To achieve sustainability on the cloud, proper consideration must be given to how virtualization is managed in terms of the following factors:

- Location of creation (which physical server) of a virtualized resource
- Load on the virtualized resource
- Load on the physical resource
- Contention for physical resources (for example, for a network card)
- Conditions that warrant migration of a virtualized resource
- Conditions that warrant powering down or making a device dormant
- Conditions that warrant clustering of virtualized entities to a server or a subset

These factors must be based on some decision-making paradigms. A live analysis of network activity should result in taking actions on the management of virtualized entities. Given the importance of green themes in ICT, it would make sense to embed autonomic decision making based on cloud activity. This would be possible if a body of knowledge is derived from the live network that could possibly be shared among peer cloud entities.

Virtualization Behavior

Virtualization is the concept of creating a level of abstraction that enables end users to share the computing power of physical machines present on a cloud (Figure 8.1). The abstraction could include hardware or software. For example, multiple operating systems could be hosted on a single server regardless of the actual hardware profile of the server. Another form of virtualization is network device virtualization for which, for example, a device such as a router could have an instance that can be visible to users.

The concept of virtualization helps save energy by using fewer energy sources to support computing resources. A number of operations on virtualized entities are possible. Once a virtualized instance is created, it could be suspended, relocated, consolidated with other virtualized instances, and migrated to other physical machines. The key concept that helps realize the "green theme" of a cloud is the ability to perform VM management actions based on changing network scenarios. The relationship between virtual entities and physical entities is straightforward. Virtualization demands processing power from the underlying hardware. This has a rippling effect on power consumption. The more the load imposed by virtualized entities on the physical entities, the more energy consumption is demanded by hardware. Therefore, a single piece of hardware could consume varying levels of energy at different times, depending on the load it is supporting. This results in a "chain reaction" locally and across the network in terms of quality of service (QoS). Therefore, studying virtualization traffic

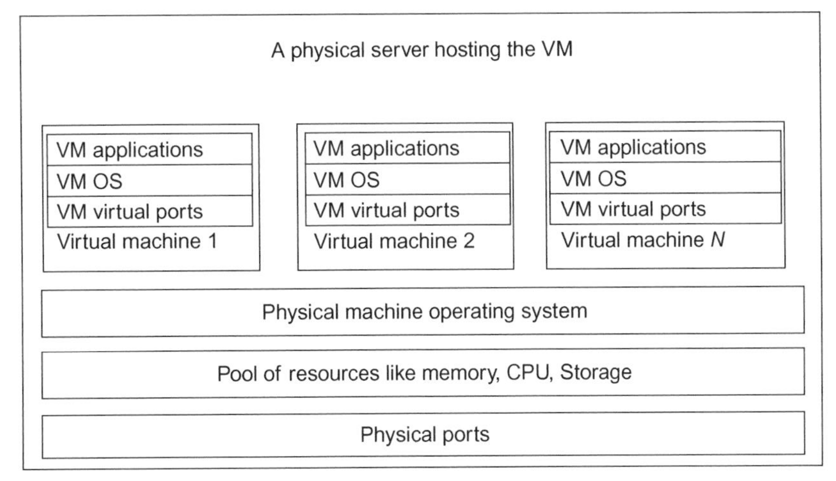

Figure 8.1
Virtualization on a single physical machine.

behavior at the granularity of each virtualized entity is one of the paramount considerations needed for managing virtualization. Ideally, it would be beneficial if a body of knowledge could be derived from virtualized instances that would help make critical runtime decisions.

Chapter Coverage

In the recent years, new plans for energy optimization for clouds based on network traffic consumption have been considered. Many disjoint pieces of research that show that virtualization management must be based on logic taken at the granularity of each virtualized entity. In this regard, there are various plans including the use of simple network management protocol (SNMP) to characterize network traffic. This chapter is intended to give the reader an idea of network characteristics and what it means in relation to green information and communication technology (ICT) (and green cloud). It will provide an overview of some of the research performed in this area and present a prototype for a model that employs SNMP for network traffic in a virtualized environment. The chapter also provides insight on how to extend network management principles for cloud-based environments regarding "greening." By considering green computing as a network management problem, it is possible to look for network attributes and statistics that could reveal opportunities for saving energy. Also, constantly gathering network data from individual virtualized entities could help create a knowledge base that feeds into critical decision making on virtualization management. The chapter also discusses the practicality of gathering very fine granular network statistics from a local physical systems and relating them to traffic contributed by virtualized entities. Furthermore, this leads on to the discussion on how virtual machines (VMs) in a data center can be managed based on information sharing between physical machines. The reader will be able to appreciate the complexity of gathering statistics in a highly intensive data center because of the volume of network traffic and distributed architecture of cloud-based services. The chapter describes research work that is still in its infancy and is likely to become significant in the future because of the need to deploy energy awareness to cloud architectures. The chapter will contribute to ideas for developing future autonomic systems for power-efficient virtualization management for cloud infrastructures.

Rationale

Relationship Between Infrastructure as a Service (IaaS) and Power

IaaS is the principle regarding where resources that are to be shared are allocated on a need-to-know basis. In a way, this is outsourcing of computing tasks to a secondary facility. Instead of deploying redundant hardware and software resources, a user can choose to subscribe to an IaaS being offered by a cloud. The infrastructure being offered on demand could be a set of servers, databases, specialist software, routers, data storage, and so on. In any case, each resource that can be shared could have multiple instances of real resources. This concept is known as *virtualization*. Although it yields to significant energy savings, it is important to remember that the activity of each virtualized instance (for example, a VM) consumes energy, depending on the processing tasks it is undertaking. In a recent study, Mouftah and Kantarci (2013) highlight the fact that the heavy use of clouds for everyday ICT is a massive contributor to energy consumption. Therefore, there is a need to further optimize cloud-based operations to save energy. This chapter is a step in that direction.

Network Processes and Power

Virtualized resources on the cloud interface with the underlying network. The load on the physical machine and the physical network is proportional to the packets being processed, sent, and received by an aggregate of virtualized instances. Hence, there is a direct link between packet transmission and power consumption. This is also evidenced by the fact that telecommunication systems, data centers, and end user processes have been proven to be the key contributors to energy consumption (Chiaraviglio and Mellia, 2013). There is a need to study packet transmission characteristics that can help refine virtualization scheduling and network management algorithms in order to reduce power consumption. This is in line with one of the best practices for creating a green data center (i.e., understanding power consumption and its impact (Guelzim and Obaidat, 2013)).

Need for Thermal-Aware Virtualization

Advancements to cloud computing algorithms have resulted in many plans requiring decisions on how to spawn, allocate, and manage virtualized resources. But there is still much work to be done in terms of considering energy use as one of the major criterion for virtualization management. It has been acknowledged that network processes contribute to power consumption. Therefore, to implement virtualization plans that take this into account, how live network statistics can be gathered from virtualized entities must be investigated. In other words, a thermally aware virtualization plan must consider the current state of network activity across the data center network (DCN) in order to manage virtualized entities.

Understanding Sustainability on the Cloud

Current State of Affairs

Government policies, the global outlook on greenhouse gases, and the drive to reduce CO_2 emissions have resulted in widespread adoption of cloud-based services. Today, it is easy to have access to high-speed computing resources and networking equipment via a cloud at a fraction of the costs. Still, recent studies have shown that the power consumption in data centers is increasing at an alarming rate. Until recently, cloud infrastructure was considered as a savior business model to alleviate the energy crisis of computing. ICTs' reliance on clouds is raising concerns about the amount of greenhouse gases being emitted by these infrastructures (Mouftah and Kantarci, 2013). It is estimated that an average corporate data center uses up to $100,000 worth of power annually. (Jingy and Xing, 2013). At the rate in which data centers are being deployed, a power crisis has been forecast. Even with sophisticated cooling systems, intelligent scheduling algorithms, and energy-efficient networking, each data center will consume a certain baseline of power. A report produced by the Climate Group claims that data centers roughly contribute 7% per annum to ICT's overall carbon footprint (The Climate Group, 2008). The report also states that with the current trends in cloud computing, approximately 122 million data centers are estimated to be in use by the year 2020. Modern data centers are in the realm of exascale computing in which they process approximately 10^{18} operations per second (Bilal et al., 2014). They point to various solutions to reduce energy consumption including traffic consolidation and link management.

It has often been argued that better hardware design, powerful processing units, better process allocation algorithms, and so on would help reduce power. However, even with improvements in parallel processing and central processing unit (CPU) scheduling, Amdhal's law proves that there will always will a portion of computing that cannot be optimized with parallelization. This means that a server will always be subject to some power consumption at all times. In terms of a data center, Gustafson's law is more relevant because it considers improved processing efficiency by adding more machines. In any case, it is deniable that the problem of power consumption is compounded as more machines are added. Certainly cloud computing is power hungry even when there is minimal processing. There is also the factor of energy being consumed in just keeping the DCN elements active. Jingy and Xing (2013) claim that between 30% and 60% of energy in data center server rooms is wasted. Also, cooling systems in most data centers run continuously regardless of their workload. In addition to servers, cooling devices, and auxiliary support systems in a DCN, there are also networking devices that consume energy. As with servers, DCNs always have a baseline power consumption that increases as packet-processing tasks are designated to them. Worse yet, there are also networked devices that consume a fixed amount of power regardless of how much they are processing. Modern data centers suffer from nonoptimized energy use because of the high amounts of energy they use at all times regardless of resource use. Ebrahimzadeh and Rahbat (2013) made a very thorough review on where the energy consumption in data centers arises. The authors break down the contributors to energy using an analytical model. The literature states that there is clearly a disparity between energy consumption and resource use because of "energy-agnostic" devices. The authors claim that servers continue to use half of the peak energy even when they are idle. This echoes the findings from an experiment that criticized the power hungriness of static network designs (Chabarek et al., 2008). The research supported the concept that redundant network infrastructure is not efficient in terms of energy consumption and that the way forward is to deploy energy-aware networks. It is therefore clear from this discussion that the common view that QoS can be guaranteed only by interconnected power-hungry network devices must change.

Because a data center's activity fluctuates throughout the day based on user demand, more needs to be done to minimize power consumption. The problem of power consumption in DCNs requires solutions from many ends. No one method is adequate to tackle this problem. Cloud architectures and cloud-based software are developing continuously, and it will be a long time before concrete green standards are put into place. Therefore, much research addresses power savings from different viewpoints. It is expected that future cloud deployments will consider amalgamating various plans to reduce power consumption.

Achieving Sustainability on the Cloud

As explained, there is no single body of knowledge that lists the steps to achieve a green data center because of the multitude of contributing factors. Currently, the process of "greening" a data center includes adding additional mechanisms for efficient power consumption. Although cooling and hardware choices are built into most cloud deployment frameworks, other areas for improvement should be considered postdeployment. Some approaches for sustainability follow:

- Heat reduction mechanisms such as fluid/dry cooling systems and computer room air conditioning (CRAC) are present in every data center. They are for the entire data center as a whole rather than individual servers.
- Deployment of hardware-optimized servers and networking equipment consume less power by employing parallel processing and different modes of operation depending on workloads. This has been made possible recently because of developments in chip-level and processor-level optimizations. The focus on green-aware hardware is expected to continue to grow rapidly in the coming years.
- Power-aware intelligent networking devices will result in fewer redundant networking devices to support the DCN. For example, recent developments in routing have created routers that are able to off-load their tasks dynamically to other routers to save energy. Modern network devices for the cloud are also able to perform more efficient port management that allows a set of ports to be shared among many network-dependant applications. This helps to keep unused ports in a dormant or power-saving state until being commissioned for use.
- There have been developments in cloud-based virtual networking algorithms such as intelligent power-aware routing techniques. For example, Chang and Wu (2010) suggest a method for routing packets in a cloud virtual network where the route for a packet is selected based on energy efficiency.

- One of the traditional approaches to green computing is using "wake-on" devices such as network interfaces. Although this plan per se has been used for a few years for LANs, it is has been extended to cloud environments and distributed networks.
- Traffic-aware green initiatives have been used with processors that could be controlled based on network traffic patterns (Christensen et al., 2004). If idle times can be predicted, processors could be suspended during periods of inactivity to save power. Obviously, processors across a large cloud data center could potentially experience several windows of idle times. The culmination of suspending processors during such periods would contribute to significant power saving in the overall network. It is important to look into how the scheduling and clustering of tasks to fewer processors as opposed to distributing them across the network would help save energy. Networking protocols can also be made to make operational decisions to control transmitters based on changing network conditions (Safwat et al., 2002).

Sustainability with VM Management

Numerous studies in the recent years have focused on trying to fully understand how the process of virtualization can be managed to adopt green considerations. One such study was conducted by Peoples et al. (2013) who identified various factors for energy consumption. One was the requirement to manage VMs to minimize energy consumption. Operations such as VM spawning, migration, setup, and so on were performed in a manner that is consistent with low energy consumption policies. Because of the fact that the tasks performed by the VMs directly contribute to the load of the underlying network, there has been much work recently on ways to express this as a metric. For example, the work carried out by Yang et al. (2014) demonstrates a method to evaluate the load for each virtual machine. This number that expresses the load can then be used as a deciding factor for migrating VMs if the present allocation of VMs is causing deterioration in the QoS. A separate management platform to make such decisions was supplied.

Green Cloud as a Network Management Problem

Importance of Virtualization Management

Some of the literature surveyed in earlier sections highlighted the fact that virtualization as a business model helps cut carbon emissions. But virtualization alone would not suffice for green computing. Ad hoc approaches to virtualization will not address the energy crisis facing the modern computing environment. What is needed in this age is efficient management of virtualization with the central goal of minimizing the number of actual hardware parts required to support these approaches. Several issues need to be addressed. For example, attention should be given to QoS changes when VMs are allocated to machines that are already hosting other VMs. Similar consideration should be given to selecting an ideal physical machine for migrating VMs when there is a need to save power.

Relationship Between Networking and Power Consumption

To appreciate why the greening of clouds should be regarded as a network management problem, it is important to acknowledge the relationship between packets and power. The transmission of packets from one location to another across the "wire" will consume energy at the sender and the receiver as will intermediate devices that route packets and power overheads required to set up connections. The knowledge of volume and frequency at which packets are transmitted in a cloud data center leads to the appreciation of the fact that simple, pure network communications contribute major energy consumption. Energy consumption is proportional to traffic rates (Meng et al., 2012). The advent of Infrastructure as an IaaS has helped cut power consumption, but the need for optimizing the reduction of power remains. One of the recent advancements includes adaptive link rate communications that consume power based on the load to be transmitted across the link rather than some predefined baseline power (Botero and Hesselbach, 2013).

Each VM can be treated as a source of network packets. Maximum energy efficiency can be gained by having fewer numbers of physical machines host the maximum number of VMs as long as doing so does not compromise on QoS. Corradi et al. (2014) report that the allocation of a VM to a physical machine might be affected by network-related constraints. When VMs are consolidated, there will be demand for various resources at the substrate level, especially networking ones (such as interfaces). Power is still consumed at times even when there is no transmission of packets, but VMs maintain connections during idle times. Therefore, network devices consume power in both idle and active states. In the active state, a device is engaged in packet processing as opposed to being idle when packet processing is minimal but the device is still consuming energy. Meng et al. (2012) report that the total energy consumption could be quantified in terms of energy use in both states. They provide an equation using the parameters that represent power consumption at different states:

$$E = P_a T_a + P_i T_i \tag{8.1}$$

In the equation, P_a and P_i represent the power consumption at device active times T_a and idle times T_i, respectively. But the equation is the minimum baseline for active and idle states. As explained, the energy use will fluctuate, depending on how busy a device is. Meng et al. (2012) refer to this as "operational frequency" (r). The authors express the power of energy consumption required to work at r as $f(r)$, and C represents constant energy consumption as follows:

$$P_a(r) = C + f(r) \tag{8.2}$$

These authors also considered other parameters to factor into power consumption of network devices in context of their research into energy-efficient routers, the details of which are not relevant to this chapter. However, the preceding explanation and Equations (8.1) and (8.2) provide valuable insight into why green computing must be considered as a network management problem. Equations (8.1) and (8.2) clearly show that energy consumption is dependent on traffic patterns. Traffic patterns affect operational frequencies (r) to which network devices are subject. This, in turn, is reflected in power consumption (E). Therefore, it can be concluded that by gaining a better understanding of network traffic, it is possible to find better greening of cloud infrastructures. Virtualization initiatives must consider network traffic as one of the deciding factors for allocation of processes to physical devices.

Need for Traffic Characterization in Virtualized Environments

The equations discussed in the previous section were in the context of single network devices. The total power consumed by a device was based on its total load/frequency of operations. But to manipulate the allocation of power to VMs based on their individual footprints to power consumption, the equations must be expanded to account individual threads of network traffic. A statistical network profile of each individual virtualized entity hosted on a physical machine will help identify least active VMs and "top-talkers." VMs can then be sent to and/or consolidated to save the power of a device.

Assuming that a physical machine has the capability to host a number of VMs, the following parameters are used:

n Number of VMs hosted at the physical machine
C Constant power consumption of a physical machine regardless of operations performed
$\boldsymbol{E_i(t)}$ Energy contribution by network activity at time t, of the ith VM, where $i=1$ to n
$\boldsymbol{B_i}$ Baseline energy required just to keep a VM and its associated connections active. This is always present for each VM. $i=1$ to n
$\boldsymbol{FP_i(t)}$ Total traffic related footprint of a VM on its substrate physical machine. $i=1$ to n
$\boldsymbol{EC(t)}$ Overall energy cost for a physical machine because of network activity of the VM at time t.

In the preceding description, $E_i(t)$ will change based on traffic-related frequency of the VM at time t. Using Equation (8.2) in the previous page, the following is possible

$$FP_i(t) = B_i + E_i(t) \tag{8.3}$$

From these equations, it is possible to model the overall energy cost accumulating on a physical machine at time t:

$$EC(t) = C + \sum_{i=1}^{n} FP_i(t) \tag{8.4}$$

In Equations (8.3) and (8.4), the granularity of network operations has not been taken into account.

In order to calculate the preceding parameters and to establish a profile for the full data center, network traffic characterization is required on an on-going basis. Various network parameters influence E_i (t). For example, energy consumed by a VM's network-related activity can be proportional to bandwidth use ($U_i(t)$) of that VM.

$$E_i(t) \propto U_i(t)$$

Hence, as an example, it will be useful to collect and monitor the use (among other attributes) for each VM. The data gathered will form a knowledge base that can feed into decision-making processes regarding migrating VMs and power management.

Role of Hypervisors in Traffic Characterization

The volume of virtualization in cloud environments and the need to share resources requires a platform that can map virtual entities to real resources. Virtualization has various levels of abstraction that are dependent on underlying resources. The task of the *hypervisor* is to enable sharing of resources among virtualized entities. Many cases of virtualization will involve virtualized instances representing a real piece of hardware as in the case of virtual networks or sharing network adapters. Hypervisors are the interface between the virtualized entities and the real-world computing entities (be it hardware or software). Hypervisors have features to support the representation of a resource for virtualization needs while isolating one virtual entity from other.

Modern hypervisors have the ability to support network administrators in monitoring the performance of resources and to evaluate how virtualized entities are using those resources. It would be ideal if hypervisors were to evaluate the network activity contribution of each VM with the aim of better placement of VMs to conserve energy. Future hypervisors must embed green elements into their operations by allowing traffic-aware virtualization management. Hypervisors must have more participation in VM consolidation and live migration based on energy conservation.

SNMP for Green Cloud Traffic Characterization

SNMP Operation in Context of Green Clouds

Because it has been recommended in the previous sections that green computing for clouds must be a network management problem, the focus now is on SNMP, which has been widely adopted as the network management protocol for corporate networks. Because of its simplistic nature and ease of deployment via various management platforms, SNMP seems to be the preferred choice. RFC1213 (IETF, 1991) broadly segregates a network into managers and managed entities. A manager or a set of managers is able to query managed devices for the most up-to-date information and statistics. Almost everything on a network can be queried by the manager. Again, because of the simplistic client/server mechanism of SNMP, any large-scale distributed network can also be managed. Indeed, SNMP was initially formed to monitor small LANs and to be an interim protocol until something permanent was standardized. However, the widespread use of SNMP has shadowed ad hoc approaches to network management.

The crux of SNMP lies in a database called the management information base (MIB), which is present in every managed entity. When a manager queries a device, a software agent interprets the request and returns the appropriate values from the MIB. Of course, the more frequent these queries are, the more accurate the management information is. An MIB holds some static values, but most objects are variables. For example, objects can reflect changes to routing tables, packet counters, error counters, time ticks, and so on. The values of these objects can be used to calculate statistics that would indicate problems on the network.

Considering green cloud in a network management context would be a significant change to how SNMP will be applied. This is because SNMP was initially designed to manage only wired networks. Typical application of SNMP involves a network manager querying objects from various network devices to calculate use, link performance, network trends, identification of bottlenecks, and so on. Applying SNMP for green computing and virtualization would seem to be a radical shift in network management paradigm. Fortunately, the abstraction created by virtualization can consider each virtual machine as an independent system. This reflects the real-world model of networks. Therefore, in network management terms, each VM can be considered as a managed entity that can respond to SNMP queries from the manager. Each VM is able to maintain an MIB that could be queried by a management interface that could be provided via the underlying hypervisor. The information gathered can feed into a VM management system. Furthermore, the information from the MIBs can be used for calculating network statistics for each VM. This can then be used to identify time slots when energy could be saved by managing VMs.

Related Work

Little research has been done on the energy use of VMs. Since the advent of cloud computing, virtualization has been seen only as an efficient mechanism to share resources without detailed considerations on how it would impact energy consumption. Although it was widely accepted that VM activities contributed to energy consumption, there have been very few initiatives to quantify and measure the power consumed on a per VM basis. Aman et al. (2010) conducted significant research that demonstrated the power metering of VMs. The authors state that the rationale for their work is the lack of accurate power management features for virtualized environments. Current green initiatives using power management techniques exist for typical LAN servers. The dearth of VM power management techniques can be explained by the complexity of cloud environments as the result of abstraction and sharing of resources. Aman et al. explain that without power metering for VMs, it is not possible to fully exploit the power-saving features brought about by cloud computing. They designed a power metering system called *Joulemeter* that considered various VM resource use concerning CPU, disk, and memory. The Joulemeter helps developers to evaluate the energy consumed by applications if they were to be hosted or ported to a cloud environment. The publication related to the joulemeter by Aman et al. is significant in two ways. First, it is one of the very few research studies that acknowledges the effect of VM power consumption and identifies the contributors to power consumption. Second, it produces a concrete method and a system to measure power consumed by each VM.

The majority of operations on the cloud are based on network applications hosted via VMs. The amounts of micro-operations performed by VMs have the potential to combine as power consumption hot spots. Hence, it is important to consider the cause of VM power consumption, rather than just measuring it. Considering it might open opportunities for VM task optimization. Many of these operations performed by VMs are network related. Therefore SNMP is propitious for collecting network-related data from each VM. Blanquicet and Christensen (2008) created a custom SNMP MIB for a green-aware network management system for traditional networks. Values obtained via the custom MIB were then used to take decisions resulting in energy conservation. Suganama et al. (2014) describe a green usage monitoring (GUM) Information base that records specific device information that is critical to energy use. They performed various experiments on collecting status information from MIBs on devices in traditional LAN environments, which was used as a learning data set to evaluate power consumption of those devices. Daitx et al. (2011) used SNMP MIBs for virtualized environments, the data from which decisions on virtualization management could be made. This was significant because the researchers reasoned that if a physical device has a virtual representation, two management planes exist. Each of these planes could be queried using SNMP regardless of whether they are virtual or physical. In other words, a virtual port of a router could be queried in the same manner as a real port. Adding strength to the suitability of using SNMP in context of virtualization, Hillbrecht and de Bona (2012) developed an MIB that recorded details of runtime attributes of VMs. SNMP has also been used in a data centers to map activity logs, which then were used to derive statistics to understand the data center's traffic profile (Benson et al., 2010). These research studies lead to the belief that any virtual entity can be treated as a real entity with regard to its behavior, and, therefore, the same data collection principles can be applied or extended.

A Model for Network Management for Green Cloud

Model Outline

The model being proposed is for efficient network data collection from virtual entities using network management principles. Specifically, the model shows how SNMP can be embedded into cloud architecture and how network management information is communicated among various elements of the network. This by no means is a concrete model but an idea that is subject to adaptations and extensions during the course of future deployments. Slight variations to this model will be performed when the concepts explained herein are tested in a lab-based environment. In general, a network management approach to green clouds must be based on the following guidelines (Voderhobli, 2015):

- Data collection from virtualized entities must be autonomic. These data must feed into a knowledge base or a learning system.
- Data collection requests must be made at a local level to avoid having network management data traverse across the network. In other words, the managed entity is a VM, and the management platform is resident on the substrate machine. Therefore, SNMP queries are sent at a local level.
- VM operations and decisions to migrate, consolidate, and so on must be based on the knowledge base derived from the MIB. A software process will act upon the information in the knowledge base to decide on when and where to migrate a VM. This software process will hereafter be referred to as an *analytical engine*.
- The optimization goal should be to use minimal physical machines to host all VMs as long as QoS is not compromised. Any machines not hosting VMs can remain dormant.

Figure 8.2 describes the relationship between various entities on a physical machine. The entity responsible for querying the virtual machines is the VM–active stats poll engine, referred to as *VASPE* henceforth. The VASPE's role is paramount in recording the statistics derived from polling in the statistics database. Basically, it sends SNMP_get queries to a designated port on the relative IPs of the VMs. The relative IPs are local ones assigned to the VMs by the host machine. Each SNMP_get query sent by the VASPE receives a response in the form of data stored in the MIB of a VM. This is then synthesized, in many cases over a period of time and stored as a statistic in the statistics database. The analytical engine "learns" the current network-related attributes and decides whether any of the VMs can be manipulated to conserve energy. It will interact with the substrate to perform VM management functions and localized operations such as putting the host system to sleep when network activity is nonexistent. Similar peer systems then interact at a network level to identify candidate VMs for migration and host servers that can accommodate the VMs. Local information can also be propagated to global databases that maintain the status information of physical machines along with use and network statistics. When the full system comes together, the plan allows for the fewest possible physical machines to host all VMs. The three elements—VASPE, statistics database, and the analytical engine—could be deployed in the hypervisor to integrate with other VM management functions.

Figure 8.3 shows the different elements in context of the full network. As the figure indicates, the global database caters to information sharing across the distributed network. From time to

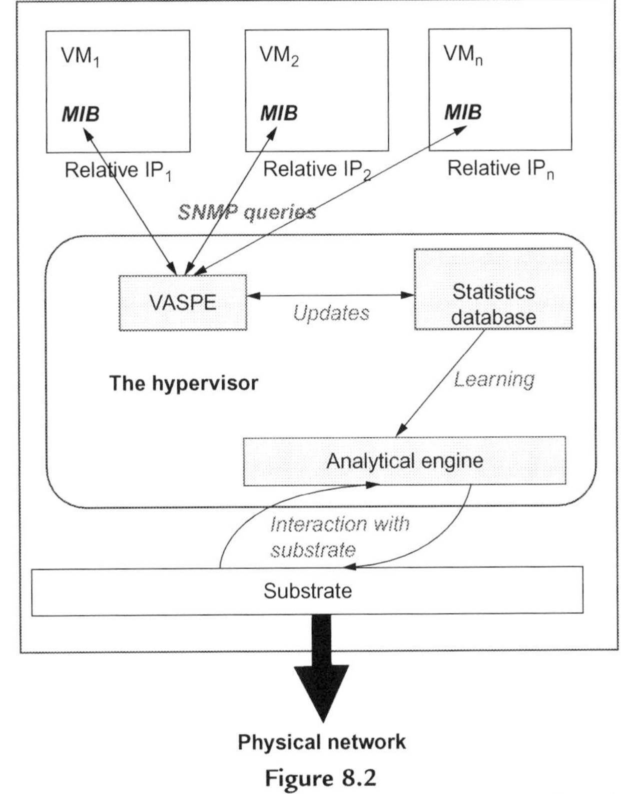

Figure 8.2
Interaction between different elements at a local machine level.

time, analytical engines on physical machines will consult the global database for various VM management functions, including electing a physical machine for hosting migrated VMs.

Gathering and Using Statistics

Statistics are not readily available from SNMP MIBs. The MIBs store data only in scalar or counter form that can be accessed by the network management platform. The data obtained by the VASPE must be used in calculations to determine statistics that are relevant for making decisions. For example, use can be one of the indicators of network traffic. Each VM use on a virtual interface can be calculated to determine how much of the total traffic at the underlying physical machine is from a specific VM. The VASPE must poll a virtual interface at regular intervals and use the information for specific calculations that will feed into the decision-making analytical engine. As an example, it is useful to consider how MIB objects *ifInOctets*, *ifOutOctets*, and *ifSpeed* can help in calculating the use. Assuming that the VASPE has been set up to poll a VM for these values at times *now* and $now+t$, the

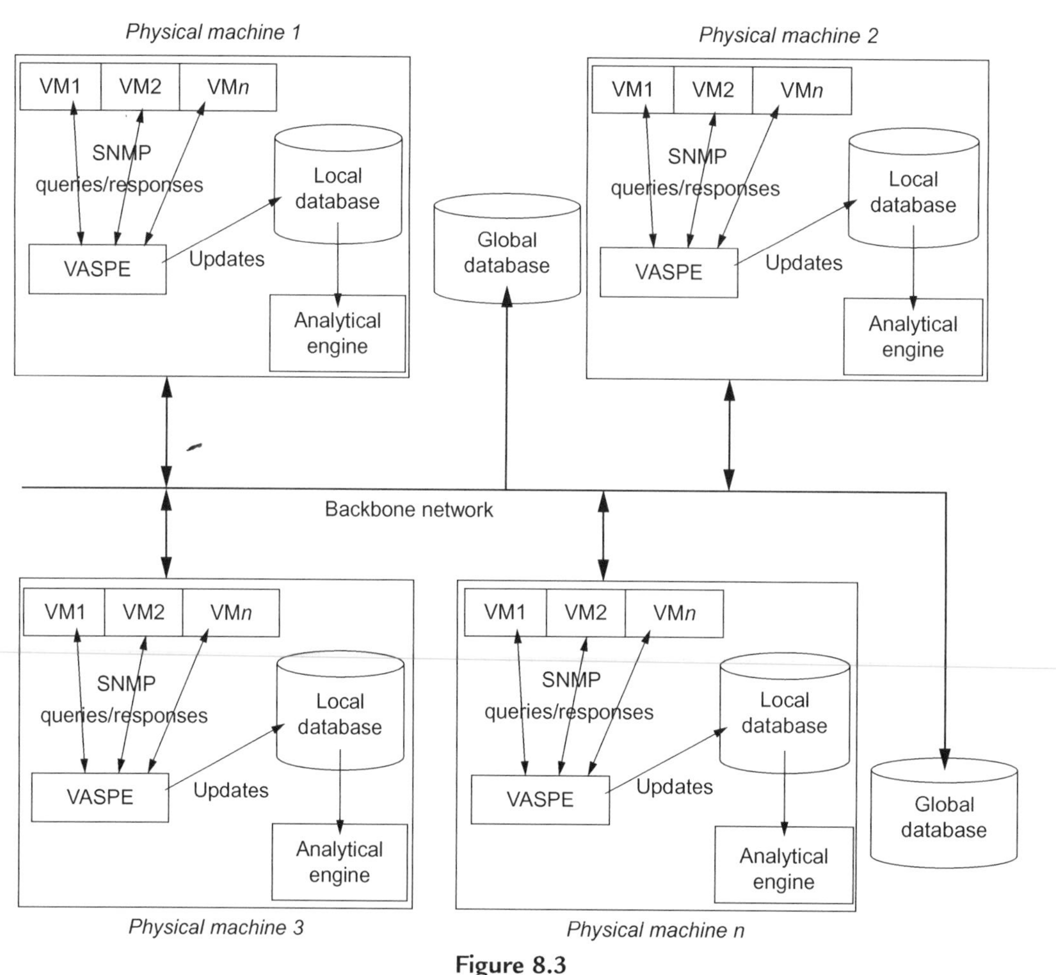

Figure 8.3
Global view of the network.

total bytes per second transmitted and received from the VM's interface can be calculated as follows (Leinwand and Conroy, 1996)

$$\text{Total bytes per second} = (\text{ifInOctets}_{\text{now}+\text{t}} - \text{ifInOctets}_{\text{now}}) + (\text{ifOutOctets}_{\text{now}+\text{t}} - \text{ifOutOctets}_{\text{now}})$$

$$\text{Time difference} = (now + t) - (now)$$

$$\text{Bytes per sec} = \text{Total bytes per second}/\text{Time difference}$$

$$\text{Utilization} = (\text{bytes per sec} \times 8)/\text{ifSpeed}$$

In the preceding, *ifInOctets* and *ifOutOctets* represent the number of bytes in octets received and sent, respectively. *ifSpeed* is the speed of the interface. In the traditional LAN sense, this is the speed of the physical interface, but in this case, it is the speed of the virtual interface.

Utilization is only one of the network attributes that characterize the traffic of a virtual machine. There are many more; some are related to interflow and intraflow traffic. Additional work must be done to investigate these as the research progresses. Statistics can be aggregated to help determine high and low thresholds to aid in decision making. For example, reaching a high threshold of use could often mean that a physical machine cannot be delegated to another VM and will not be a candidate for future VM allocations until use value drops.

By using *use* as a reference, typical sequence of operations using statistics could be:

- Each virtual machine present on a physical machine VASPE sends a SNMP_get request to a virtual port on a relative IP.
- VASPE must repeat the operation over a period of time to gather statistics. For example, to calculate use, it must request the number of bytes in/out at two different times.
- Statistics calculated will be fed into a local database, which is also in the hypervisor. This could be used to calculate a physical machine's overall use. This information can be stored in a global database that could be accessed by other physical machines. Hence, a medium of communicating information is relevant for the full network.

If $maxVM_p$ is the total number of VMs on a physical machine p, *instance* is an instance of a VM on the machine, and $LDbase_p$ represents the local database of physical machine p, then the following routine is needed for each physical machine across the network to feed individual VM statistics into corresponding local databases.

1: **while** ($maxVM_p$! $=$ 0) **do**

2: $inst = maxVM_p$;

3: **while** (($instance$! $=$ 0)) **do**

4: snmp_get $(inOctets_{instance}, outOctets_{instance})_{\mathrm{now}}$;

5: snmp_get $(inOctets_{instance}, outOctets_{instance})_{\mathrm{now}\,+\,1}$;

6: calculate $utilisation_{instance}$;

7: writeToDatabase $(utilisation_{instance}) \rightarrow localDbase_p$;

8: **end while**

9: $maxVM_p$–;

10: **end while**

This assumes, for example, that use is calculated at one-second intervals. There are implications of frequent SNMP polling, the discussion of which is not covered in this chapter. This process is performed across all physical machines, each of which holds a database that records the use for its corresponding VMs. Furthermore, the total current use for each physical machine is recorded on the global database. With the information from the local and global databases, the following operations can be performed:

- If use was a deciding factor to check whether a VM must be migrated, then a VM with the least contributing use can be migrated to a physical machine that can accommodate it.
- The analytical engine can select a physical machine to host the migrated VM. Alternatively, a VM with heavy use can be relocated to a physical machine with low overall use. Overall use can be queried from the global database. As explained earlier, because traffic is a contributor to power, clustering VMs based on use will free some physical machines.
- If a physical machine is no longer needed to sustain VMs, a hardware interrupt initiated via the analytical engine can put the machine to sleep. The global database is then updated to reflect this. Over the course of time, if a newly spawned VM does not have room on a physical machine, one of the dormant ones can be delegated to host it. This is also true when a VM changes its profile from low use to high, causing the host physical machine to be overwhelmed with a high use threshold.

With thresholds being set based on SNMP derived statistics, it can also be ensured that VM management considers the QoS.

Conclusions and Future Work

This chapter is one of the many publications that relate to SNMP-based traffic characterization for green clouds. It is still in its preliminary stage and thus is a theoretical outline for the proposed model. This chapter presented some ideas to inculcate the use of SNMP-based network management principles for green clouds. Currently, analyzing network-related information from a large data center is very problematic. It is hoped that a system using a SNMP-based traffic method can help alleviate this problem. The plans described in this chapter represent a novel approach to creating a distributed system that constantly monitors a network to find windows of opportunity for saving power using traffic characteristics. SNMP was traditionally used with LANs and WANs. The research related to the proposition described in this chapter demonstrates a significant move in using SNMP for green clouds.

For now, the concept looks ideal on paper. But future work will entail rigorous experimentation to test the theory's validity in busy cloud data centers. Currently, we are in the process of setting up an isolated emulation environment to build the first prototype of the proposed system. The experiments will address various concerns regarding the frequency of SNMP polling because network management traffic itself could be detrimental to the performance of the network. Therefore, there should be investigations on how SNMP-querying intervals should be balanced to produce accurate statistics while maintaining a low traffic footprint. These investigations will be included in SNMP overheads. We should also look into hardware overheads as the management entities; that is, VASPE, the analytical engine, and local databases will rely on the host physical machine. This could potentially cause some resource depletion, the effect of which will propagate to the VMs.

References

Aman, K., Feng, Z., Jie, L., Nupur, K., Arka, B., 2010. Virtual machine power metering and provisioning. In: Proceedings of ACM Symposium on Cloud Computing (SOCC).

Blanquicet, F., Christensen, K., 2008. Managing energy use in a network with a new SNMP power state MIB. In: Proceedings of 33rd IEEE Conference on Local Computer Networks, LCN 2008, pp. 509–511.

Benson, T., Anand, A., Akella, A., Zhang, M., 2010. Understanding data centre traffic CHARACTERIStics. ACM SIGCOMM Comput. Commun. Rev. 40 (1), 92–99.

Botero, J.F., Hesselbach, X., 2013. Greener Networking in a network virtualisation environment. Comput. Networks 57, 2021–2029.

Bilal, K., Saif Ur Rehman, M., Osman, K., Abdul, H., Enrique, A., Vidura, W., Rizwana, I., Sarjan, S., Debjyoti, D., Mazhar, A., Usman Shahid, K., Assad, A., Nauman, J., Samee, K., 2014. A taxonomy and survey on Green Data Center Networks. Future Generat. Comput. Syst. 36, 189–208.

Christensen, K., Gunaratne, C., Nordman, B., George, A., 2004. The next frontier for communications networks: power management. Comput. Commun. 27, 1758–1770.

Corradi, A., Fanelli, M., Foschini, L., 2014. VM consolidation: a real case based on OpenStack Cloud. Future Generat. Comput. Syst. 32, 118–127.

Chiaraviglio, L., Mellia, M., 2013. Energy-efficient management of campus PCs. In: Jinsong, W., Sundeep, R., Honggang, Z. (Eds.), Green Communications: Theoretical Fundamentals, Algorithms, and Applications. CRC Press.

Chabarek, J., Sommers, J., Barford, P., Estan, C., Tsiang, D., Wright, S., 2008. Power awareness in network design and routing. In: Proceedings of INFOCOM 2008—The 27th Conference on Computer Communications.

Chang, R.-S., Wu, C.-M., 2010. Green virtual networks for cloud computing. In: 5th International ICST Conference on Communications and Networking in China, pp. 1–7.

Daitx, F., Esteves, R.P., Granville, L.Z., 2011. On the use of SNMP as a management interface for virtual networks. In: The Proceedings of the IFIP/IEEE Intl. Symposium on Integrated Network Management.

Ebrahimzadeh, A., Rahbat, A.G., 2013. Greening data centres. ICT for Sustainability, Green Networking and Communications.

Guelzim, T., Obaidat, M., 2013. Green computing and communication architecture. In: Handbook of Green Information and Communication Systems. Elsevier, pp. 209–227.

Hussein, T.M., Burak, K., 2013. Energy-efficient cloud computing: a green migration of Traditional IT. In: Handbook of Green Information and Communication systems.

Hillbrecht, R., de Bona, L.C.E., 2012. A SNMP-based virtual machines management interface. In: Proceedings of IEEE Fifth International Conference on Utility and Cloud Computing UCC-2012, pp. 279–286.

IETF, 1991. RFC 1213 Management Information Base. [Internet] Available from: http://www.ietf.org/rfc/rfc1213.txt (Last accessed 7th June 2014).

Jingy, Z., Xing, F., 2013. Cloud computing—a greener future for IT. In: Green Communications, pp. 221–243.

Leinwand, A., Conroy, F.K., 1996. Network Management—A Practical Perspective, second ed. Addison-Wesley.

Mouftah, H., Kantarci, B., 2013. Energy-efficient cloud computing: a green migration of traditional IT. In: Handbook of Green Information and Communication Systems. Elsevier, pp. 295–329.

Meng, W., Wang Y., Hu, C., He, K., Li, J., Liu, B., 2012. Greening the internet using multi-frequency scaling scheme. In: Proceedings of 26th IEEE International Conference on Advanced Information Networking and Applications.

Peoples, C., Parr, G., McClean, S., Scotney, B.W., Morrow, P.J., 2013. Performance evaluation of green data centre management supporting sustainable growth of the Internet of things. Simulat. Model. Pract. Theory 34, 221–242.

Safawat, A., Hassanein, H., Moufta, H., 2002. A MAC-based performance study of energy aware routing schemes in wireless ad-hoc networks. In: Proceedings of Global Telecommunications Conference, 2002. GLOBECOM '02. IEEE.

Sun, H.S., Lee, S.E., 2006. Case study of data centres' energy performance. Energy Build. 38 (5), 4078–4094.

Suganama, T., Nakamura, N., Izumi, S., Tsunoda, H., Matsuda, M., Ohta, K., 2014. Green Usage Monitoring Information Base. [internet] Available: http://tools.ietf.org/html/draft-suganuma-greenmib-03.

The Climate Group, 2008. SMART 2020: Enabling the Low Carbon Economy in the Information Age.

Voderhobli, K., 2015. A SNMP based traffic characterisation paradigm for green-aware networks. In: Delivery and Adoption of Cloud Computing Services in Contemporary Organizations, IGI Global (due to be published in 2015).

Yang, C., Liu, J., Huang, K., Jiang, F., 2014. A method for managing green power of a virtual machine cluster in cloud. Future Generat. Comput. Syst. 37, 26–36.

Future Solutions

CHAPTER 9

Energy Harvesting and the Internet of Things

Christian DeFeo*
Premier Farnell Inc., Leeds, UK
**Corresponding author: cdefeo@premierfarnell.com*

Energy Harvesting: Intelligence and Efficiency

Energy harvesting, also known as *energy scavenging*, is a technology that takes advantage of ambient power sources including light, heat, and motion. The technology is not completely new: Many people are familiar with it in their solar-powered calculators, which were initially available at the end of the 1970s.

The solar-powered calculator and the latest energy-harvesting technologies prove a simple, if eloquent, point: We have around us a great deal of power that is often not used, for example, light, both natural and artificial, that merely illuminates and heat that streams through our homes via hot water pipes, radiators, and furnaces that diffuses outward. Nor are kinetic energy and magnetism tapped to their full potential. Energy permeates our world, yet somehow our exploitation of it is strangely limited; we are currently mostly confined to tapping into concentrated stores of energy such as fossil fuels or batteries.

Far from being a particularly green technology, batteries generate a particularly large amount of waste. They require nonrenewable resources, not only for the chemicals for their production but also for the metals that must be mined to create the casings, the factories that assemble the units, the fossil fuels that are burned in their transportation, and the processes that are involved in their disposal and recycling. And even then, not every last part of the battery can be recycled, thus increasing the stock of potentially toxic waste in our landfills.

Additionally, a battery-powered appliance represents a great deal of energy profligacy: In a standard remote control, batteries more often than not sit unused, their power draining out as time creeps on. The same holds true for many other battery-powered devices whose idle time is longer than the period they are actually used, but because of the constraints of conventional engineering, the power available must be provided constantly.

The development of solar-powered calculators highlights another issue: In order for these devices to be made viable, LED to LCD displays were used to make switches. This created no appreciable difference in functionality; however, substantial amounts of energy were saved via this change. The thought process that led to this substitution is critical to the success of green technologies and a key component in creating a successful energy-harvesting application. The issue involves not only substituting solar cells for a normal battery but also applies to the fact that the devices they power must make intelligent substitutions and find efficiencies to reduce consumption. If power is generally perceived as plentiful and available, the impetus to build solutions in this manner is reduced. It is far simpler for an engineer to focus on the application itself rather than what powers it; the current drive for additional functionality, as evidenced by the smartphone revolution, may lead to a commercial scenario in which energy is given less consideration in favor of providing further utility rather than repurposing current levels of utility in a more efficient way.

Nevertheless, energy-harvesting technologies are beginning to be part of our everyday lives. For example, the French firm Arveni has created a television remote control for Philips that is powered by motion; insofar as a remote and the user experience is concerned, all is well. Remote controls require only a quick burst of power to fire off their signals to the television. Meanwhile, when the remote is not needed, there is no power steadily bleeding away. The leading-edge sensors of EnOcean's new range of products can be used in Internet of Things (IoT) applications that rely on energy harvesting to power them.

However, energy harvesting has yet to go mainstream; beyond the aforementioned considerations of additional functionality as opposed to repurposed functionality, energy-harvesting technology in and of itself presents a problem to engineers. In the development of most applications, power is taken as a constant, which is how engineering has been taught for decades. The amount of power required certainly can vary, but its availability is taken as a given. Energy harvesting offers no such assurance: Sometimes there is not enough light to generate power or sufficient heat, or enough kinetic energy to power a solution.

This inconsistency requires engineers to think about a contingency: What does one do in a situation when no power sources are available? How can an application take best advantage of a situation when energy is abundant? Should one think in terms of a power "budget" and then "spend" it accordingly? Consideration of these challenges has wider implications than building, for example, an efficient remote control or calculator: More mainstream renewable energy sources must contend with this fundamental variance in their availability. By addressing these challenges in relatively small applications, fundamental principles for larger solutions may also be uncovered.

The element14 community (www.element14.com) is an online resource for electronics engineers. At the present time, it has more than 230,000 registered users spanning the globe; these engineers represent every skill level ranging from 30-year veterans of the electronics industry living in the western United States to recent graduates of engineering institutes in India.

Among the many activities that the community undertakes, it sponsors design challenges that feature new technologies; this has the dual purpose of marketing products and embedding education. In February 2013, element14, in collaboration with Würth Elektronik, Linear Technology and Energy Micro (now part of Silicon Labs), launched a design challenge featuring the Energy-Harvesting Solution to Go kit.

The kit has the following components:

- EFM32 Giant Gecko Starter kit (EFM32GG-STK3700) that provides an energy-harvesting sensitive microcontroller necessary for coping with inconsistent power availability.
- A multisource energy harvester that could harvest energy from four different sources: solar, heat, motion, and magnetism (induction).
- Piezo transducers that generate energy through motion and vibration.

The community used gamification tactics to encourage participation: in exchange for working on the projects, engineers were eligible to win an unspecified grand prize. The recruitment period began on February 25, 2013, and closed on March 18. There were 75 applicants.

The proposals suggested that energy harvesting had a wide range of potential applications including the use of the kit to create a home brew monitor and as part of a system to monitor the outbreak of forest fires. Seven applications were eventually selected. The chosen projects were as follows:

- A remote control that uses multiple energy harvesters (e.g., kinetic energy) rather only one source
- A system to recycle energy used in reptile enclosures
- A carbon monoxide alarm that does not require batteries
- A home brew monitor
- A forest fire monitoring system
- A bike light
- A Reptile Tank Energy Recycler that makes most efficient use of heat and light

As these selections indicate, the intent was to feature projects in which intelligent energy solution substitutions could be made in existing applications, such as the carbon monoxide alarm, or to find a means to make existing energy expenditure more efficient. Time constraints also dictated a relatively limited scope: The challenge ran from March until August 2013 and relied on the engineers who participated in their spare time.

From the very beginning of the challenge, it became clear that energy harvesting requires a difficult adjustment in engineering approach. The main difficulty lay in developing effective energy-harvesting and expenditure strategies. How does one take advantage of energy when it is available. How does one adjust the solution when it is not available?

The solutions easiest to develop were ones that relied on energy being more or less available for a predictable period. This was the case with the home brew monitoring system: The process of fermentation generates its own heat. Active fermentation processes can raise the temperature of beer by 10-15 °F.[1] In this instance, the Peltier module was used to power a timer that could alert the user to when the brewing process had completed. It was a relatively straightforward matter for the engineer who worked at Griffith University in Australia to develop this application and complete it in less than two months.

The carbon monoxide alarm was a greater challenge; this was developed by an engineer in Poland who noticed that a low battery light on the carbon monoxide alarm in his kitchen was blinking. He queried what might have happened had he not noticed that the blinking light; relying on a battery in that instance was not only wasteful but also potentially dangerous.

The engineer experimented with a number of approaches, but conventional gas sensors required too much power for a successful application to be built. element14 helped him acquire an electrochemical sensor from a Japanese manufacturer that had the benefit of reducing the power requirements.

However, in order to gather enough energy, he eventually had to use the Peltier module; in this instance, he fastened the harvester to a hot water pipe. The concept was to have a carbon monoxide alarm fixed to a wall behind which hot water pipes run. As they radiate heat, the alarm makes use of the diffused energy. The prototype was successful; as the efficiency of energy harvesting continues to improve, no doubt it will become much more embedded in our portable devices.

The IoT, "Hyped" and "Hidden": A Green Technology

The *IoT* is a recently coined term used to describe a series of technologies that enable machine-to-machine communication and machine-to-human interaction via Internet protocols. It is a technology that can be said to be both hyped and hidden: the elements that are "hyped" are markedly visible in the devices and appliances that are making their way into our homes, offices, and pockets. The "hidden" applications are presently being incorporated into our public infrastructure, power grids, and factories. Whether hyped or hidden, the term implies a significant paradigm shift in how the Internet is used.

When the predecessor to the modern Internet, ARPANET, was first conceived by the US Department of Defense, it was to ensure that communications could be maintained between human beings in the case of a nuclear attack. A similar person-to-person focus was the intent of the original proposals for the World Wide Web: One researcher would create an information resource that would then be accessible to other researchers (specifically at CERN), acting as a latter-day version of the Victorian book *Enquire Within Upon Everything*.

[1] Stika, Jon. "Controlling Fermentation Temperature: Techniques" Brew Your Own, March/April 2009. http://byo.com/stories/item/1869-controlling-fermentation-temperature-techniques (accessed 31.05.14).

In parallel, the development of machine-to-machine applications took place prior to the popularization of the Internet. In May 1996, a document was published by the Consultative Committee for Space Data Systems outlining a proposal for packet telemetry services.[2] This standard was created so that satellites and probes could communicate with facilities on Earth and stream sensor data so that the environment could be assessed. Furthermore, the protocol enabled making rational decisions about where the probe should go next or what tests it should perform.

Although in a less long distance but altogether more pervasive sense, the IoT intends to achieve functionally similar purposes for a remote space probe. Using such technological applications, it should be possible to monitor and intervene in scenarios without being physically present; it should also be possible for monitoring and intervention to occur without any active interaction with a human being. Massive reductions in the cost of microcontrollers and modules that add additional functionality mean that the creation of these applications has moved from the purview of a technological elite (e.g., space laboratories) to the individual hobbyist.

For example, it is entirely possible for an individual inventor to create an umbrella whose handle embeds a small inexpensive microcontroller, which can communicate via WIFI to an online weather service. If the weather service predicts inclement weather, the umbrella's microcontroller can receive the information and then cause an LED to light up in a particular color, for example, blue for rain or white for snow. The owner of the umbrella then has a simple visual cue as to what weather to expect that day and knows to take an umbrella.

It is not necessarily obvious that this IoT application is an exemplar of green technology. Many IoT applications will embed novelty as well as convenience; it is likely that they will constitute a significant portion of the hyped solutions. As with many IoT technologies, both hyped and hidden, the savings are small and individual, yet they become vast when replicated throughout a population.

These savings become even more obvious in another potentially hyped home application. An engineer in Belgium, Frederick Vandenbosch, created an IoT alarm clock at the prompting of and with the support of the element14 online community.[3] In its first iteration, the alarm clock accessed online train timetable information and if a train is cancelled or delayed, it enables the user to sleep for longer. Provided the appliance is eventually mass manufactured and achieves widespread use, it will help prevent overcrowding on trains; it will also allow journeys to be more efficiently planned (Figure 9.1).

---

[2] "Packet Telemetry Services: Recommendation for Space Data System Standards" Consultative Committee for Space Data Systems, May 1996. http://public.ccsds.org/publications/archive/103x0b1s.pdf (accessed 31.05.14).

[3] Vandenbosch, Frederick. "IoT Alarm Clock—Part 4" element14. http://www.element14.com/community/groups/internet-of-things/blog/2014/07/02/iot-alarm-clock-part-4 (accessed 02.08.14).

Figure 9.1
Internet of things alarm clock from element14.com.

Environmentally beneficial efficiencies will become more obvious when traffic reports can be embedded in the next iteration of the alarm clock. A clock that notes particularly heavy traffic and actively intervenes to prevent the user from unnecessarily being in the traffic would reduce pressure on the roads as well as lower carbon emissions created by idling motors. Once again, on an individual scale, this offers a marginal saving on both fuel and carbon; however, if the appliance should be commonly used in an entire country, the benefits are magnified. For example, according to the Canadian government (Natural Resources Canada, 2013).

> ...if Canadian motorists avoided idling for just three minutes every day of the year, CO_2 emissions could be reduced by 1.4 million tonnes annually. This would be equal to saving 630 million litres of fuel and equivalent to taking 320,000 cars off of the road for the entire year.[4]

The benefits would be even greater should such an appliance be used on a global scale.

This example, however, pales in comparison to the implications if businesses and governments should use IoT technologies in more hidden ways. At the Consumer Electronics Show held in January 2014, John Chambers, the CEO of Cisco, estimated that the "internet of everything" (IoE) could yield $19 trillion in savings—a conservative estimate. These figures, however, are not only from efficiencies in production and affect the environment. He cited the following examples:

[4] "Emission impacts resulting from vehicle idling," Natural Resources Canada, November 25, 2013. http://www.nrcan.gc.ca/energy/efficiency/communities-infrastructure/transportation/cars-light-trucks/idling/4415 (accessed 25.08.14).

1. *A smarter trash can*: Sensors embedded in a trash can detect how full it is, what materials it contains, and whether fumes emanating from the trash are particularly noxious. Their use can lead to more efficient waste management and more effective recycling.
2. *Smart street lights*: They can detect when foot traffic beneath them is the greatest and thus can lower their output when there are fewer people. Similar adjustments can also be made for weather conditions. Among the data that Chambers cited, "70% of the world's energy is consumed in cities, adding that 40% of total government energy spending in Europe is dedicated to street lighting. If these lights are connected to a smart network, he said, the cost could be reduced up to 80%."[5]
3. *Smart building management*: It should be possible to monitor and detect what is happening in a building and make the right decision by using a mobile phone application.

Cisco has cited other examples, such as the use of smart power grids that can more efficiently deploy energy than conventional ones. It should not be necessary to allocate the same amount of power to a residential neighborhood at 3 am as at 9 pm; the energy saved could instead be more efficiently used by enterprises and services that require all-night availability.

Cisco has also proposed that IoE technologies actually serve to increase energy consumption when required, albeit in more efficient ways. A recently created promotional video "The Storm" presents a graph reporting data from a shoreline battered by inclement weather (Figure 9.2). This power in this storm is explained to increase the output from offshore wind turbines. This increased output allows the energy company to offer electricity at reduced prices,

Figure 9.2
Still from Cisco promotional video "The Storm."

[5] Endler, Michael. "CES 2014: Cisco's Internet of Everything Vision," Information Week, January 11, 2014. http://www.informationweek.com/strategic-cio/executive-insights-and-innovation/ces-2014-ciscos-internet-of-everything-vision/d/d-id/1113407 (accessed 31.05.14).

which allows smart homes to increase their power consumption to maintain warmth and comfort.

Cisco's work in this area is neither merely theoretical nor confined to marketing videos. According to a paper published by Cisco in 2013, the city of Amsterdam, The Netherlands, has no less than 47 IoE projects in train, including smart energy grid systems, street lighting, parking applications, and building management.[6]

The company and the city collaborated with Philips to create IoT street lighting applications in the Westergasfabriek zone of Amsterdam. Lighting is an ubiquitous element of metropolitan areas, indeed, it's so ubiquitous that it's easy to forget its cost and environmental implications: Cisco estimates that 19% of electricity consumption is expended on lighting[7] It is also estimated that one-third of the world's roads are still lighted by technology dating back to the 1960s. Philips (2014) has calculated that the installation of new lighting solutions can save up to €10 billion per annum.[8]

Philips and Cisco illustrated the benefits of IoE-enabled lighting as shown in Figure 9.3.

Traditional lighting operations

Physical failure inspection

- A scouting team drive during night to visually spot failures

Paper based mapping & archiving

- Use of paper maps and files to manage the maintenance of the lighting stock

Undifferentiated lighting levels

- Lights burn uniformly throughout the night

Estimation based metering

- As multiple entities are connected to the grid, the energy consumption is roughly estimated by the utility

Intelligent lighting operations

Remote monitoring

- The lighting failures are automatically reported by the system, saving time and costs

Smart asset management

- The digital system smartly plans and routes the maintenance works to minimize street blockages

Smart dimming & scene setting

- Lights are dimmed during low traffic hours to save energy or enhanced in problematic neighborhoods to improve safety

Intelligent energy metering & billing

- A smart meter accurately calculates the energy consumption taking into account the varying rates and automatically bills all entities

Figure 9.3

Cisco and Philips intelligent and traditional lighting comparison.

[6] Bradley, Joseph, Lauren Buckalew, Jeff Loucks, James Macaulay, 2014. "Internet of Everything in the Public Sector: Generating Value in an Era of Change Top 10 Insights." Cisco Systems.

[7] Mitchell, Shane, Nicola Villa, Martin Stewart-Weeks, Anne Lange, 2014. "The Internet of Everything for Cities: Connecting People, Process, Data, and Things To Improve the 'Livability' of Cities and Communities." Cisco Systems, pp. 4-5.

[8] "The LED Lighting Revolution" Philips, May 2012.

IoT water management would result in similar results.

The Cisco approach suggests that efficiency is at the core of what makes the IoT most effective and valuable. Because efficiency in terms of energy and resource deployment is also a green priority, it can be said that the IoT is a green technology. What is more, IoT technologies allow for green interventions at both a macro level (e.g., infrastructure planning) and a micro level (e.g., encouraging an individual to stay at home when there is traffic congestion).

However, IoT has another environmentally friendly dimension, namely, the ability to collect data that has hitherto been problematic for individuals and cities to acquire on a real-time basis. Cisco is aware of this dimension and in collaboration with element14, Texas Instruments, Sierra Wireless, Wuerth Elektronik and the Eclipse Foundation deployed a design challenge on September 22, 2014 to create an Internet of Things-enabled pollution sensor.

The intent of this challenge is to create an inexpensive device that can be used by both governments and individuals to monitor air quality on a regular basis. Governments can use the device to adjust traffic flow or perhaps alter policies regarding the use of motor vehicles in certain areas. In this instance, however, technology also empowers activism. With real-time data about the state of the environment, it becomes no longer a matter of supposition or sporadically accrued data that governments need to change course. This information can be made widely available on a global basis. This democratizes data and thus encourages the discussion regarding environmental policy between states and their citizens.

As with air quality, similar devices could be built to detect the soundness of structures using mesh networks of sensors. From a government's perspective, hiring a structural engineer to examine all the bridges in any given area, for example, is expensive. These examinations might occur on a regular basis, but they do not provide real-time data.

On August 1, 2007, the I-35W Mississippi River Bridge, located in Minneapolis, Minnesota, United States, collapsed during the evening rush hour, killing 13 people.[9] The bridge was examined on an annual basis by the Minnesota Department of Transportation; it was one among approximately 75,000 that had been deemed unsafe. No system of sensors could make up for the lack of proper maintenance, but it certainly could have given real-time data regarding structural changes that would have indicated a collapse was imminent. Had that been known, the personal, social, and financial costs related to the collapse might have been avoided. Had this information been in the hands of activists, it could have been used for environmentally positive ends, for example, calling for a change in how traffic in Minneapolis was handled, perhaps by making additional rail facilities available. Such sensors also have conservation benefits: For example, old structures could achieve just-in-time maintenance rather than be replaced entirely.

[9] http://en.wikipedia.org/wiki/I-35W_Mississippi_River_bridge

Conclusions

The IoT and Energy Harvesting can be described as "perfect" green technologies: unlike many other "green" technologies, the end user is not necessarily called upon to make positive choices to save the environment, rather, environmental efficiencies as imbued by these technologies provide subtle rewards. The user of the Internet of Things alarm clock is granted more time in bed. The user of the Internet of Things umbrella has greater convenience as a benefit. The user of the Energy Harvesting empowered remote control need not run out of battery power again. Individuals, governments (both local and national) as well as businesses save money. What is more, these solutions are becoming ubiquitous: it may be a supreme irony that one of the key technologies which allows us to preserve the planet is one that is least noticeable.

Further Reading

Anon., May 1996. Packet Telemetry Services: Recommendation for Space Data System Standards. Consultative Committee for Space Data Systems. http://public.ccsds.org/publications/archive/103x0b1s.pdf (accessed May 31, 2014.).

Anon., May 2012. The LED Lighting Revolution. Philips.

Anon., November 25, 2013. Emission impacts resulting from vehicle idling. Nat. Resour. Canada. http://www.nrcan.gc.ca/energy/efficiency/communities-infrastructure/transportation/cars-light-trucks/idling/4415 (accessed 25.08.14.).

Bradley, J., Buckalew, L., Loucks, J., Macaulay, J., 2014. Internet of Everything in the Public Sector: Generating Value in an Era of Change Top 10 Insights. Cisco Systems.

Endler, M., January 11, 2014. CES 2014: cisco's internet of everything vision. Information Week. http://www.informationweek.com/strategic-cio/executive-insights-and-innovation/ces-2014-ciscos-internet-of-everything-vision/d/d-id/1113407(accessed 31.05.14.).

Mitchell, S., Villa, N., Stewart-Weeks, M., Lange, A., 2014. The Internet of Everything for Cities: Connecting People, Process, Data, and Things to Improve the 'Livability' of Cities and Communities. Cisco Systems, pp. 4–5.

Stika, J., March / April 2009. Controlling fermentation temperature: techniques. Brew Your Own. http://byo.com/stories/item/1869-controlling-fermentation-temperature-techniques(accessed 31.05.14.).

Vandenbosch, F., IoT Alarm Clock - Part 4. element14. http://www.element14.com/community/groups/internet-of-things/blog/2014/07/02/iot-alarm-clock–part-4 (accessed 02.08.14.).

Wikipedia, I-35W Mississippi River Bridge. http://en.wikipedia.org/wiki/I-35W_Mississippi_River_bridge (accessed 31.05.14.).

CHAPTER 10

3D Printing and Sustainable Product Development

Steve Wilkinson*, Nick Cope
Leeds Beckett University, Leeds, UK
**Corresponding author: s.wilkinson@leedsbeckett.ac.uk*

Introduction

During the last few years, there has been an exponential growth in the use of 3D printing additive manufacturing processes and, in particular, significant media interest in predicting the next industrial revolution delivered by these "new" printing technologies. The technology has been presented as new by the media, but, in fact, the first commercial 3D printer was created in 1984 when Chuck Hull, working for the US company 3D Systems Corp, developed the stereolithography (STL) process using lasers to selectively cure liquid photopolymers. At the same time, Hull defined the STL file format that is now the de facto standard for the exchange and printing of 3D models. The development of this first 3D printer established the slicing and filling algorithms present in many devices today. During this period, alternative technologies (i.e., selective laser sintering (SLS) and direct metal laser sintering and extrusion) were also under development. However, during this time, none of the technologies was referred to as 3D printing or even additive manufacturing.

The media interest in 3D printing was triggered by the open sourcing of the designs for extrusion-based plastic printers, notably the Rep-Rap system in 2005. This system originated at the University of Bath, created by a team led by Dr. Adrian Bowyer, senior lecturer in mechanical engineering, and funded by the UK Engineering and Physical Sciences Research Council. The initial aim of this project was to create the means for developing self-replicating machines (Figure 10.1).

These open-source designs have sparked a growing market of low-cost 3D printers that use a variety of technologies. In the hands of the open-source community, this has facilitated the move away from thinking of 3D printing as rapid prototyping (RP) and toward the concept of 3D printing being the final manufacturing process.

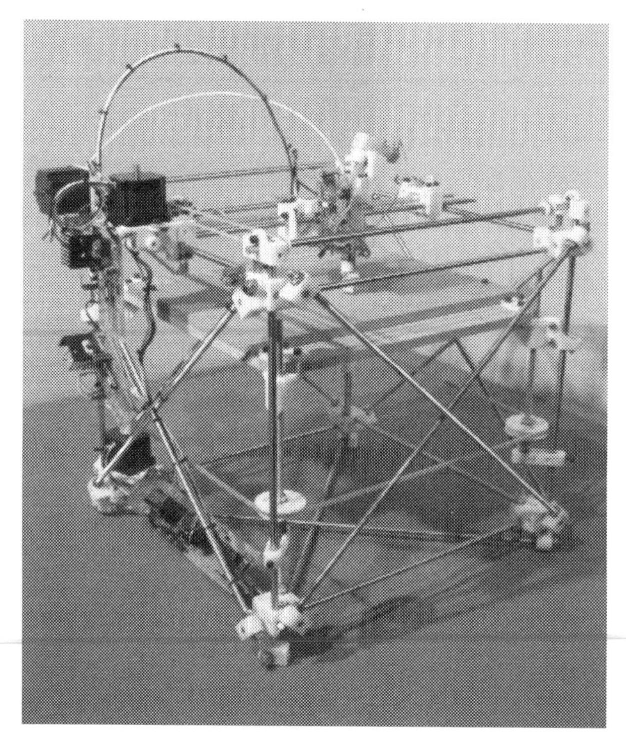

Figure 10.1
REP-RAP 3D printer.

3D Printing Design Pipeline

All 3D printing processes use layer-based technologies to build up a 3D object from slices through the 3D CAD design. The product design needs to be represented in an appropriate 3D CAD system that captures the intended solid object unambiguously (it should be noted that not everything is printable; for example, a 3D model in a CAD system may not be "printable" if it does not conform to the rules and constraints of the 3D printer system and its associated software). 3D software such as 3D Studio Max, SolidWorks, ProENGINEER, and Blender may be used to generate the appropriate 3D printable data (Figure 10.2).

In Web-to-print 3D systems, the geometry needs to be designed and uploaded using a CAD system or parametric design tool. Bespoke parametric customer CAD interfaces are used to define a parametric product such as the cell cycle from a nervous system that can generate parametric 3D printable jewelry (Figure 10.3).

The geometry is then imported directly from 3D CAD geometry into 3D print preprocessing software, which checks the design for printability by simulating the sliced printing process; such software is normally matched to a given range of 3D printers. For example, the Catalyst software environment imports STL files and generates the low-level data to drive the printer,

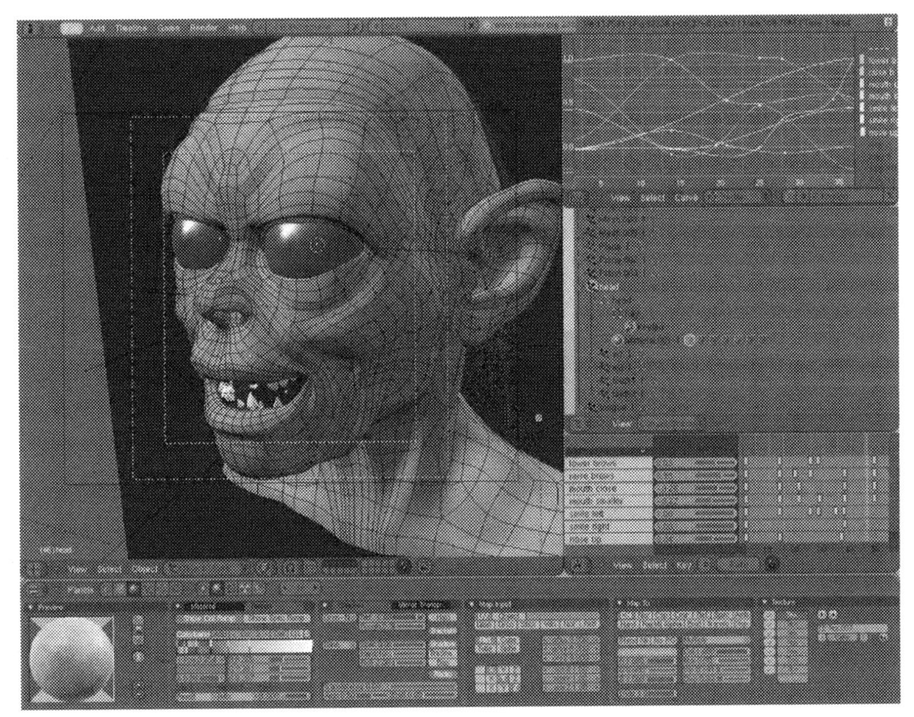

Figure 10.2
The 3D modeling package blender.

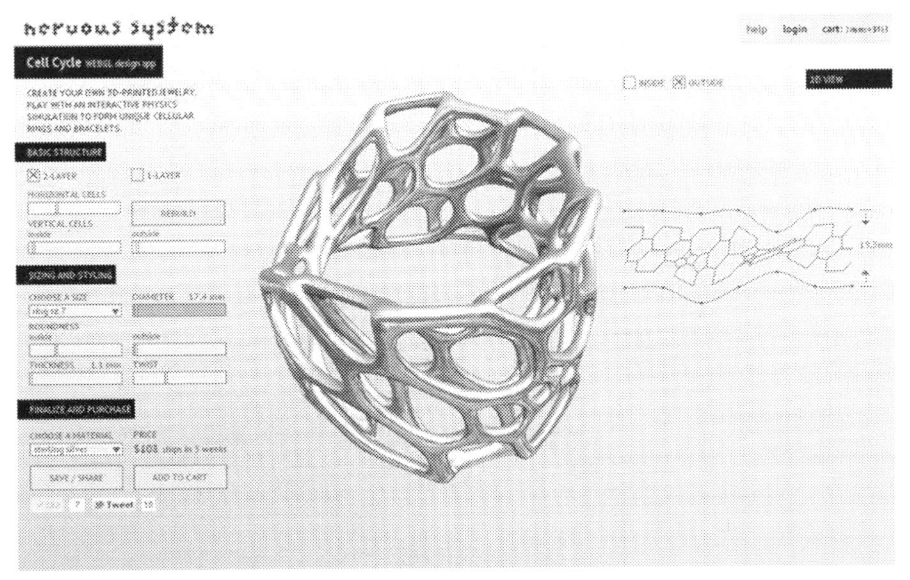

Figure 10.3
The cell cycle parametric design tool from nervous systems hosted on Shapeways' 3D printing site.

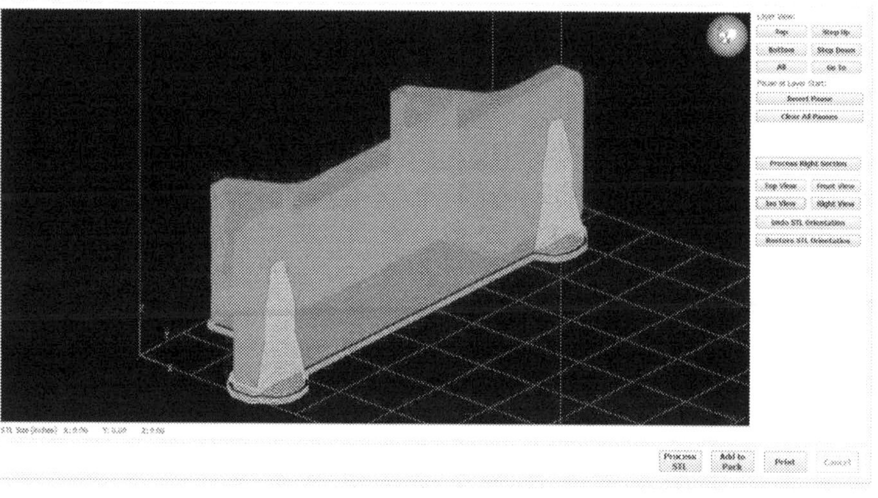

Figure 10.4
Catalyst 3D printing software.

allowing decisions regarding orientation, resolution, and support materials to be made by the designer (Figure 10.4).

There are, obviously, many advantages in being able to produce physical objects quickly and relatively inexpensively as a final product or a prototype. At the prototyping stage, product models that do the following can be handled by both the designer and the client:

- Produce visual models for market research, publicity, packaging, and so on
- Reduce time to market for a new product
- Generate customer goodwill through improved quality
- Expand product range
- Reduce the cost and fear of failure
- Improve design communication and help eliminate design mistakes
- Convert 3D CAD images into accurate physical models at a fraction of the cost of traditional methods because there are no tooling costs

Using 3D printing requires the selection of the appropriate system and underlying technologies. Web-based 3D print companies typically offer via the Internet access to a range of printers appropriate for different materials, physical performance, finish, and design.

The Shapeways material catalog includes the following (using appropriate printers):

- Strong and flexible plastic: A starter material with easy design rules
- Metallic plastic: A brittle nylon plastic filled with aluminum dust
- Detail plastic: An acrylic-based polymer that can print fine details

- Frosted detail plastic: An ultraviolet (UV)-cured acrylic plastic that prints fine details
- Steel: Suitable for jewelry and mechanically durable pieces
- Sterling silver: Available in three levels of polish
- Brass: Both gold-plated and polished brass have a smooth glossy sheen, whereas raw brass has a rough matte finish
- Bronze: Has a unique coloring of rose gold with a subtle marble effect
- Elasto plastic: An ultraflexible off-white plastic with a rough texture
- Full-color sandstone: Gypsum printable with color textures on the model surface that feels like matte clay
- Ceramics: Food-safe glazed ceramics available in glossy and satin finishes and a variety of colors
- Gold: Solid hand-polished gold
- Castable wax: High-resolution wax for investment casting
- Platinum: Solid polished to a beautiful smooth sheen
- Full-color plastic: Material printed on a ProJet machine

The choice of materials and printing system depends on the quantities involved, the complexity and the application of the product, the purpose for which it will be used, and any further manufacturing processes required. Examples follow:

- Final manufactured part/product
- Promotional models (functional and nonfunctional)
- Patterns for castings
- Electrodes for electro discharge machining of dies

The underlying technologies and materials must match the desired function of the product; the materials/printers combinations have identified design rules for any printable design. In some cases, it is necessary to ensure that the printed component matches or exceeds the performance of conventionally manufactured parts by using the grain and internal structure of a printed material for particular engineering requirements.

The values of the parameters limiting what is practical to print depend on the properties of the selected material and the operational parameters of the printer.

Nearly all 3D printing machines process geometry by importing STL files. The STL format is an ASCII or binary file used in manufacturing. It is a list of the triangular surfaces that describe a computer-generated solid model. This is the standard input for most 3D printers and is one of the basic file types also used in other manufacturing tools such as desktop Computer Numerical Control (CNC) milling.

CAD data are processed by slicing the computer model into layers, each being typically 0.10-0.25 mm thick. The machine then uses this data to construct the model layer by layer, each being bonded to the previous until a solid object is formed. Because of this laminated method

Stepped construction

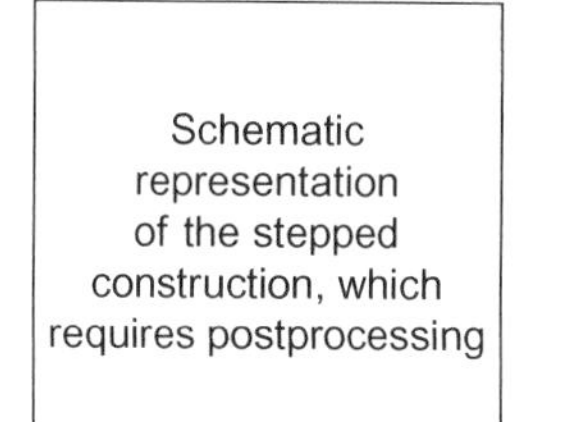

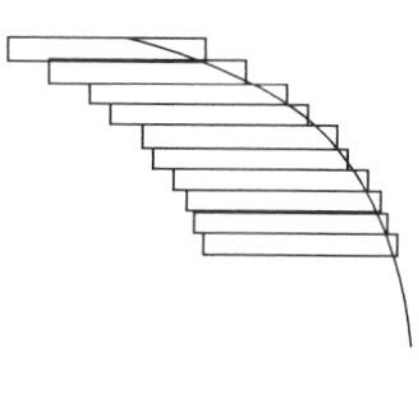

Figure 10.5
Stepped construction.

of construction, a stepped surface is developed on curved faces (Figure 10.5). The removal of this stepped surface is essential if maximum advantage of the process is to be realized.

A range of processes is used in different 3D printers including these:

1. Stereolithography apparatus (SLA)
2. Selective laser sintering (SLS)
3. Solid ground curing (SGC)
4. Laminated object manufacturing (LOM)
5. Fused deposition modeling (FDM)
6. Powder-based inkjet 3D printing

The Underlying Printing Processes

Stereolithography Apparatus

The STL file of the component is sliced according to the device's software. Each slice is then etched onto the surface of a photosensitive UV curable resin using a "swinging" laser. The resin is cured where the laser beam strikes the surface. Each layer is typically 0.13 mm thick. At the end of each pass, which covers the entire surface of that layer, the platform descends to allow liquid resin to flow over the previously cured layer. A recoating blade passes over the surface to ensure that a consistent layer thickness is achieved and that no air is trapped between the layers.

Models that have an overhang are produced on "stilts." That is, the software anticipates these problems and produces a platform on which the model is produced. To prevent the model from sticking to the build platform, a lattice base is automatically created, thus reducing the contact area of resin to platen. This technique is shown in Figure 10.6. On completion, the model is carefully removed and washed in a solvent to remove uncured resin and placed in a UV oven to ensure that all resin is cured.

Schematic diagram of the stereolithography (SLA) apparatus

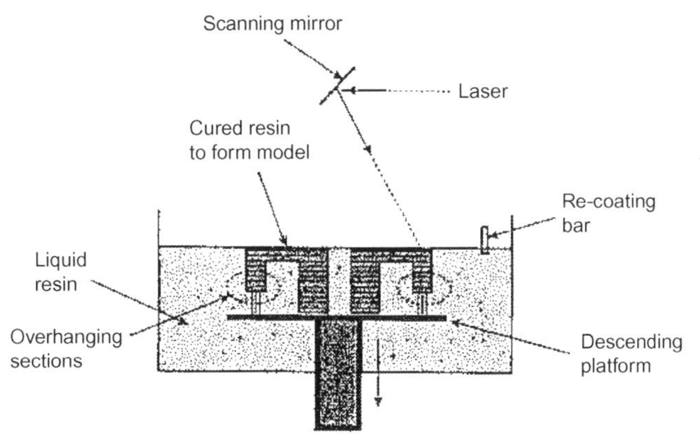

Figure 10.6
Stereolithographic printing.

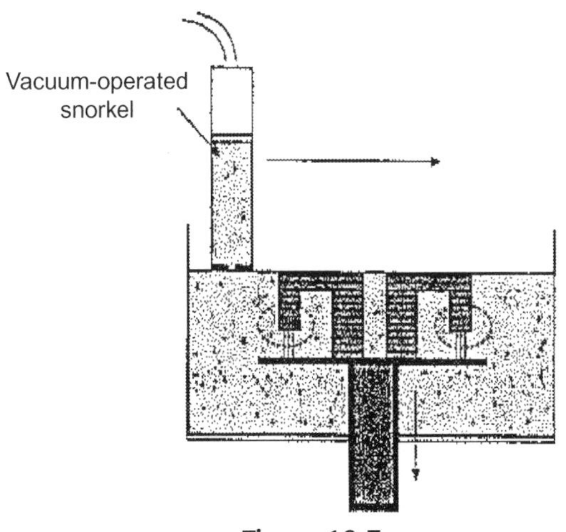

Figure 10.7
Adapted SLA process with snorkel.

A recent improvement to this technique has been the development of a *snorkel* as shown in Figure 10.7. This is filled at the start and end of each pass. As the snorkel passes over the model, a controlled amount of resin is released, ensuring that the entire surface is evenly coated. This technique ensures that resin cavities are not formed in areas that did not become cured by the laser. For example, internal cavities that became filled with uncured resin are likely to destroy the model because of the internal expansion of the uncured resin upon curing. Also this method allows different resins of varying viscosity to be used.

More recent developments have been prompted by problems caused by the expansion of the model when it is used as a disposable pattern (such as the wax pattern in the lost wax process). Where the resin model is produced to form solid walls, expansion during the "burning-out" stage weakens the ceramic shell and can cause failure in the firing and/or casting stages.

An engineering company that specializes in SLA systems has developed a machine and software that together allow the model to be constructed in the form of a honeycomb. The honeycomb structure collapses in on itself during "burning-out," thus avoiding the problems of expansion. Each pocket of the honeycomb structure is connected to its neighbor by a small hole and this allows for the uncured resin to be drained out of the prototype model prior to use.

Selective Laser Sintering

SLA and SLS processes are very similar except for the materials used and the form they take. Whereas SLA uses a liquid UV curable resin, the SLS process uses a powdered material. The use of powdered material is one of its major advantages because, theoretically, any powder that can be fused is capable of being transformed into a model and can be used as a true prototype component. Currently, the range of materials includes:

- Nylon
- Glass-filled nylon
- Polycarbonate
- Waxes
- Metal, which requires both a postsintering operation and copper infiltration cycle before the part can be used
- Sand bonded with heat-cured resin, for example casting patterns, that can be used in reverse engineering

As with other RP systems, the laser scans each slice of the component. In the case of SLS, the powder is sintered by the action of the laser scan. As each layer is sintered, the platform lowers the cured layer of powder, and a roller spreads another layer of powdered material over the previous layer. This is shown in Figure 10.8. During the build cycle, any nonsintered powder helps to support the model, thus avoiding the need for stilt supports in the liquid resin system of the SLA process. When the prototype is removed from the bath of powder, it requires a light brushing to remove any loosely attached material.

Solid Ground Curing

SGC is a physical imaging technology used to produce accurate, durable prototypes. Models are built in a solid environment, eliminating curling, warping, support structures, and any need for final curing. The process consists of:

- A computer analysis of a CAD file processes the object as a stack of "slices."

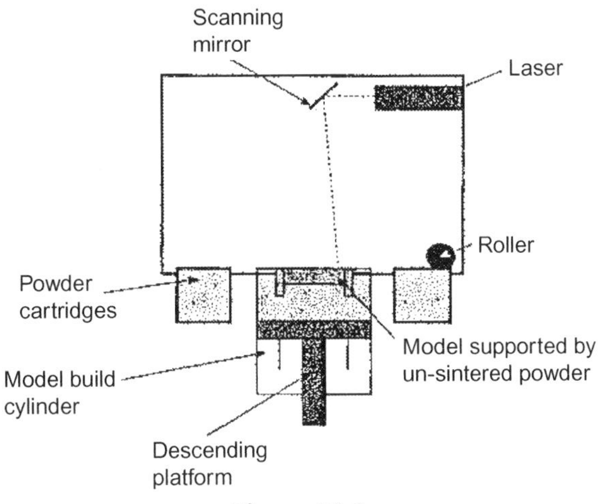

Figure 10.8
Selective laser sintering.

- The image of each working slice is "printed" on a glass photomask using an electrostatic process similar to laser printing.
- That part of the slice representing solid material remains transparent.
- This photomask is used next to produce the component slice by UV projection onto a photoreactive polymer. The UV light cures the polymer resin.
- The unaffected resin (still liquid) is removed by vacuuming.
- Liquid wax is spread across the work area as the process proceeds, filling the cavities previously occupied by the unexposed liquid polymer resin.
- A chilling plate hardens the wax, and the entire layer, wax, and polymer are now solid.
- The layer is milled level to remove any excess material.
- To build the next layer of the prototype, another layer of photoreactive polymer is laid on the previous work surface and spread evenly.
- An UV floodlight is again projected through the next photomask onto the newly spread layer of liquid polymer. Exposed resin again corresponding to the solid cross-section of the part within that slice polymerizes and hardens.
- The process is repeated slice by slice, each layer adhering to the previous one, until the object is finished.
- The wax is removed by melting or rinsing, revealing the finished prototype. (Alternately, it can be left in place for shipping or security purposes.)

The SGC method has 10-15 times the throughput of other methods based on photoreactive polymers. Any geometric shape can be created in any orientation. Parts can generally be made overnight, often in batches, and need no extra curing after they emerge. The use of wax means that supports do not have to be added to overhangs. Additionally, a build session can be interrupted, for instance, to expedite another project, and erroneous layers can be erased.

Laminated Object Manufacturing

As the name implies, the LOM process laminates thin sheets of film (paper or plastic), and the laser has only to cut the periphery of each layer and not the whole surface as in SLA.

The build material (paper with a thermo-setting resin glue on its underside) is stretched from a supply roller across an anvil or platform to a take-up roller on the other side. A heated roller passes over the paper, bonding it to the platform or previous layer, as shown in Figure 10.9. A laser focused to penetrate through one thickness of paper cuts the profile of that layer. The excess paper around and inside the model is etched into small squares to facilitate its removal.

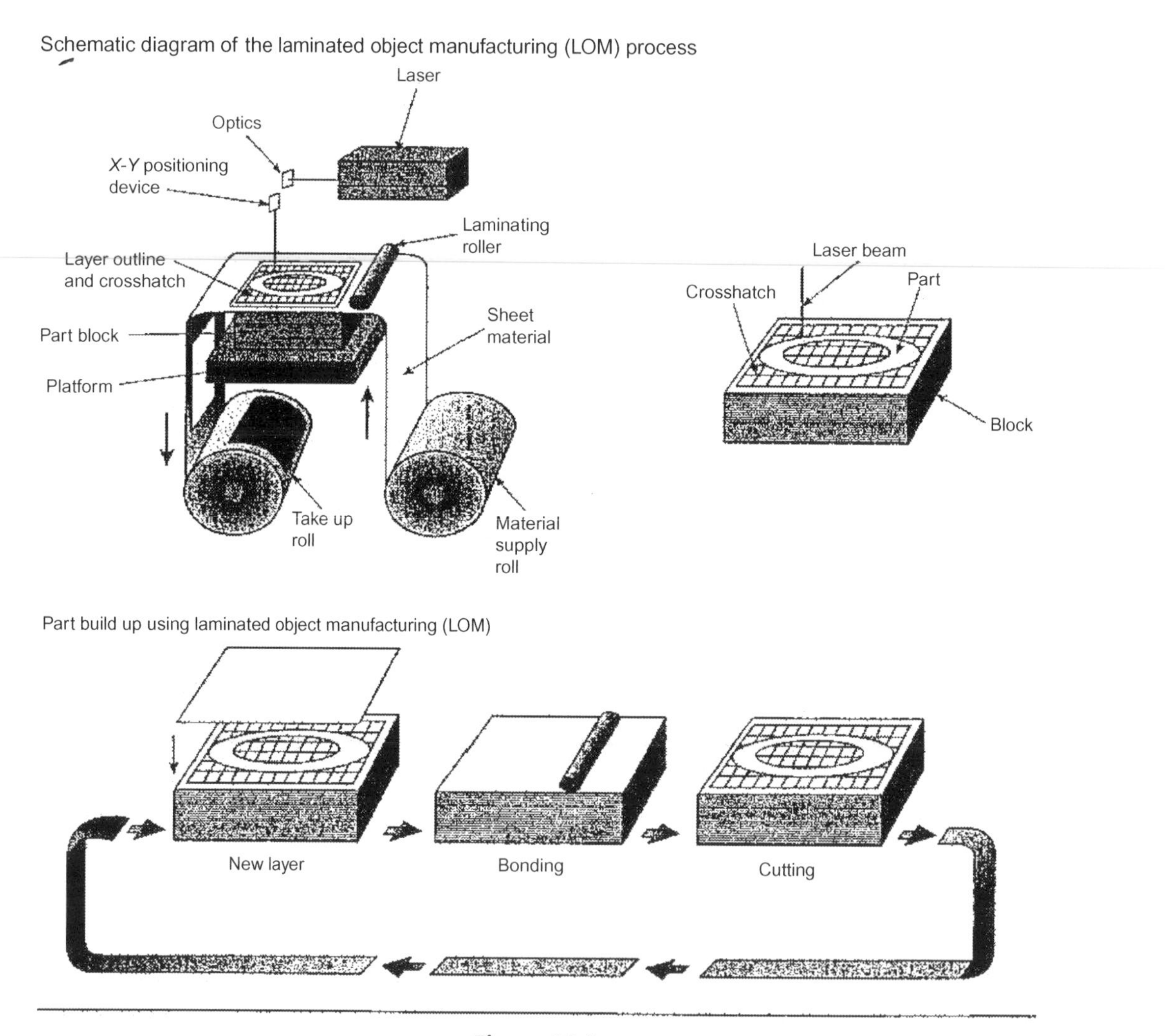

Figure 10.9
Laminated object printing.

Prior to its removal, this surplus material provides support for the developing model during the build process. The complete process of continuously gluing and cutting layer by layer until the model is complete is shown in Figure 10.9.

To reduce the build time, double or even triple layers can be cut at one time, increasing the size of the steps on curved surfaces and the postprocessing necessary to smooth those surfaces. LOM objects are durable, multilayered structures that can be machined, sanded, polished, coated, and painted. They can be used as precise patterns for secondary tooling processes such as rubber molding, sand casting, and direct investment casting.

Fused Deposition Modeling

The materials suitable for the FDM process include:

- Acrylic-butadiene-styrene (ABS), also known as *high-impact polystyrene*
- Medical ABS
- Investment casting wax
- Low- and high-density polyethylene
- Polypropylene

A thermopolymer is extruded from a traveling head having a single, fine nozzle. The head travels in the *X*-axis while the table or platform travels in the *Y*-axis and descends at predetermined increments in the *Z*-axis. On leaving the nozzle the thermopolymer adheres and hardens to the previous layer as shown in Figure 10.10.

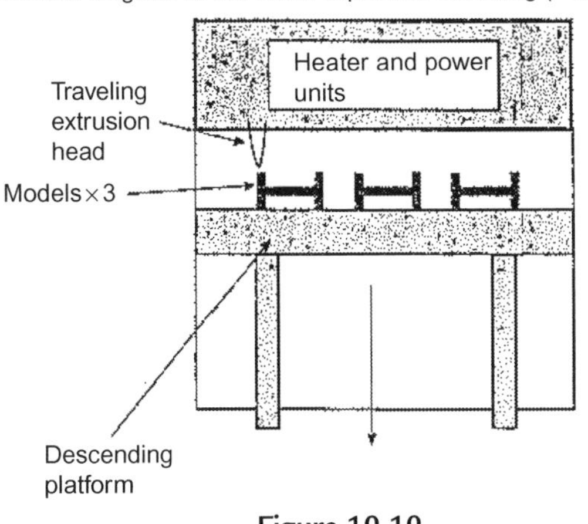

Figure 10.10
The FDM process.

Inkjet Technology: Powder-Based Printers

The FDM process sprays fine droplets of glue onto specific points on a powder bed. This causes the powder to stick together in the right places (Figure 10.11). Then another layer of power is deposited, rolled flat, and glued. The finished product emerges when the layers are complete and the unglued powder that provides support while the structure is growing is blown away.

The disadvantage with all of these processes is that the object can be made of only a single material: metal, plastic, or whatever. The real power will come when 3D printing can combine materials allowing us to create or recreate anything we desire.

Current technology also allows ink droplets to seep into the air gaps between the powder granules, giving color though the finished object. This is a step closer to inserting other materials into the voids to create a composite material. Once perfected, this could lead to a number of innovations, for example, recreating printed circuit boards. One could print a new motherboard directly from the Internet. Recent research has shown that features as small as 0.2 mm can be printed. Other benefits include the ability to arrange the alignment of particles to increase strength in certain directions where it is required. Other features include using cavities to save weight, depositing the material where strength is required, and omitting particles where it is not required.

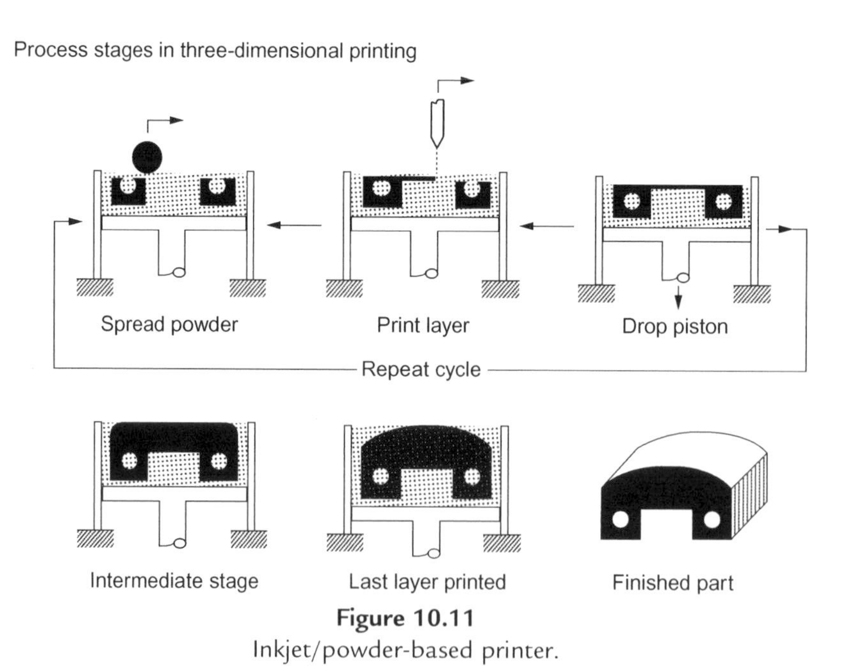

Figure 10.11
Inkjet/powder-based printer.

Hybrid Systems: Integrating 3D Printing with Subtractive Machining Technology

The hybrid approach is the integration of 3D printing technology with advanced CNC machining; the control software of such hybrid machines, however, uses the same input data as the material addition (3D printing) techniques. Typically, parametric software automates the generation of cutter paths for the removal of material. Hybrid machines maintain the flexibility of 3D printing with advantages of subtractive machines. The main advantages of material subtraction are:

- Speed of production
- Strength of materials used

However, the main disadvantages are:

- Limitation to models with no cavities.
- Inability to machine re-entrant angles, that is, the shape must be able to be machined with a vertical cutter. If the shape curves underneath, however, the product can be turned over and the other side machined.

Most RP systems are based on incremental build techniques as in laminated manufacturing technology. Conversely, new software that is currently available, such as DeskProto, is a *decremental* approach. This new system starts with a solid block of material (any material can be used) and removes as much as is needed to create the prototype. The decremental build technique fulfills the three criteria for RP.

1. The software can import STL files, which are standard files for any RP process.
2. It builds using an automatic process and, in this, differs from other Computer Aided Manufacture (CAM) systems in that it is easy to use and functions almost automatically.
3. It creates a prototype within a short time span. The actual milling time with DeskProto can be very short indeed. Even more important is the short preparation time required, for example:

- Milling can be done without an operator
- Translation from CAD geometry to CNC tool paths is very quick

Short delivery time (i.e., the complete system of software and milling machine) is ideal for in-house use. DeskProto Company has tried to create a "black box" approach (i.e., a highly automated system).

The hybrid approach is not new to RP. In hybrid rapid manufacturing, a metallic object is produced through a combination of additive and subtractive processes. The additive process relies on speed while ensuring the desired material integrity. The RP object is only near net because geometric quality is relatively poor as the result of the steps involved in building in layers. Only following fine-cut CNC machining process ensures the required geometric finish. Some stress relief may be required between processes, especially if high-powered laser welding is used as in the laser-engineered net shaping (LENS) approach.

The LENS technique uses powdered metal fed into an argon-shielded laser. The control is a full five-axis system that can deposit a variety of metals onto a surface, either to build up as layers into a full metal shape to then be finished by machine. Figure 10.12 shows the LENS system being used to build from new or repair worn components.

The emergence of personal fabricators utilizing both 3D printing and subtractive machining is following the same path as the development of 3D printers with entrepreneurial startups producing low-cost home/office machines. An example of such a startup is the FABtotum home fabricator Factotum, 2013, combining 3D scanning, 3D printing, and multiaxis CNC machining with a cost less than £1000. See Figure 10.13.

Direct metal deposition (DMD) is a similar process to LENS and was developed at Michigan University and Precision Optical Manufacturing and adopted by Trumpf, the German laser

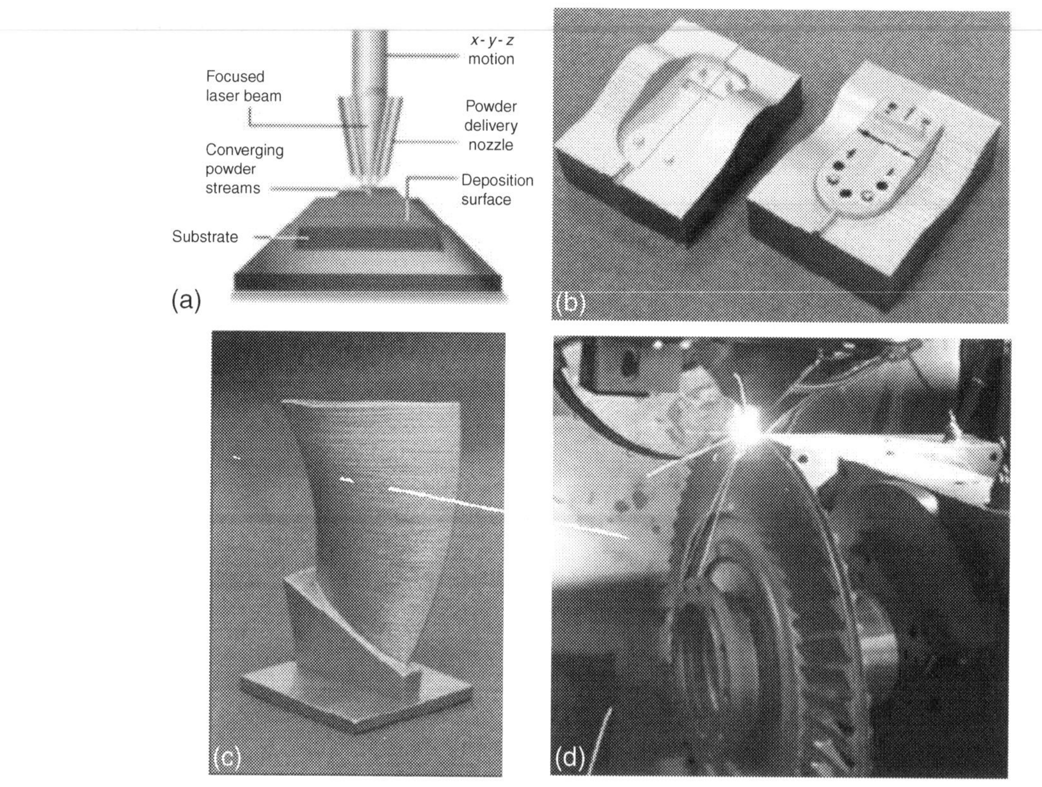

Figure 10.12
Laser-engineered net shaping. (a) LENS process, (b) a pair of dies made using LENS, (c) a turbine blade made using LENS, (d) LENS used for refurbishing a turbine.

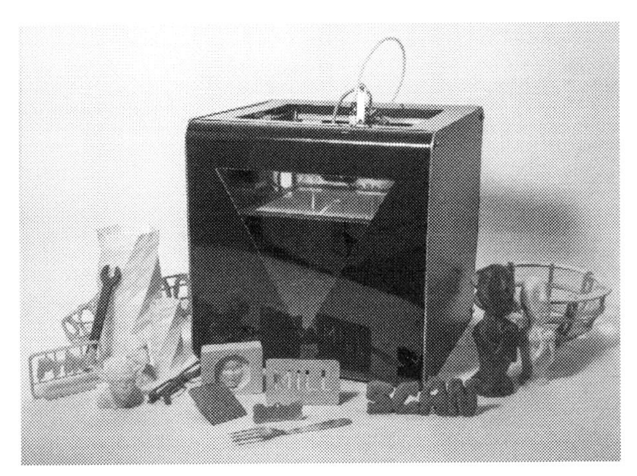

Figure 10.13
The FABtotum personal fabricator.

company. This system uses optical feedback to ensure the integrity of deposition. This machine in Figure 10.14 is known as TrumaForm DMD 505 and has a very high-powder utilization rate. It is mounted onto a full six-axis robot so that a freeform surface can be produced because the head is normal to the part surface.

Arc hybrid layered manufacturing is a process developed at IIT Bombay (Karunakaran et al., 2009). The advantage of this process is that the Metal Inert Gas (MIG) welding equipment deposits the layers in a similar fashion to the previously described systems. It can be retrofitted to any existing CNC machine. An example of an injection molding tool near-net shape and the finished part are in Figure 10.15.

Environmental Considerations

The media have promoted 3D printing, in particular its ability to lower the environmental impact of manufacturing processes. Many of the proposed environmental reasons for adopting 3D printing and related technologies are outlined in the Factory of the Future report (2013) and listed here:

- The green agenda (reducing energy, water, waste)
- Changes in communications practices and social networks
- Changes in demands for proximity to customers
- Changes in the costs of resources
- Changes in the costs of transportation
- Changes in performance metrics (e.g., increased emphasis on quality)
- Improvement to lead times and customer satisfaction
- Less constraint because of machine use and batch efficiencies

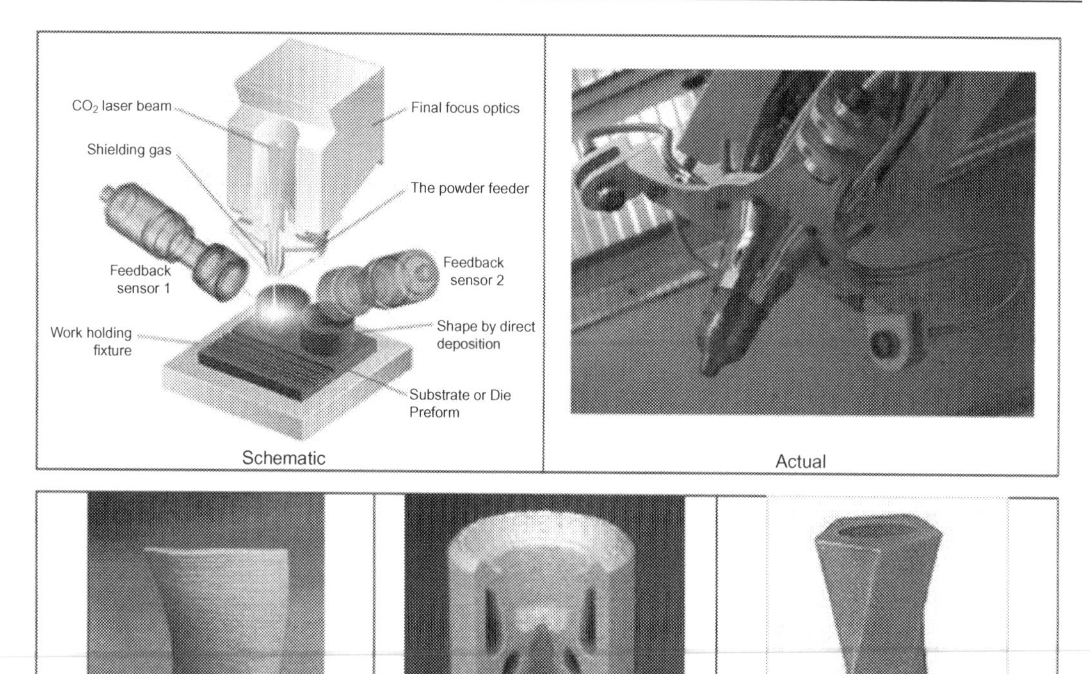

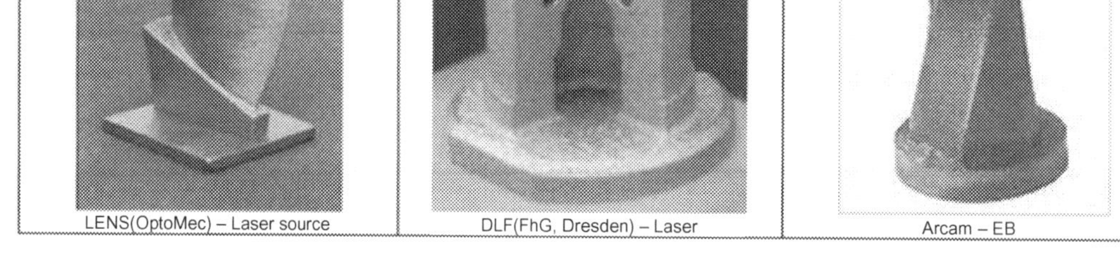

Figure 10.14
Laser deposition head.

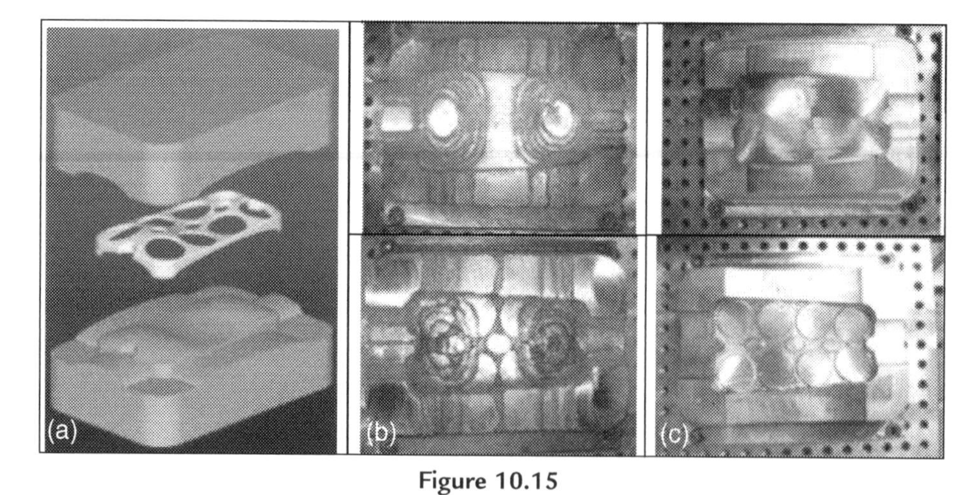

Figure 10.15
Near net molds. (a) CAD models of Egg template and its injection molds, (b) near-net molds, (c) finished molds.

There are a number of examples of how 3D printing has allowed a new range of ultrapersonalized products to be designed and manufactured. These technologies have generated a wave of democratization of the manufacturing processes allowing consumers to be designers and makers of their own products. The Makerbot Thingiverse Website that is a community-based repository for sharing 3D printable models has seen an exponential growth from 2008 with more than 100,000 objects uploaded (Thingiverse, 2014).

This growing repository of parts and wide access to printers for the maker/repair communities has facilitated the use of 3D printing and the repair and replacement of parts required for a wide range of products. The ability to print and repair rather than scrap a product helps to reduce waste and the need for more of Earth's resources, both in terms of raw material and energy.

Using hybrid technology in manufacture/remanufacture has a number of advantages, such as the ability to repair high-value components including jet turbine blades, that impact the environment in a number of ways

- Reduction of the need for energy and materials because there is no requirement to make new components from the production of raw materials to the manufacturing processes involved.
- Localization of production (i.e., no need to transport raw materials, unfinished components)
- No need to input energy and resources into scrapping a high-value component

The production of new components by this process means that there is less wasted material and energy because only a finished light cut is required for produced shapes; it is not necessary to rough cut the majority of material.

The Factory 2.0 Philosophy

3D printing can be viewed as an advanced manufacturing system that enables any individual to become a designer and maker of his or her own products. Online communities discussed in Factory 2.0 (2011) show new ways of collaborating online in order to bring manufacturing back to the United States. Since the liberalization of patents on this technology, the number of 3D printing machines and facilities has grown; it is now possible for small companies or individuals to purchase such machines. The proximity of production to consumer means that transportation costs can be drastically reduced and lead times can be cut dramatically with obvious benefits to the environment. Fox (2010a,b) states that "Web 2.0 combined with advanced manufacturing technologies (AMTs) could revitalize manufacturing, generate new employment, and reduce environmental impacts."

Fox goes on to discuss how the new AMTs enable many new ways of combining materials and embedding functionality. As a result, previously difficult trade-offs, such as geometric complexity versus production time and cost is now addressed. The latest ideas in 3D printing as previously stated can help revitalize manufacturing and generate new employment, especially when combined with the power of the Web. It can lead to sharing new product ideas, especially through social networks such as Facebook, Flickr, and YouTube, thus rapidly propagating new ideas around the world. Important companies such as Electrolux, Lego, and Philips have used this idea through competitions and so on to produce new product designs via customer input (i.e., gives customers what they want more accurately).

New Web-based manufacturing companies, such as Fabjectory, Figure-Prints, Ponoko, and Shapeways, currently use Web technologies to allow designers to manufacture products in a variety of materials, set up online shops with products that are manufactured on demand, and generate parametric products that customers can modify and print.

Agile Manufacturing

Agile manufacturing processes are facilitated by 3D printing technologies that enable quick response orders and are the production of a variety of parts in small batch sizes or individual components, avoiding the need to reconfigure or set up traditional machine tools. The cost benefits include not having to use capital to provide storage of real parts because each one is made to order. The Factory of the Future (2013) states that 3D printing and other technologies such as robotics is also a "game changer" of the future. These technologies will enable reconfigurable factories that can make single customized part production a reality and use of less energy than mass production (i.e., making just what you want when you want it).

Recycling

As noted, the use of 3D printing to repair and recycle broken products by making 3D printable spare part components is growing. The other dimension of recycling involves the materials used for 3D printing, typically spools of filament.

Some of base material used in 3D printed products can be processed back into fresh-coiled modeling filament or used as recycled material. Machines developed by Coca-Cola and Will.i.am's that recycle plastic bottles as filament. Any type of recyclable plastic from food wrappers and drinks bottles to water pipes and Lego bricks can be turned into filament. These machine grind up the plastic, melt it down, and then extrude it as a material that can be fed into a 3D printer (Figure 10.16).

The Ekocycle 3D printer was created in collaboration of Coca-Cola and Will.i.am, who is the chief creative officer of 3D Systems. Such recycling machines are less expensive alternatives to the spools of filament used by 3D printers.

"As humans we create hundreds of millions of tons of waste every year," says Will.i.am (2014). "A lot of it is cheap PET plastic." To address the waste problem, mechanical engineering student Tyler McNaney created the Filabot recycling machine (Figure 10.17). As well as waste

Figure 10.16
The Ekocycle 3D printer.

plastic, the machine allows users to recycle their failed or unwanted printed objects. "It is a one stop shop for all the filament you could ever need" (McNaney, 2014).

The dissolvable support material can also be made from organic or recyclable material and hence is easily sourced and thus has less impact on the environment.

Organic Materials

The possibility of using new raw materials derived from plants is good for the environment. For example, the ASUS uses bamboo pulp to make a plastic substitute to use in the company's laptop computers, which are eminently recyclable and use only sustainable crops such as bamboo rather than precious fossil fuels (Figure 10.18, Ochi, 2012).

Bamboo has an immense tensile strength that rivals that of many alloys (Asus, 2014). Asus has produced a highly resilient notebook. It was tested under arduous conditions and is the first notebook to have survived the unforgiving conditions of snow-capped Qomolangma Peak (8848 m). Asus has shown that bamboo has a renewal rate that no other plant can match, growing 60 cm in 24 h and regenerating itself upon harvesting without being replanting (Figure 10.19).

The major use of pulp material is for casting, so its use is not technically 3D printing; however, steps have been taken to use such materials as part of the filaments used by extrusion printers.

Figure 10.17
The Filabot recycling machine.

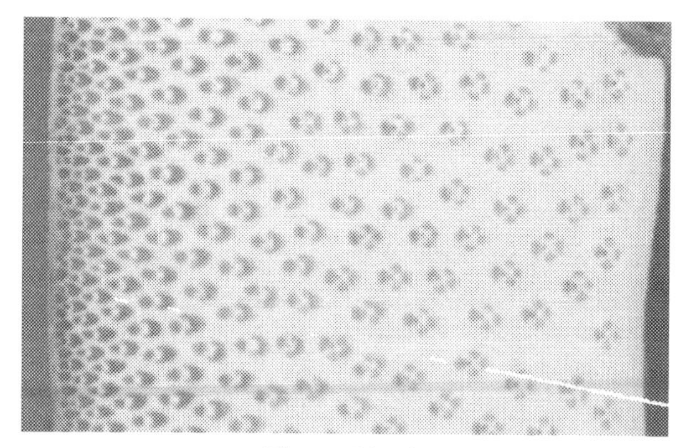

Figure 10.18
Bamboo pulp. Ochi (2012).

One recycled material now used is made of a printable wood now available in the form of the experimental material Laywood, which can be extruded through the nozzle of a 3D printer. This material is a mix of recycled wood fibers and polymer binders. An added advantage is that Laywood emulates wood with a grain but does not warp, twist, or shrink because of its internal structure (Laywood, 2013).

Figure 10.19
Asus bamboo notebook.

Factory of the Future: The Next Steps

An ambitious example of the use of 3D printing cited in an article by Factory of the Future (2013) is the Urbee car1, designed and manufactured by Kor Ecologic using a Stratasys additive manufacturing machine. The article discusses the Stratasys2, which was used to print all the vehicle's exterior components using FDM, which eliminated the need for tooling, machining, and handiwork.

Many of the advantages from 3D printing have been noted. These also include direct-write additive manufacturing machines that can incorporate the functionality of circuits, sensors, and controls into assemblies during their production. Another innovative product relates to health care; the production of customized prosthetics and made-to-measure implants can be produced directly from MRI scans if regulations permit. Current research on the use of the combination of metallic and synthetic/macro and nano structures that are ideal for body part replacements is being studied. With an aging population, this will reduce the impact of the cost to the environment and provide a healthier population. However, the weakness along the direction of build (similar to the grain of wood) is a problem, but it can be addressed by using mass parallel printing to produce complex structures (Figure 10.20).

The Red Eye3 company based in the United States uses deposition machines to produce approximately 5000 individual parts from an initial request in about 2 weeks. The growth rate for the business is 30% annually, primarily from increased demand.

Figure 10.20
The Urbee car.

Conclusions

The use of additive (3D print) rather than subtractive manufacturing has a significant environmental impact by reducing the need for consumable materials, minimizing tooling, manufacturing topologically optimized components, and producing consolidated assemblies.

Additive manufacture enables large reductions in energy consumption from traditional process such as casting, which must go through large-scale heating and cooling cycles. The ability to reduce the number of components though consolidated assemblies reduces the energy required to assemble and transport separate parts to different geographical locations.

The democratization of the technology can end the design and manufacture of products, particularly of spare parts for recycling and repair. And the open-source community is generating new ways to recycle print materials and use sustainable materials such as wood and bamboo fibers.

Some of the limitations of 3D printing such as speed of manufacture are being addressed with the use of hybrid technologies within the domestic market and high-value manufacturing.

References

Asus, 2014. http://www.asus.com/Notebooks_Ultrabooks/U6V_Bamboo/.
Deskproto, 2014. http://deskproto.com/products/essentials.htm.
Factory of the Future, 2013. Foresight. Government Office for Science, UK.
Fox, S., 2010a. VTT, The Technical Research Centre of Finland.
Fox, S., 2010b. Dr Stephen Fox. http://eandt.theiet.org/magazine/2010/08/post-industrial-world.cfm.
Karunakaran, K.P., et al., 2009. Hybrid rapid manufacturing of metallic objects. In: 14èmes Assises Européennes du Prototypage & Fabrication Rapide, 24-25 Juin 2009, Paris.
Laywood, 2013. http://www.3ders.org/articles/20120925-test-printing-of-new-laywood-filament.html.
http://www.dezeen.com/2013/01/14/filabot-3d-printing-recycling-tyler-mcnaney/.

Ochi, S., 2012. Fabrication of press-molded products using bamboo powder. J. Mater. Sci. Res. 1927-0585.1 (1), 156, E-ISSN 1927-0593.
Thingiverse, 2014. http://www.thingiverse.com/about.
http://www.dezeen.com/2014/07/02/coca-cola-will-i-am-3d-printer-recycled-plastic-bottles/.

SECTION V

Case Studies

CHAPTER 11

Automated Demand Response, Smart Grid Technologies, and Sustainable Energy Solutions

Azad Camyab
Pearlstone Energy Ltd., London, UK

Abbreviations

ADR automated demand response
ATSP ADR technology solution provider
BFC Bracknell Forest Council
BMS building management system
CCGT combined cycle gas turbine
DNO distribution network operator
DR demand response
DRA demand response aggregator
DRAS demand response automation server
DSBR demand-side balancing reserve
DSR demand-side response
DUoS distribution use of system charges
EMR electricity market reform
FIT feed-in tariff
Gensets generating sets
HBS honeywell building solutions
LAN local area network
LCNF low carbon network fund
LCPD large combustion plant directive
NG National Grid
Ofgem Office of Gas and Electricity Markets
STOR short-term operating reserve
TNUoS transmission network use of system

TSO transmission system operator
TVV Thames Valley Vision
VPP virtual power plant

Background

The Challenge of Maintaining Equilibrium in the UK Electricity System

Energy has become front-page news, and the debate about its future has never been so important. Understanding what that future might look like is crucial if we are to meet the long-term challenge of providing safe, reliable, and secure energy in a sustainable and affordable way. Of course, we cannot be certain how the energy future will evolve. Factors such as environmental legislation, energy costs, and economic developments will all have a major impact on the future energy landscape.

Generation of thermal electricity in the United Kingdom is coming to an end. A large amount of coal- and oil-fuelled generation has and is retiring because of age and environmental legislation, including the European Union (EU) large combustion plant directive (LCPD). Furthermore, challenging economics for gas-fuelled generation has resulted in few new power plants being built and many, including units only two years old, being mothballed until the business case for their operation becomes more favorable.

The resulting decline in capacity and generator availability has led to very tight capacity margins—the difference between electricity supply and demand levels that can make the role of the National Grid (NG) (the UK's transmission system operator [TSO]) in matching generation and demand quite challenging. Indeed, spare electric power production in the electricity system is predicted to fall to only 2% by 2015, increasing the risks of blackouts across the United Kingdom should an unforeseen operational issue arise or the United Kingdom experience a cold winter.

The white paper on electricity market reform (EMR) published by the Department of Energy and Climate Change (DECC) in July 2011 proposed a number of changes that the government estimates will save the UK economy £9 billion between 2010 and 2030 while ensuring decarbonization and the continuity of electricity supplies.

One of the key assumptions in the EMR document is that planned low carbon generation and gas capacity will be addressed as planned on schedule, and the proposed package of price incentives is designed to ensure that this happens.

However, a number of factors could undermine this forecast such as:

- Nonprice constraints causing delays in the construction of new nuclear and wind power plants (e.g., financial constraints, problems with new technology, equipment shortages, and planning permission delays)

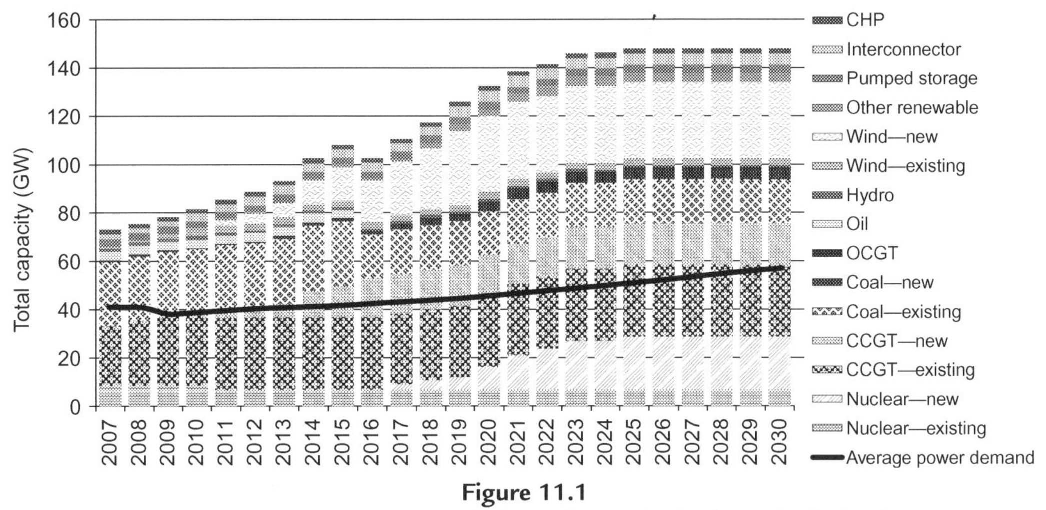

Figure 11.1

Existing, under construction, and planned capacity in the United Kingdom.

- Delays in resolving exactly what the structure, level, and duration of the feed-in tariffs (FITs) and capacity payments will be (including measures to improve electricity market liquidity) that could result in investment decisions and construction being delayed
- General uncertainty over the long-term future of UK gas production

See Figure 11.1, for the current forward expectation of total available fuel type generation to 2030 as prepared by the NG.

The information in Figure 11.1 assumes that 1.5 GW of nuclear generation capacity comes offline in 2011 and 8.4 GW of coal and 3.6 GW of oil capacity comes offline in 2016 as required under the LCPD. The figure paints an optimistic picture of total capacity relative to average electricity demand.

However, data in Figure 11.2 indicate the total average output in comparison to average demand assuming the impact of delays in new nuclear construction, a lack of investment in combined cycle gas turbine (CCGT) and wind power generation, and the drop in availability and output of aging fossil fuel plants.

To get from total capacity to average output, some assumptions on average load factor going forward have been made. Because most existing coal and gas sources are nearing 20 years old, it is assumed that the average load factor of the existing coal and gas fleet will drop by 2% per annum post-2015 (i.e., gas 66% in 2015 to 34% by 2030, coal 43% in 2015 to 11% by 2030) when accounting for more flexible operation, plants coming offline, and or drop in efficiency.

The results reveal a significant potential gap between average forecast demand and average available power generation.

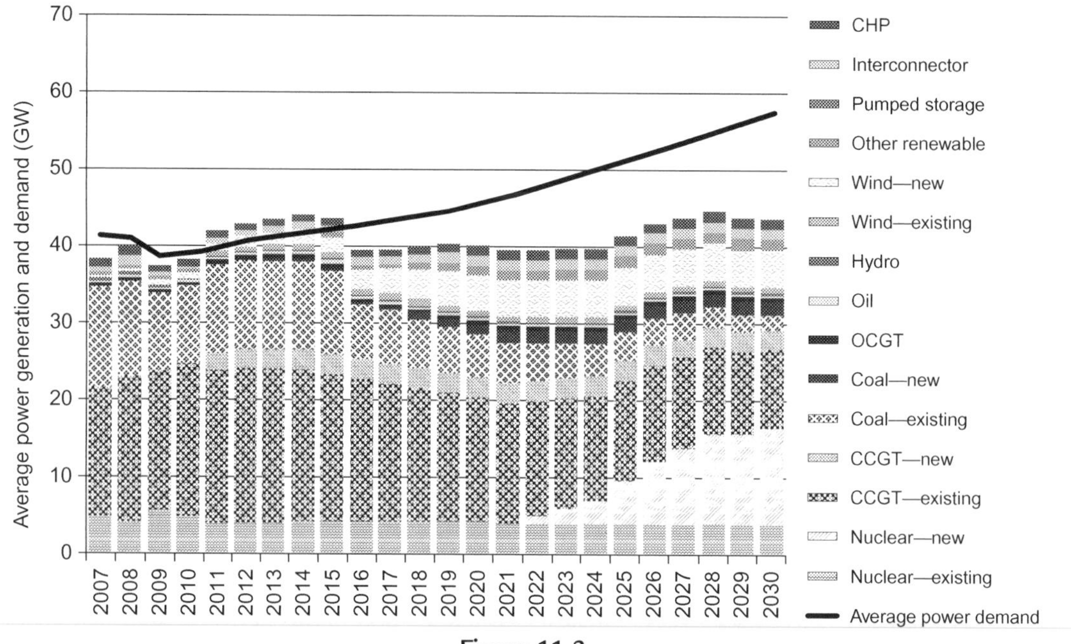

Figure 11.2
Average demand compared to average power generation.

This analysis is based on the following assumptions:

- No new CCGT built other than that currently under construction
- Only 50% of planned wind capacity built
- New nuclear delayed by five years, ramp up in 2022
- All existing nuclear capacity to benefit from continuing life extension throughout the period
- Load factors derived from historical power generation data (coal = 43%, CCGT = 66%, nuclear = 58%, and wind = 30%)

Such a scenario would have severe consequences not only for domestic users but also for businesses for which blackouts would cause major disruption and have a dramatic effect on productivity.

Furthermore, large power stations can quit suddenly or be delayed coming back online after maintenance shutdowns. Interconnector swings based on market prices can cause predicted availability of electricity from neighboring countries to disappear and in fact go in the opposite direction. Unforeseen forced outages such as the fire incidents in the second half of 2014 at Didcot B, a gas-fired power station in Oxfordshire, and the coal-fired plants at Ironbridge and Ferrybridge are good examples of how significant capacity can suddenly be lost in a short period of time, causing major imbalance in generated electricity.

Additionally, a significant number of old Magnox nuclear power stations are reaching the end of their lives and already operate on fewer cells than were originally designed. At the same time, solar- and wind-based generation can be notoriously erratic and intermittent, causing them to be unreliable in times of heavy demand.

Since 2004 in the United Kingdom, generated electricity from offshore wind energy has seen an annual growth rate of 54%. Generation from offshore wind alone increased by 68% during Quarter 1 (q1) of 2013, yet this rate will need to accelerate much faster over the next few years to meet the UK government's committed 2020 renewable energy target. However, generation from offshore wind cannot be guaranteed for periods of imbalance and therefore, at present, it cannot be regarded as a reliable source for periods of extreme demand that the NG is increasingly required to meet beginning in winter 2014.

As part of its remittance to satisfy the demand for electricity in the United Kingdom, the NG is required to maintain a "reserve" to ensure that it keeps electricity generation and demand balanced. This reserve is achieved by keeping a large number of part-loaded, fossil-fueled power stations on standby. This means these units are constantly running without generating electricity in case they are needed; that is an expensive and wasteful exercise, which generates millions of tons of carbon emissions every year.

Despite this, the UKs electricity system is currently sized to meet demand peaks that occur only infrequently. This leads to generating plant and transmission and distribution networks being underused for much of the time. However, increased use of electrical appliances, gadgets, IT systems, and electric vehicles means that demand peaks are rising. To meet them, even though for only be for a few hours at certain times of the year, distribution network operators (DNOs) simply install larger substations and dig up the roads to put in more cabling, all at huge disruption and capital expenditure to customers and ultimately the UK taxpayer.

Buildings and Flexible Load

Commercial and industrial buildings and facilities across the United Kingdom use electricity for different things in different ways. But they do not need it at full demand all the time. As a result, buildings and some facilities are able to reduce or turn off the electricity to various devices for short periods of time without negatively affecting the building's performance or comfort conditions for its occupants. This is known as the building's *flexible load*. Importantly, by turning load down or off, as opposed to alternatively turning generation on, no fossil fuel is used so there are no fuel costs, and no carbon is emitted.

A fast-acting, clean flexible load is a unique asset that currently is virtually untapped in the United Kingdom. This is a valuable commodity that can potentially meet the pressures of peak electricity demand while giving businesses security about their power supply needs. It is also vitally important in working toward a more sustainable and carbon-neutral solution by using what is already easily available in the system rather than the economically and environmentally costly alternative of constructing new power stations that may prove not to be commercially viable and so lie idle for significant periods of time.

The main facilitator for enabling DR to be fully operational is through third-party service provider companies such as Pearlstone Energy, a demand response aggregator (DRA) operating in the United Kingdom. The DRA will be able to access this flexible load in an economic, reliable, and practical way and on an aggregated basis by using, for example, the proven automated demand response (ADR) platform provided by ADR technology solution providers (ATSPs) such as Honeywell. The DRA creates value by providing "total energy solutions," and, managing sustainable energy demand and power generation to maximize revenue for its customers in the United Kingdom (and global) power markets. The DRA will use an ADR platform to enable buildings to participate in the NG's range of balancing services, such as the short-term operating reserve (STOR) program, as well as other demand response (DR) programs and initiatives for which they receive payment. However, programs such as STOR require reserve to be delivered in blocks of many megawatts, the flexible load potential from individual buildings and facilities is often too small to meet these requirements and so is of no value on their own. The DRA creates value for building owners by using the ADR technology to access the flexible load from each facility and provide this as an aggregated block to the program operators. In this way commercial, public, and industrial buildings as well as standby generators can earn revenues from offering their flexible load using ADR in a reliable, fast, cost effective, and automated fashion.

And the potential of this largely untapped resource is significant. In July 2012, Element Energy and De Montford University were commissioned by the Office of Gas and Electricity Markets (Ofgem) to explore and size the load flexibility in UK nondomestic buildings. Excluding industrial facilities, potential levels of between 1.2 and 4.4 GW of DR were identified, so the overall potential from commercial, public, retail, health care, and industrial facilities will be much higher.

Currently, the NG procures the nonsynchronized requirement for balancing UK electrical grid's frequency primarily by contracting for STOR via a competitive tender process from a range of service providers in the form of standby generation and/or demand reduction. The STOR programme was expanded in 2013 to include DR from sources with variable baselines (variable baseline units) such as commercial and industrial buildings. Against this background, the proven ADR platform provided by ATSPs, such as Honeywell, that is used across the world provides the opportunity for building owners with variable electrical use profiles to obtain payment for the provision of "clean" flexible electrical load in their buildings for the first time in Europe. Despite current low STOR prices, both ATSPs and DRAs have embarked on partnerships using ADRs to enable urban and industrial communities to access new revenues because of the belief as noted in the first part of this chapter that the need for response will increase, and as a result, prices for DR are forecast to increase from 2015.

Through the RIIO-ED1 process, Ofgem's next price control period for UK-based DNOs requires them to use demand-side response (DSR) solutions from 2015. As proved by a range of successful projects funded by its significant investment in the Low Carbon Network Fund (LCNF) between 2010 and 2015, Ofgem is convinced that DSRs can be smarter and more cost effective than traditional reinforcement methods to resolve network constraints and defer or reduce the £36 billion of investment Ofgem has forecast as being required by 2020 to keep the distribution networks fit for their purpose. The potential for DSRs here is to reduce or shift peak demands on nearly constrained networks that happen on a regular basis in certain months of the year, typically from November to February, in the late afternoon for 2-4 h rather than simply install larger expensive distribution assets.

Furthermore, with EU aide approval, capacity market auctions commenced in 2014, embracing new and existing generation capacity including DR and permanent reductions in electricity demand. Participants will be able to bid both one and four years ahead in the auctions, giving them flexibility and confidence to invest in the future. A buffer has sensibly been incorporated for the transition period for demand-side reduction within the capacity market.

In response to the very low capacity margins that are imminent, the NG recently launched a new emergency reserve service paving the way for the growth of DR. Demand-side balancing reserve (DSBR) offers a new opportunity for the demand side to participate in the provision of demand- and supply-balancing services. The NG has tendered for an amount of demand reduction and standby generation capability to be available at peak times on nonholiday weekdays during the winter for a setup payment of between £5 and 10 per kW per year. Usage payments for delivery will range from £500/MWh to £15,000/MWh. The idea is that in addition to providing much-needed backup to prevent potential blackouts, this will promote significant growth in the provision of demand-side services before DR participation in the capacity market.

All this change and subsequent development marks the first time that the United Kingdom has moved away from an energy policy of providing electricity supply to meet peak demand whenever it is. This policy has proved to be unsustainable, inefficient, insecure, expensive, and impractical—more so as energy and plant construction prices rise and we are constrained by the need to reduce carbon emissions. The rise and significant growth of DR service provision is therefore an inevitable consequence.

To provide a practical solution to the problem of peak demand supply, DRA companies such as Pearlstone Energy have taken the opportunity to offer their customers the prospect of building capacity now in order to be in a lead market position to provide DR from 2014 and significantly when the potential "capacity crunch" hits the United Kingdom. The benefits are twofold because this strategy of developing the market for DSR provision will also meet the requirements of fast-peak demand relief for constrained DNO-owned networks.

The Opportunity

Within a few years, the impact of DR on the UK capacity market will be manifold. It will enable any property owner to exploit her or his building's untapped energy as a revenue source, provide electricity to the grid when needed, and ensure that all power use is optimized more efficiently and sustainably. In effect, the DRA will enable property owners and asset managers to access their assets' untapped revenue sources while meeting energy-saving demands, cutting costs, and contributing to the reduction of carbon emissions.

The Flexible Load in any Building has Potential Value

In any building environment, the daily comfort and working conditions require commercial and industrial buildings to use electricity to power a range of devices such as lighting, heating, cooling, ventilating, air-conditioning systems, pumps, fans, and motors.

However, in many cases, it is possible to reduce use or simply turn off electricity to these devices for short lengths of time without negatively affecting the building's performance or comfort conditions. This is known as the building's "flexible load." It is the role of the DRA using an ADR platform to draw on flexible loads across many sites in a coordinated and automated fashion in order to create a virtual power plant (VPP) of negative watts, or negawatts. This "negative load" can present significant value to participating organizations.

The value proposition for the participating businesses is that the DRA is able to optimize and monetize this aggregated flexible load and share the revenue streams based on the level of load offered by an individual building. The DRA accesses this flexible load in an economic, reliable, and practical way without any impact on building performance or comfort for the occupants by using, for example, a proven ADR solution. Many gigawatts of "negative load" are already being successfully drawn on across many hundreds of buildings by ATSPs such as Honeywell's ADR technology to manage grid stress in the United States, China, India, and the United Kingdom.

The Solution

The DRA provides comprehensive value-adding DR services for its customers and offers them the ability to earn revenue from the clean, flexible electrical load in their buildings. The unique association with a recognized ATSP differentiates the provision from other competing DRAs because of the fact that ATSP provides the DRA with the infrastructure to do this.

The DRA's use of ATSP's ADR solutions delivers aggregated DR (both load shed and standby generation) to the NG's STOR program and other new balancing services, sharing the resulting payments with customers and providing them with a source of new revenue.
It also does the following:

1. Delivers DR for DNOs as fast relief for nearly constrained primary networks.

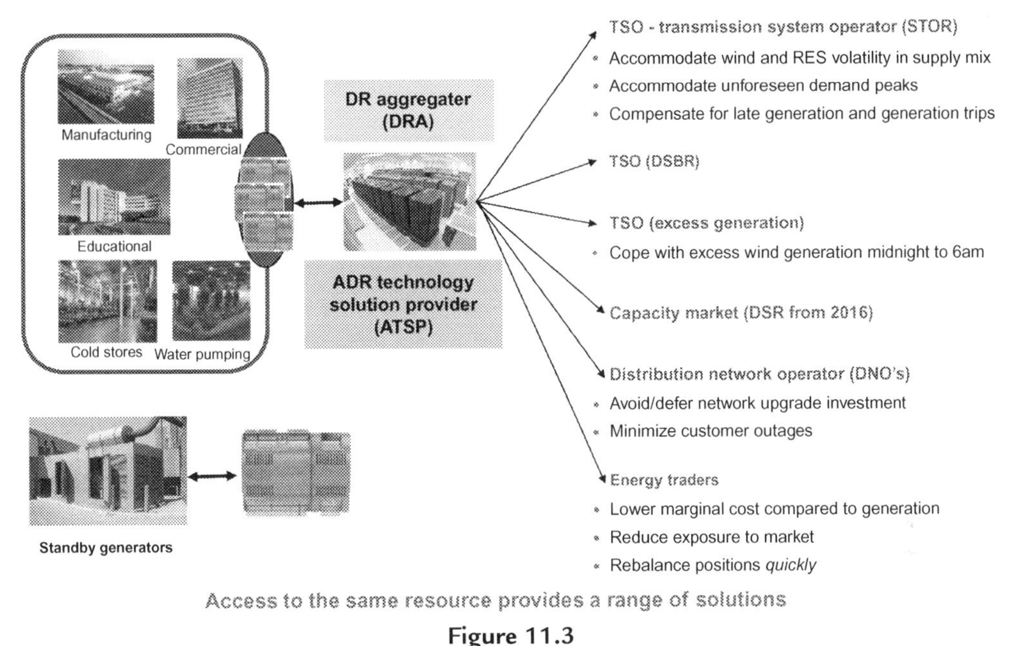

Figure 11.3
Combined value proposition to ADR customers.

2. Enables traders with access to DR to provide potential arbitrage benefits between day-ahead and hour-ahead markets and help balance their energy portfolios and avoid imbalance charges.

This is described in Figure 11.3.

ADR Value Streams

ATSP's ADR solution enables DRA companies such as Pearlstone Energy to access and aggregate DR from both standby generation (existing and greenfield) and load reduction from commercial and industrial buildings to create new income on a sustainable, predictable basis.

The resulting value streams for the DRA include:

- STOR revenues
- DSBR and capacity mechanism revenues
- DNO payments
- Opportunity to differentiate offering from those of competitors
- Opportunity to provide customers a new revenue stream by enabling them to sell the energy they do not need at certain times
- Ability to adjust customer load shape and potentially provide lower kWh charges
- Proactive management of network demand charges (Triad, distribution use of system [DUoS]) for customers using the demand-limiting functionality of ADR and receipt of management service payments

- Offering of variable baseline customers a new service using a system tested and approved by the NG

The resulting value streams for building owners are:

- No cost to install
- Ability to obtain a new source of revenue for little or no impact on the building operation or occupants
- Ability to channel the DR payments into a "green fund" to pay for energy efficiency or carbon-reducing measures that currently do not meet internal capital investment criteria
- Ability to obtain a new tool to support almost real-time visualization of site energy consumption and awareness of energy use
- Ability to reduce kWh charges by offering site-load flexibility

A summary of the value proposition by the DRA to the building owner is shown in Figure 11.4.

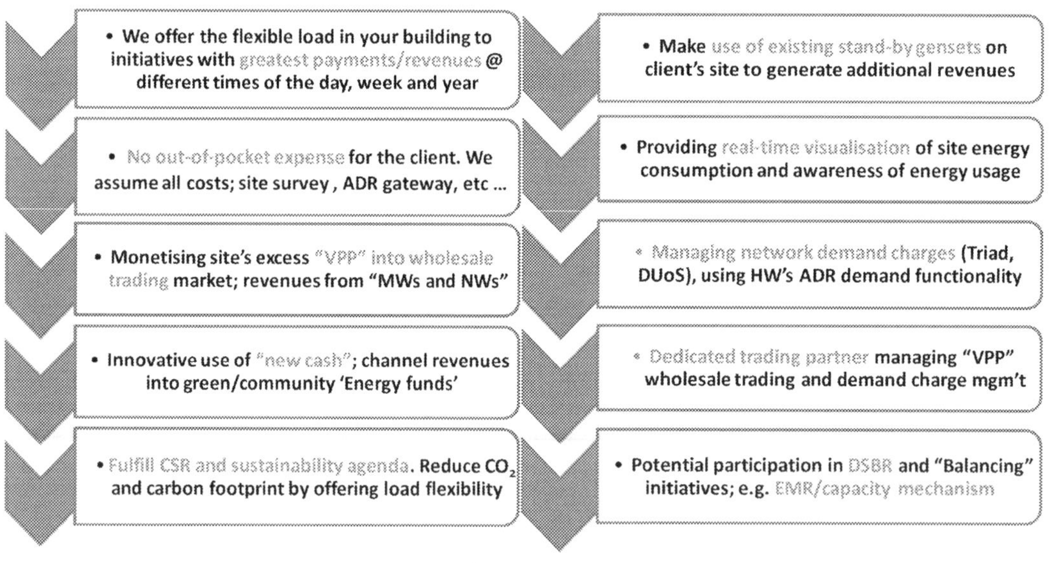

Figure 11.4
Summary of the value proposition to building owners.

The resulting value streams for the DNO include:

- Avoiding an investment of multimillion pounds of customers' money in large assets to meet peak loads that is in effect for the most time a stranded asset

- Obtaining a flexible solution that can be scaled to meet expansion needs or canceled if no longer required (e.g., if a large site on the network closes, removing the constraint situation)
- Obtaining a dynamic view of energy consumption of all participating buildings in the DR program

The resulting value streams for the TSOs are:

- Obtaining a clean, reliable, and fast response to balance the grid at times of stress without emitting carbon emissions. This would contribute to the NG's remittance to develop market mechanisms that contribute to a low carbon economy
- Accessing a huge, largely untapped source of response (i.e., flexible load in commercial and industrial buildings)

ADR Solution Platform

ADR Solution Overview

See Figure 11.5.

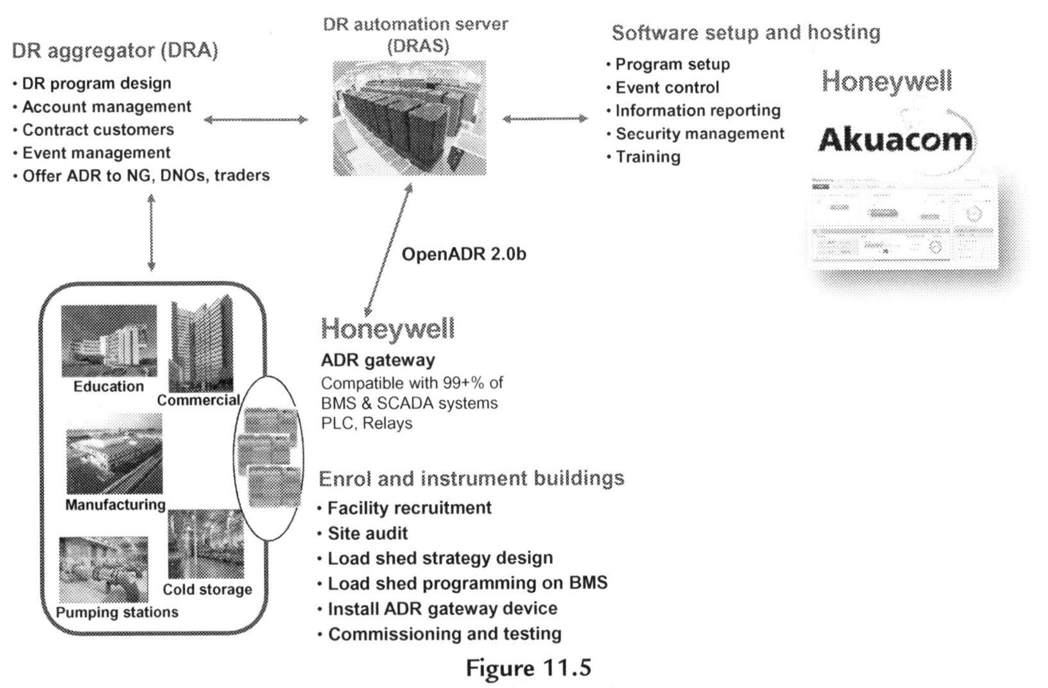

Figure 11.5
ADR turnkey solution.

ADR Solution Description

A full, turnkey ADR solution delivers aggregated DR to the NG's STOR and other "balancing service" programs and initiatives, sharing the resulting payments with participating customers and providing them a source of new revenue. This solution allows for fast-peak demand relief from almost constrained networks and for a backup for single-feeder primary networks that could suffer from peak demand constraints if there were a primary feeder fault be paid for the periods they require backup. The latter solution is made possible by ATSP's auditing, instrumenting, and connecting its buildings to those that sign up with the DRA that receive their DR service for which the ATSP receives a fee per building.

An ATSP such as Honeywell licenses the technology infrastructure to the DRA; setting up the DRA's DR program(s) on their demand response automation server (DRAS), adding each building to the DRAS, managing the internet portal for the DRA, managing the software (updates and security), reporting performance, and providing initial training and ongoing support. The ATSP also enables energy traders with access to DR to provide arbitrage benefits from day-ahead and hour-ahead markets.

The relationship between the DRA and ATSP typically works on the premise of establishing a new business to provide ADR (load shed and standby generation) as a service to DNOs, TSOs, and traders and establishing individual service contracts with them. Suitable ATSP customers are also offered the DR service on a national basis. The DRA initially contracts with its customers to meet the NG's minimum load reduction delivery volume of 3 MW or 3 MW of standby generation. The DRA's customer-building portfolio is then managed using the ADR system to deliver DR to the various DR program operators when they call for it. The NG tenders for these services in auction tender rounds under the STOR program.

Background to STOR and DR Initiatives

The NG STOR program responds to minute-by-minute changes in electric supply by calling on facilities to shed electrical load during these critical times. The NG in effect pays organizations to remove themselves or reduce demand from an overloaded grid. It is one of the best solutions to providing a reliable and stable supply of electricity back to the grid while protecting against rolling blackouts and brownouts. By participating in the STOR Program, organizations know that they are doing their part to safeguard the grid—and their business—against the unexpected.

At certain times of the day, the NG needs reserve power in the form of either generation or demand reduction to be able to deal with actual demand that is greater than forecast demand and with plant breakdowns. When it believes it is economic to do so, the NG procures part of this requirement in advance through STOR.

The need for DR varies across the year, different days of the week, and various times of the day. The NG splits a year into six seasons that include both "work" days (including Saturdays) and "nonwork" days (Sundays and most bank holidays) and specifies the periods in each day that require DR.

Technical Requirements for Providing STOR

According to the NG (see Figure 11.6), a STOR facility provider (as aggregated generators, demand reducers, or a single unit) must be able to:

- Offer a minimum of 3 MW generation or steady demand reduction (this can be from more than one site).
- Deliver full MW within 240 min or less after receiving instructions from the NG. However, the NG normally needs it much sooner and from experience, generator sets or demand-reduction sites incapable of delivering full capacity within 10 min would find it hard to be

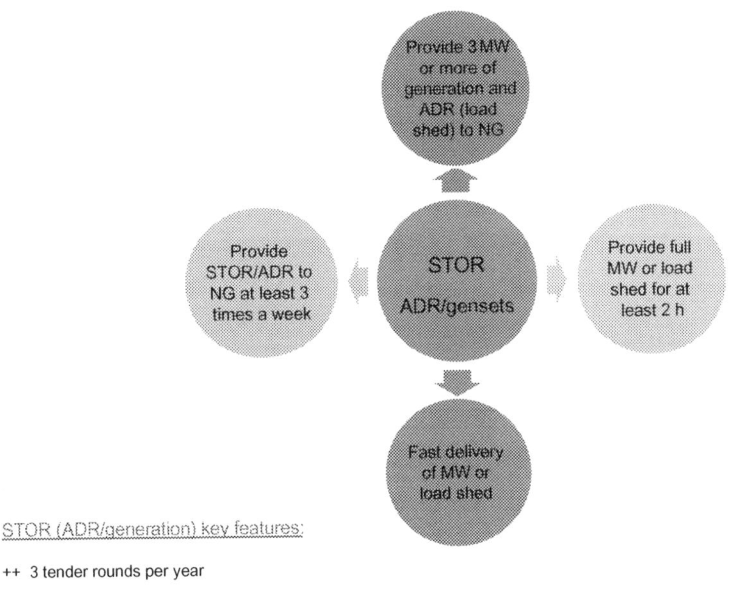

Figure 11.6
Overview of the National Grid's STOR program in the United Kingdom.

included into the NG's STOR program. The ability to respond fast is a key element in securing these contracts as well as:

- Providing full MW for at least 2 h when instructed
- Having a recovery period after provision of reserve of not more than 1200 min (20 h)
- Being able to provide STOR at least three times a week

Payments from NG

There are two forms of payment that the NG will make as part of the STOR service:

- *Availability payments (£/MW/h)*: DR service providers are paid to make their unit/site available for STOR service within an availability window.
- *Utilization payments (£/MWh)*: DR service providers are paid for the DR delivered as instructed by the NG.

Summary of ADR Project Goals and ADR Project Phases

The DRA measures ADR project success by being able to:

- Collaborate with the building owner to assess the facility's energy demand
- Gain building owner approval for automated load-shed strategies
- Integrate the automated load-shed strategies with existing building control systems
- Connect the building to access flexible load without disruption

The ADR project phases are as follows:

Audit facility: To perform a DR-focused energy audit walkthrough of joint DRA/ATSP activity with the facility staff or management representatives to fully understand the facility's equipment, controls, schedules, and critical business requirements.

Develop shed strategies: To identify and quantify electrical load-shed potential within the facility and then develop and present the demand-reduction opportunities to the facility staff for approval.

Implement customer approved shed strategies: To develop project milestones and timelines involved with connecting the building(s).

Meter data: To retrieve the building's historical energy consumption data necessary for a facility assessment by the ATSP's energy team.

Install OpenADR gateway: To install an OpenADR gateway in close proximity to the central building management system (BMS). The OpenADR gateway is mounted in accordance with local codes and standards and is approximately the size of a small laptop computer. The OpenADR gateway will be connected to the local area network and communicate via the

Internet to the DRA's remote (cloud-based) DRAS software to access event start/stop signals and information.

Integrate OpenADR gateway with BMS: To coordinate the integration of the OpenADR gateway with the BMS. The preferred deployment method is for shed strategies to be programmed into the building's existing BMS. The OpenADR gateway will poll (outbound) into the DRAS software every minute, looking for new OpenADR signals that have been set up by the DRA operators following receipt of a dispatch signal from the NG or other DR program operator. Upon identifying a new signal, the OpenADR gateway, in turn, signals the BMS system via digital output (DO) to implement a preprogrammed load-shed strategy. The DRA works closely with the facility's BMS integrator to test the deployment and ensure that shed strategies are functioning as agreed by the customer before offering DR to the NG for STOR and other DR initiatives.

Designing Reliable Load-Shed Strategies

Each building owner has unique business requirements to support his or her brand and operational requirements. Technical specialists from the ATSP and the DRA work closely with the owner to perform the audits necessary to identify and design appropriate load-shed strategies, design the overall solution, and provide testing and training. The shed strategies generally focus on delivering the maximum available kilowatt reduction across the building's various electricity-using plants without impacting the comfort of the building's occupants or the building's performance.

Each load curtailment strategy is coauthored and preapproved with the building owner to ensure that it is reliable before registering it onto DR programs.

For most DR programs, the load-shed target is approximately 10-20% of the building's maximum electricity demand through automation and control. This can often be more than 30% for well-instrumented buildings.

Figure 11.7 shows the results of a DR event held for a single commercial building in the United Kingdom toward the end of 2012.

- The (- - -) line represents the building's demand profile determined by an electrical consumption baseline algorithm that measures the demand over the previous 10 days on a rolling average, taking into account demand variation such as weekends and public holidays.
- The (——) line shows the actual demand profile on the event day taken directly from the building's main (fiscal) electricity meter.
- The (....) line is the load profile resulting from the difference between the forecast and actual profiles ((- - -) minus (——))

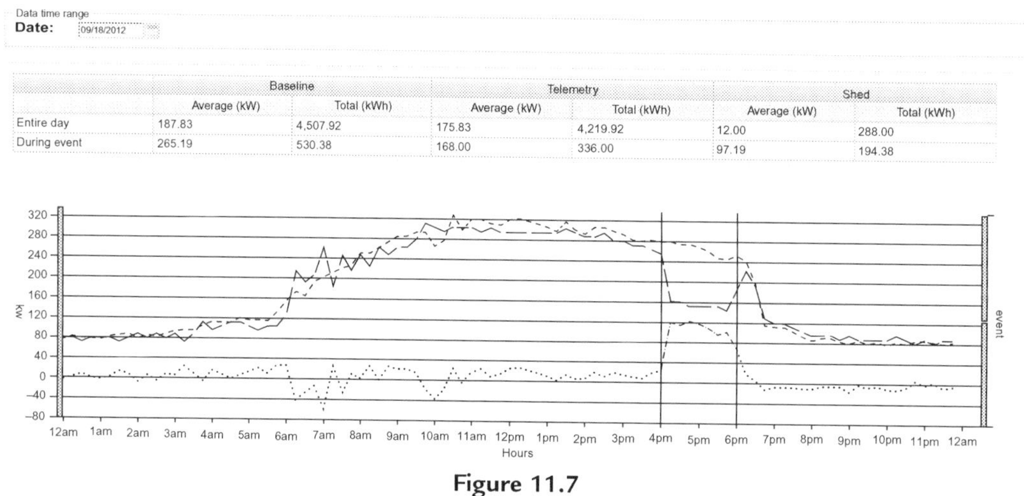

Data time range
Date: 09/18/2012

	Baseline		Telemetry		Shed	
	Average (kW)	Total (kWh)	Average (kW)	Total (kWh)	Average (kW)	Total (kWh)
Entire day	187.83	4,507.92	175.83	4,219.92	12.00	288.00
During event	265.19	530.38	168.00	336.00	97.19	194.38

Figure 11.7
Typical load profile and load-shedding.

- The (-·-·-) line represents the amount of load reduced during a DR event and the area below it is the virtual metered data provided to a DR program operator for settlement purposes.

For the building shown in Figure 11.7, 36% of the total daily kilowatt reduction was delivered with no impact on the building's operation, safety of the occupants, or the building's general internal environment. In fact, none of the building occupants noticed any change.

Among the load devices controlled, the key Heating, Ventilating, and Air Conditioning (HVAC) plant included within the load-shedding strategy to deliver these results were as follows:

- Main chillers: off
- Chilled water pumps: off
- Main air-handling units: inverter controlled to 20 Hz
- Conference Air Handling Unit (AHU): off
- Boilers: off
- Heating pumps: off

Turnkey Implementation

The building is connected to a DR programmed as a turnkey solution by the DRA including equipment installation, software implementation, training, and support. All load-shed strategies identified from the site audits are loaded onto the DRA's remote DRAS.

ATSP then provides a portal for the DRA for real-time monitoring and management of the load curtailment events. ATSP also provides the facility managers of the enrolled building a portal to monitor the building's electrical load consumption in real time and view their performance during curtailment events.

Solution Elements

Cyber Security

Cyber security is one of the important concerns in the evolving smart grid space. The system needs to detect, prevent, and communicate system security threats including cyber security threats and terrorism and then recover from them.

ATSP's ADR solution such as Honeywell's has gone through a thorough evaluation to ensure that the critical infrastructure cyber security standards will be followed for all the elements in all of their DR projects. Functional requirements of cyber security are fully appraised and considered at three levels, including physical, server, and customer security and continue to do so on an ongoing basis.

Architectural Approach

The solution offered by ATSPs provides a number of flexible and scalable architectural options for achieving the ADR programmer's objectives. Where feasible, ATSPs such as Honeywell's will demonstrate and test these different architectures to use as a basis for subsequent deployments and programmed development. The Honeywell ADR system operates as described in Figure 11.8 when signaling an event requiring load shedding to take place.

Automating Market Anticipation

The DRAS brings full automation to DR by implementing automated market participation for energy providers and consumers. The DRAS functionality supports programs that range from simple curtailment to complex time of use, dynamic pricing, and customer bidding. The DRAS accepts DR events and tariff information from DR program operators and turns these into standardized OpenADR signals that are identified by an OpenADR gateway located on each building in the ADR program. The gateway sees this as an event and triggers an appropriate action based on the set of rules defined for the event and preapproved by the building owner.

The DRAS software provides the two-way messaging infrastructure to broadcast price and reliability signals directly from the DR program operator to preprogrammed load control devices at a customer site, enabling highly predictable load reductions to be automated in direct collaboration with the customer (Figure 11.9).

Open Standard

Automation of DR programs is widely accepted as an effective industry solution for shifting and shedding electric loads. Unfortunately, many of the industry solutions available today are not standardized, creating problems for utilities, aggregators, and regulators. The OpenADR Alliance was formed to accelerate the development, adoption, and

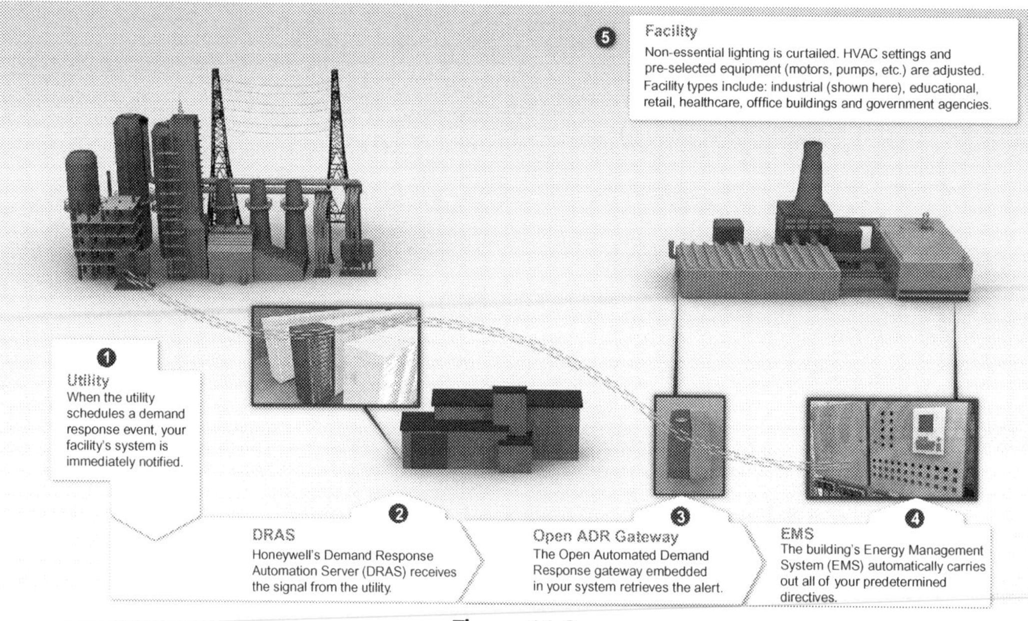

Figure 11.8
ADR operational architecture.

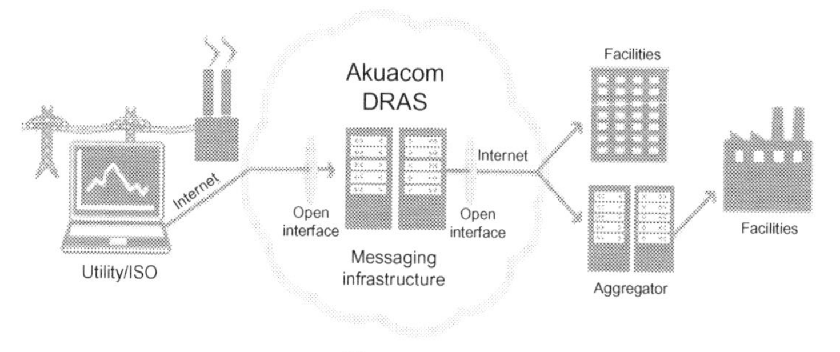

Figure 11.9
DRAS operating system.

compliance of OpenADR standards throughout the energy industry that allow any facility using OpenADR-compliant hardware to understand the DRA's DR messages (or those of any other aggregator using OpenADR) and respond in an automated fashion. OpenADR is a communications data model built on Internet standards including extensible markup language (XML). Much of the complexity of the DR program is managed by the DRAS and translated into simple signals for the facility's existing BMS.

OpenADR leverages existing BMSs and open protocol communication standards to help lower the cost of delivering ADR. For more information, see http://www.openadr.org/.

OpenADR Gateway

Overview: The OpenADR gateway is an embedded controller/server platform designed for remote monitoring and control applications. The unit combines integrated control, supervision, data logging, alarming, scheduling, and network management functions, integrated Input/Output (IO) with Internet connectivity and Web-serving capabilities in a small, compact platform. The OpenADR gateway supports a range of protocols including LonWorks®, BACnet®, Modbus, and oBIX as well as numerous other building automation protocols and Internet standards.

ADR System Architecture

See Figure 11.10.

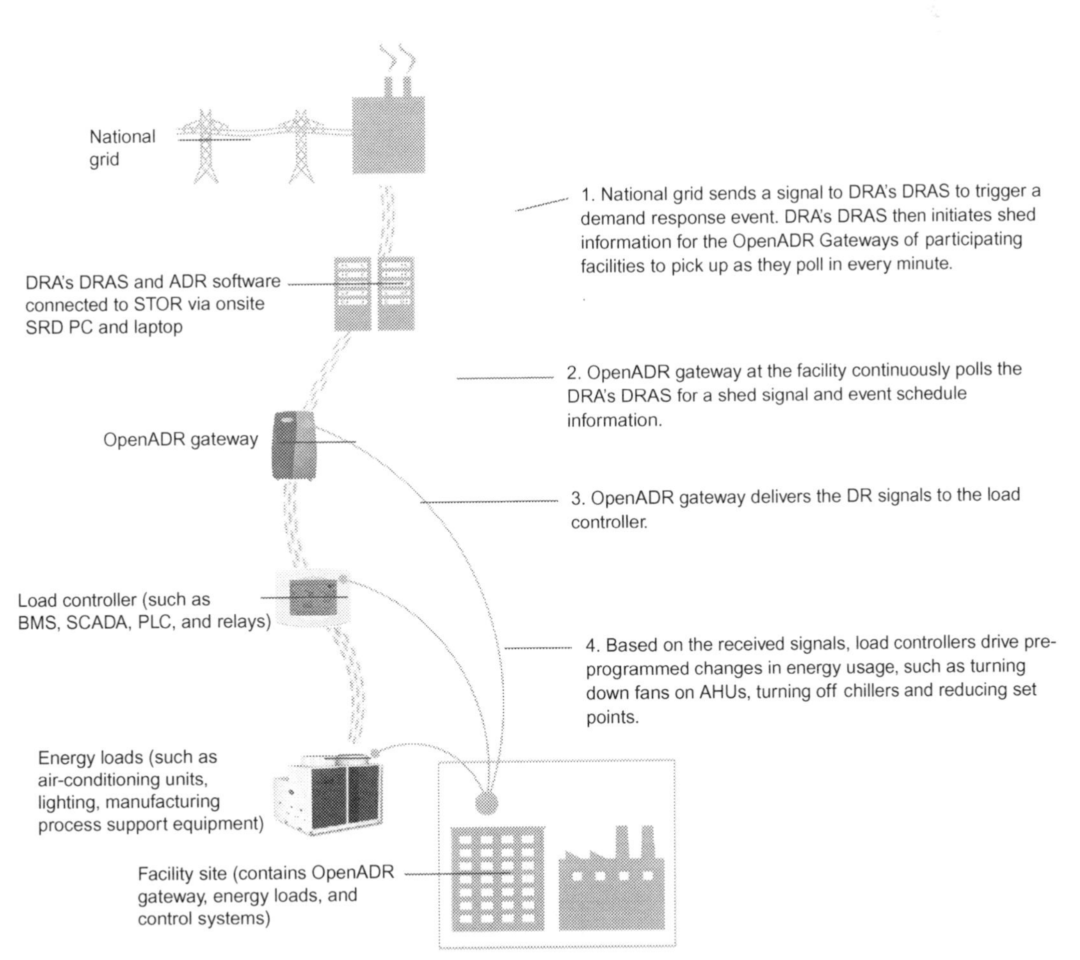

Figure 11.10
An overview of the ADR system architecture.

Flexible Method for Scheduling

ATSP's ADR platform can signal load-shedding events on an emergency or kilowatt target (or price) basis. The DRA can use the same platform to implement multiple DR programs (although they cannot be used at the same time). This ensures that as the program expands, different types of program offerings can be made without changing the hardware infrastructure.

Hardware

For a typical DR project, ATSP and the DRA utilize an OpenADR gateway device to communicate with each building including generator sets. The gateway is built using open standard signaling protocols enabling it to "talk" to more than 99% of BMS, Programmable Logic Controller and supervisory control and data acquisition systems without expensive integration costs.

The OpenADR application is typically configured for each individual site to account for differences in site equipment and to ensure compliance with both the building owner's business needs and DR program requirements.

The configuration process is streamlined and repeatable across different buildings with minimal effort.

The configuration for controlling standby generators and collecting meter data is shown in Figure 11.11.

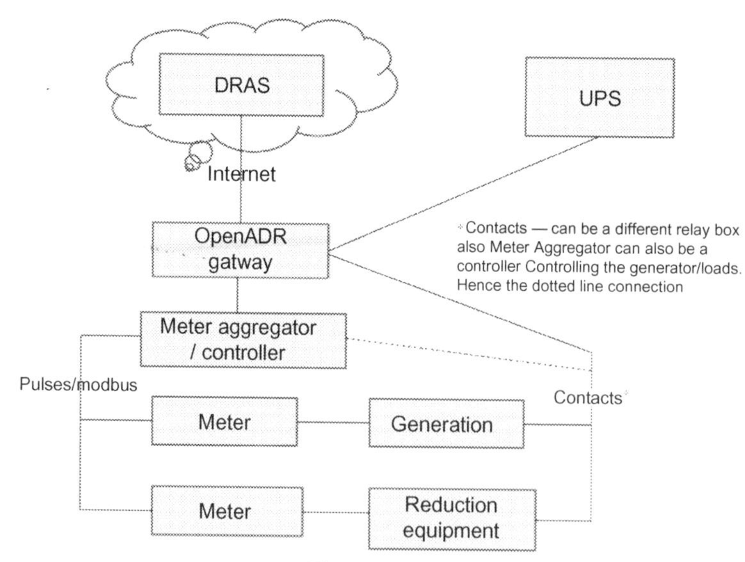

Figure 11.11
Smart functionality and DR decision support capabilities.

Demand Forecasting

The ADR software is continually being upgraded with additional functionality driven by customer needs. The ADR software also includes functionality to help accurately predict the demand by each commercial and industrial building as the DRA adds facilities to its overall DR program. Dynamic learning functionality equipped with adaptive forecasting engines learn the response to a DR event from previous DR events using data from electricity meters deployed in the DRA customers' premises (typically 15 min of sampling). This allows the DRA to make individual customer demand predictions under various applied DR strategies (e.g., time of day, day type, weather conditions) and therefore supports decision making such as achievable load reduction amount and timing.

In a day-ahead mode, the ADR platform predicts the entire next day demand (13-37 h ahead). In a day-of mode, it predicts afternoon and evening demand having the morning data (3-9 h ahead). Currently in development is a real-time mode for a few hours or even minutes ahead (Figure 11.12).

DR Event Optimization

Optimizing the forecasted baseline load that needs to be reduced to meet the target demand curve is made manually using approximate combinations of the DR availability of individual participants. Additional enhancements include a new optimization algorithm that accurately predicts the effect of a specific DR event (including lead and rebound effects) by using a combination of forecasts for individual loads in customers' facilities. This will allow the DRA to schedule optimum timing of DR events as well as optimize customers' facilities and loads to involve and to match the load reduction targets of specific events. The optimization is

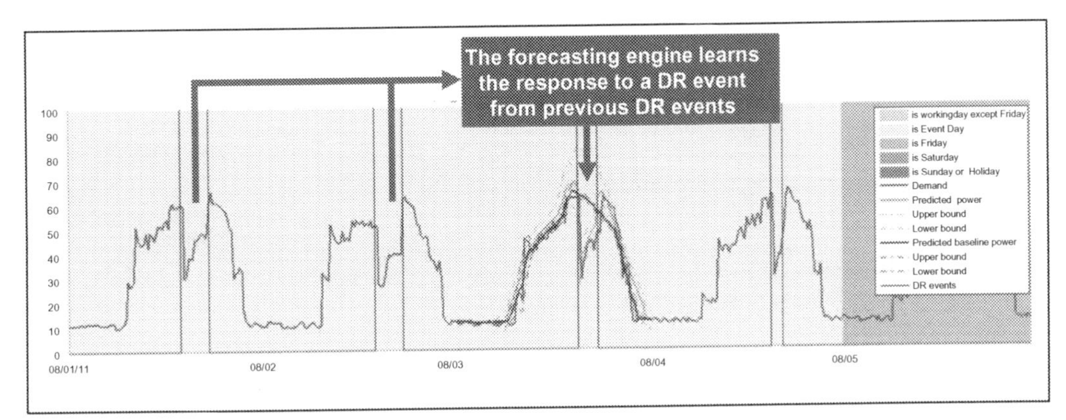

Figure 11.12
Dynamic learning functionality for accurate demand forecasting.

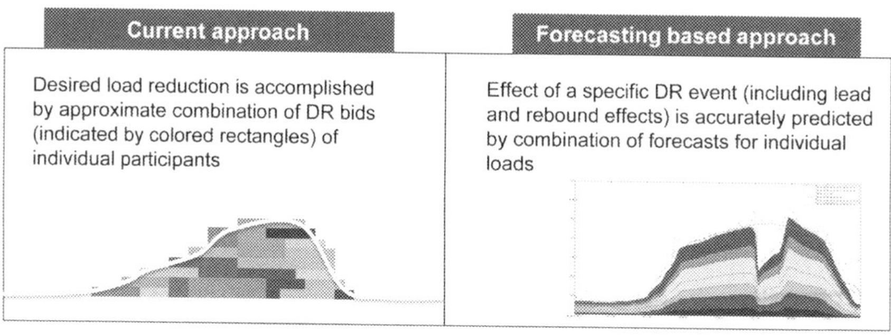

Figure 11.13
Demand response event optimization.

implemented as a decision engine that proactively runs the demand-forecasting module and analyzes the impact it will have and ultimately selects loads to meet requirements (Figure 11.13).

Intuitive Web Interface

The DRAS has browser-based interfaces for the DRA operator as well as each of the connected facilities. The DRA operator interface allows a user to view the status of the system, determine that facilities are online, and associate facilities with programs in general or for specific events.

The interface also supports setting up programs and triggering events. In addition, there are Web service interfaces into the DRAS for each of the functions.

The facility manager at each building has a view into the DRAS called their "My Site" interface. This allows the facility manager to configure the building's program settings and monitor each building's DRAS customers and account information from anywhere in the world. See Figure 11.14.

Demand Response Automation Server

The DRAS with OpenADR-compliant software offers the following functions, tools, and services:

- Enables virtual aggregation of DR over multiple sites
- Provides a software-as-a-service (SaaS) platform that meets industry standard authentication and security requirements
- Secures communications via certificate encryptions

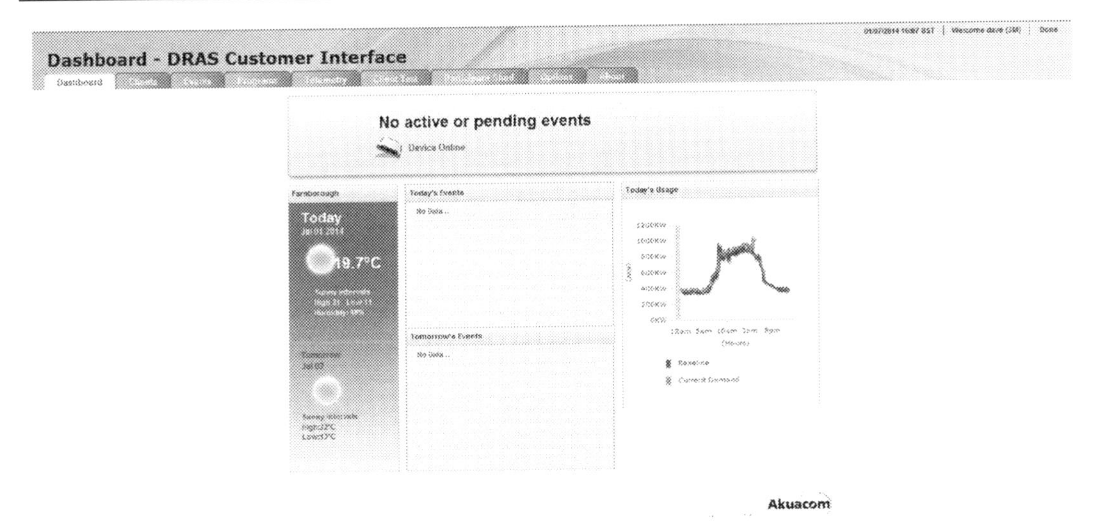

Figure 11.14
Intuitive Web interface.

- Provides a standards-based open interface that allows third-party applications to integrate with the system and offer services over the IP network
- Has dynamic target tracking capability
- Provides a Web-based management framework
- Allows administrators to control and manage devices over the IP network
- Provides tools to create automated policies to control energy consumption
- Provides user security through role-based access control
- Configures the interfaces and protocols on the appliance
- Creates and edits schedules, alarms, logs, and trends
- Backs up and restores configurations
- Provides a graphical programming tool that allows users with a working knowledge of controls and automation to translate their knowledge into efficient and powerful control logic
- Supports network protocols such as SNMP (simple network management protocol) and SOAP (simple object access protocol)/XML
- Empowers building owners with Web-based dashboards and tools to see their real-time energy use and have greater control over their electricity costs

DRAS Baselining

The NG and facilities need baselines for use data so that there is something to compare to when assessing demand shed. Baselines are a simple form of forecast, predicting a day's use with simple calculations based on the use measured on previous days. In addition to using historical data from previous days, some baselines are also adjusted based on measurements made in the morning of the same day. These adjustments are called *morning adjustments (MAs).*

To form the baseline, the DRA's DRAS receive an upload of the facility's electrical meter data once per minute and record it in the database. The typical baseline calculation takes meter data from previous nonevent workdays as defined by the DRA in collaboration with the NG and averages that data over time. So, a 10/10 baseline averages readings for a certain number (e.g., 10) of previous nonweekend, nonholiday, nonevent days and populate a baseline curve with these averages. The *n*/*m* baseline model used is the average hourly load shape of the "*n*" highest consumption days within "*m*" selected similar days.

The current day use is then compared to the baseline; the difference is the shed amount used during the DR event. The use data can also be compared to the calculated baseline before the event to determine the shed potential. Current use data that are well below baseline use are unlikely to meet the required provision levels for a specific NG STOR window.

Another variable in the baseline calculation is the "*MA*" factor. The DRA has the ability to apply a MA that shifts the baseline based on the use levels before the event begins. If the use is unusually high in the morning because the temperature is unusually high, for example, then the baseline can be adjusted to calculated shed against this higher baseline. The ultimate goal of the baseline is to project what energy would be used if no event is called.

Morning Adjustment

The MA is used to make a modification to the baseline to accommodate those situations in which the weather or other factors are significantly different from the previous few days. In these cases, the calculated n/m baseline might not accurately predict the normal (i.e., nonshedding) use behavior. For example, if the event day is going to be extremely hot but the past two weeks have been much cooler (the average high temperature was about 15° but the expected high on event day is 25°), then the baseline will not accurately predict the actual use for event day.

The MA calculation uses that day's actual use data reported in the morning time or hours before the event starts. The calculation constructs an average offset from the baseline to that day's actual data and then reconstructs the baseline using that offset. Thus, the new baseline for the day is a better predictor for the rest of the day (especially the usual peak time event periods in the afternoon) than the old baseline.

Virtual Meter Data

The DRA's DRAS provides the baseline data for individual sites or as an aggregated figure across multiple sites. It also provides the minute-by-minute telemetry of actual electricity use. The ADR software subtracts one graph from the other to determine the electrical load shed during a DR event in a highly accurate manner and provides this as virtual meter data for the

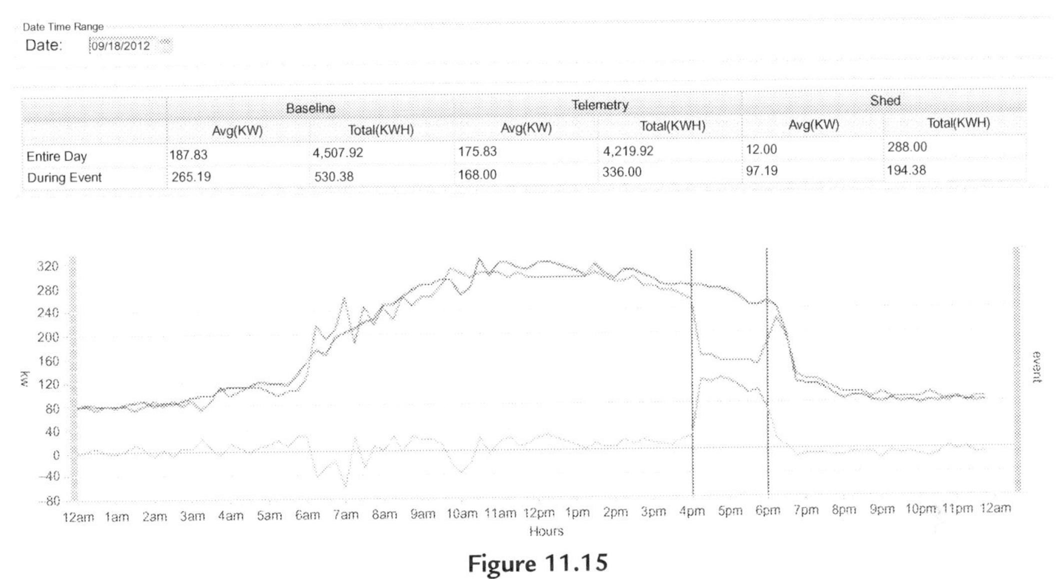

Date Time Range
Date: 09/18/2012

	Baseline		Telemetry		Shed	
	Avg(KW)	Total(KWH)	Avg(KW)	Total(KWH)	Avg(KW)	Total(KWH)
Entire Day	187.83	4,507.92	175.83	4,219.92	12.00	288.00
During Event	265.19	530.38	168.00	336.00	97.19	194.38

Figure 11.15
Illustration of a screen shot of metered data.

event. This meter data is shown numerically for a 2-h event between 4 pm and 6 pm in a table and in a graph as illustrated in the screen shot in Figure 11.15.

Case Study 1: Thames Valley Vision Project, United Kingdom (Auto DR Element)

Utility: DNO: Scottish and Southern Energy Power Distribution
Project Description: Thames Valley Vision (TVV) is a Scottish and Southern Energy Power Distribution (SSEPD) smart grid project run in partnership with other local, national and international businesses. The £30 million project is part of a £500 million program funded by the LCNF run by Ofgem, the UK energy regulator. TVV is based in Bracknell (in Berkshire, Southern England) because the local network is due for major system upgrade before 2020. SEPD is seeking to identify more cost-effective solutions than traditional network reinforcement that will require new and larger substations to be built and require disruption by digging up roads to lay more cables. Lessons learned at Bracknell and the Thames Valley, which are home to many large companies typical of much of Britain's network, can quickly be applied nationwide. To meet the LCNF objectives of facilitating low carbon solutions by trying new technologies and practices, TVV collects new data and develops sophisticated modeling to facilitate the operation and planning of networks to avoid costly system reinforcement.
Project Partners: SSEPD, Honeywell, GE, DNV Kema, EA Technology, University of Reading, Bracknell Forest Council (BFC)

***Key Utility Objectives*:**

- Prove (to Ofgem) that demand on a constrained primary substation can be reduced/moved when peak demand might exceed its limits
- Show that, as a result, required asset upgrade plans and investment can be deferred
- Enable GE's new "intelligent network" system to automatically reduce network load when it senses that a substation will be overloaded at peak times by instigating Honeywell's ADR system
- Produce a range of unique "learning outcomes" that can be shared with other DNOs to facilitate successful implementation of an ADR solution (e.g., commercial arrangements with building owners, engagement and enrolment, load-shed strategies, event frequency, and duration)
- Publicize the TVV project's progress and successes to raise awareness and strengthen SSEPD's reputation

***How ADR Element of the TVV Project Supports These Objectives*:**

- Honeywell has provided SSEPD with its ADR solution to shed load in buildings connected to SSEPD's constrained 50 megawatt (MW) primary substation in Bracknell at times of peak demand (typically 4-6 pm).
- Honeywell will catalog all aspects of the project building process and load-shed events and publish these as learning outcomes.
- Honeywell will work with GE to ensure that their two systems can seamlessly communicate with each other.

Date Agreement Signed with Honeywell: February 3, 2012
Term: 5 years

Honeywell Building Solutions Scope: Phase 1

- Number of buildings: up to 30
- To be connected by end June 2015
- DRAS
  - To be cloud based
  - To be US based
  - To be integrated with GE's new PowerOn Fusion distributed energy management system by 2015

***Key Success Factors for SSEPD*:**

- By demonstrating to Ofgem that 30 large electricity-using facilities connected to the Bracknell primary network can be convinced to participate in a DR program
- By demonstrating to Ofgem that sufficient load on can be shed to relieve peak demand on the Bracknell primary substation and that this can be maintained for a period of time into the future that makes it economically more attractive than traditional reinforcement

- By ultimately developing a DNO-specific cost-benefit analysis of ADR that shows it to be a valid alternative solution to traditional network reinforcement
- By receiving positive feedback from organizations participating in the DR program
- By demonstrating consistent and reliable performance of Honeywell's ADR technology
- By demonstrating the effectiveness of multiple customer engagement activities by a DNO
- By understanding the different aspects of ADR adoption by owners of various buildings including commercial, public, health care, retail, and educational

***SSEPD's Definition of Success for Honeywell's Role in this Project*:**

- Enroll, instrument, and commission 30 (relevant) buildings by end June 2015
- Prove the concept and technology
- Develop the ADR solution as the TVV project progresses (e.g., integrate with GE's network control system)
- Demonstrate that organizations can be encouraged to enroll in the ADR program initially *without* financial incentives (i.e., by being a "good neighbor")
- Produce the learning outcomes

Current Project Status:

***Tier 1 Pilot*: Three organizations, three buildings completed 2011. These are:**

- Honeywell House, Bracknell and Wokingham College, BFC Time Square
- Pilot testing program conducted 2011-2012
- Pilot testing report by Imperial College submitted to and approved by Ofgem

***Tier 2 Deployment*:**

- Six additional buildings connected at the end of January 2013
- Aggregated load shed to date: 592 kW
- Installed by the end of June 2015: 39 buildings with aggregated load shed of more than 1.2 MW

Why is ADR Important?

ADR is a smart grid technology that can be employed by a DNO to reduce electricity demand peaks on their supply networks.

The recent advances in computer-based technologies and the proliferation of electrical appliances in our homes and offices along with population growth have introduced new and larger loads on local electricity networks that were originally specified and built many years ago when demand was much simpler. These new loads are now causing networks to approach their maximum capabilities, posing expensive and disruptive infrastructure upgrade decisions for DNOs to make concerning the building of larger substations and the laying of new, higher capacity cables in the roads.

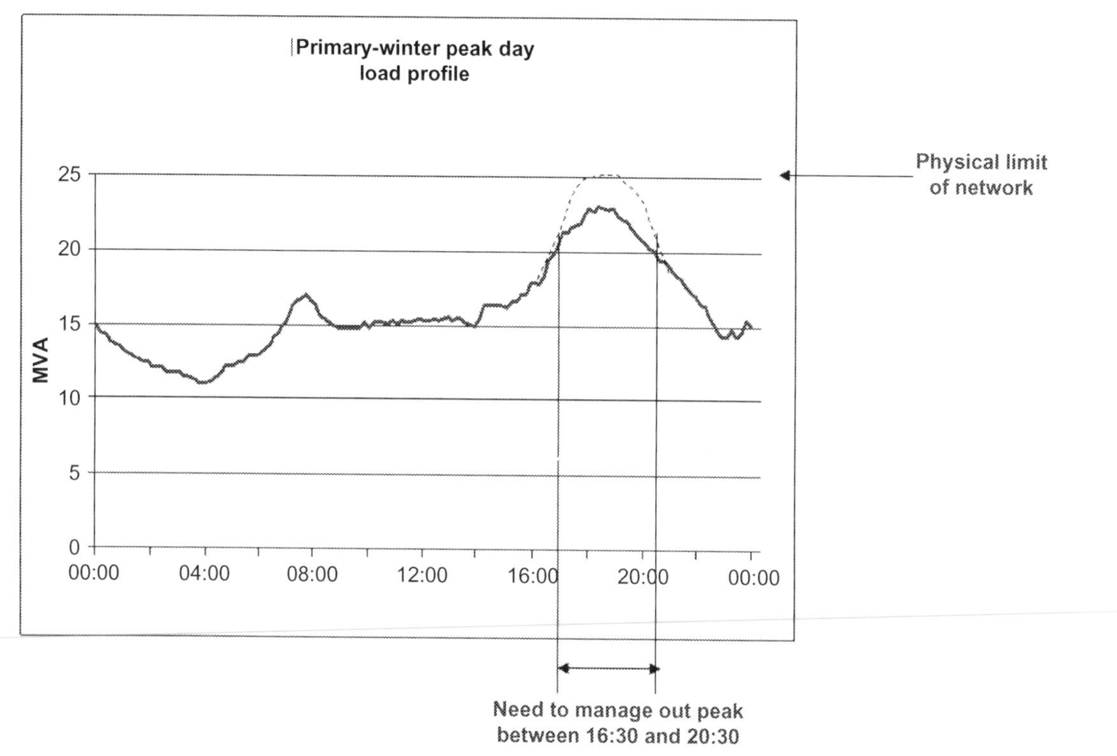

Figure 11.16

Load profile of a primary substation supplying a local community.

The diagram in Figure 11.16 shows the load profile of a primary substation that is supplying electricity to a local community with a typical mix of commercial, retail, educational, and residential buildings.

Obvious within the graph in Figure 11.16 is a significant peak in demand between 16:30 and 20:30. This peak is a result of the overlap between the demand required by businesses during operational hours and the demand required by homes in our communities. The risk in the future from the introduction of further new electricity-consuming technologies, such as electric vehicles and heat pumps, is that this demand peak will increase and exceed the physical limit of the network, resulting in supply interruptions.

It is important to note that the DNOs would always prevent demand shortages and ensure security of supply by investing today in larger capacity plant and infrastructure. This would be at the cost of the DNO and would have to be passed on to the consumer, potentially increasing the cost of energy in the long run.

ADR is a technology that postpones the need to install new capacity. ADR enables the DNO to reduce and move the electrical load on the network during peak periods, ensuring a secure supply without the need for expensive and disruptive new infrastructure upgrades.

How Does ADR Work?

When SSEPD identifies that a peak demand event is going to happen, it can employ ADR to reduce electricity consumption across buildings on their network in a concerted fashion. To enable this, an OpenADR gateway device installed in each building enrolled on the ADR program identifies an action signal via the Internet on SSEPD's cloud-based DRAS. This initiates an electricity load-shedding strategy programmed into each building's management system and preagreed electricity use areas are turned down and off. Only areas that have no or minimal impact on the building occupants are accessed. These areas include, for example, adjusting frequencies on air-handling units and turning off lights in unoccupied offices and pumps in ornamental lakes. At all times, the building owner retains complete control by having to confirm or deny participation prior to any event (Figure 11.17).

The result of a load-shedding event on a commercial building with a maximum demand of 290 kW is shown in Figure 11.18. You can clearly identify the load shed between 4 and 5 pm where the shedding strategy was able to reduce the demand by 45 kW. Information on the electricity use of each building along with the consumption profile during an event is accessed via a Web portal to provide a complete audit trail.

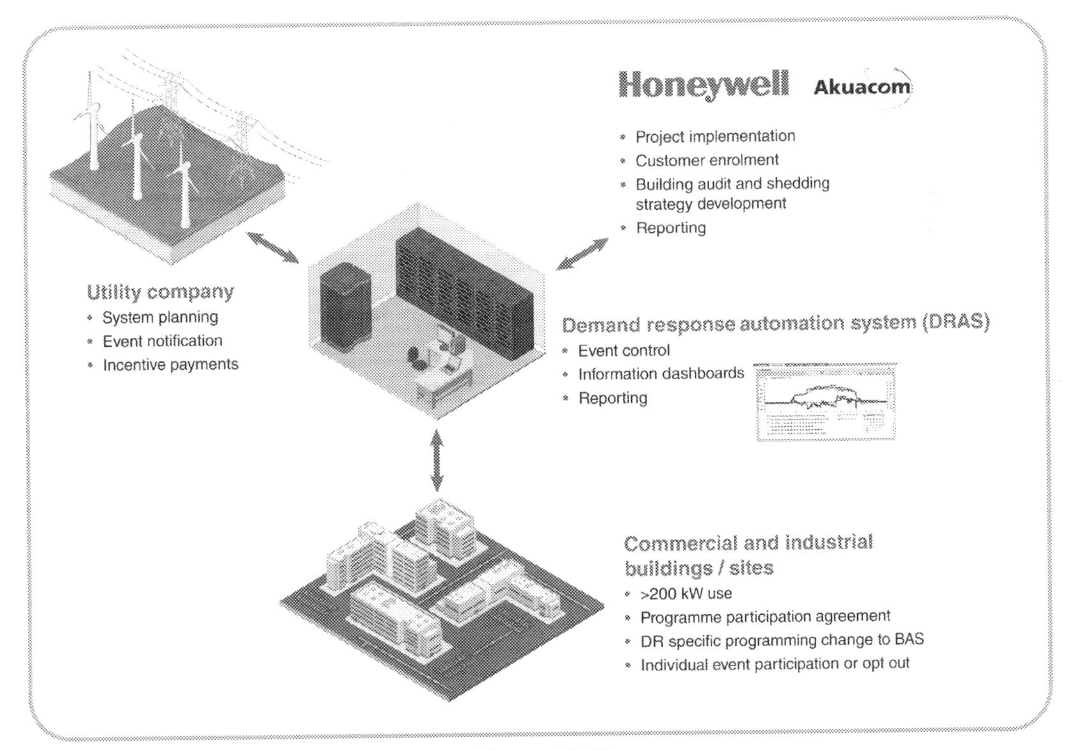

Figure 11.17
OpenADR gateway and signal from DRAS.

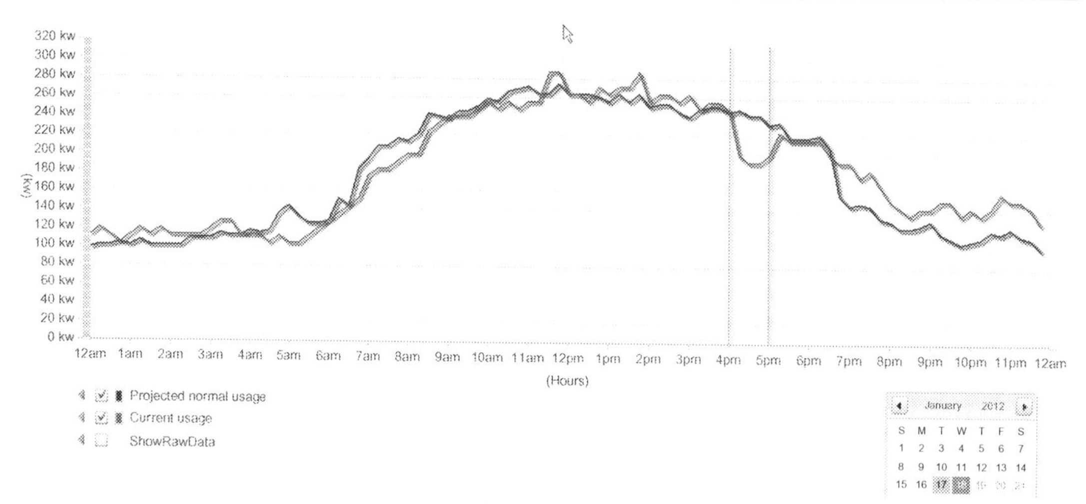

Figure 11.18
Typical load-shedding event on a commercial building.

What are the Benefits?

- Increased understanding of each building's electricity use (every 15 min on 24/7 basis)
- Enhanced ability to forecast electricity consumption to improve cost control
- Increased comfort conditions for building occupants
- Reduced green-based electricity: no fossil fuel-based additional generation
- Enhanced reputation by participating in an innovative, high-profile, publicly profiled Ofgem-funded project
- Publicized major government initiative proving DR as a vital technology for the future
- Recognized contribution to ensure energy security for the local area
- Recognized contribution to defer major network upgrade and the resulting local road work and disruption
- Initiated generation of a new revenue stream
- Prevented impact on business operations in the building
- Recognized installation and commissioning of proven technology in the building at *no cost*

SSEPD TVV Project SDRC Report

Honeywell Input and Perspective
October 2014
DNO—Commercial Customer ADR Engagement Strategy

Early DSR Adopters

Early adopters on the TVV ADR project included project partner organizations such as BFC and Honeywell, each of which wanted to lead the project by example: BFC to encourage organizations in its area to participate and Honeywell to demonstrate the technology.

Early adopters also included Bracknell and Wokingham (B and W) College, which was eager to use the public nature of the TVV project to publicize its participation and raise awareness of the college through press coverage, the Web, and case studies (written and video). There were other benefits the College identified that encouraged its early involvement. Despite being a new building, the College's BMS and energy system had never been commissioned adequately, meaning that heating and cooling varied widely across the college and energy consumption was much higher than it should have been. By accessing the expertise of the Honeywell ADR engineers, B and W College was able to optimize the system and greatly improve its energy efficiency and building operations. In addition, using ADR, the College ran a number of load reduction events on separate parts of the building to identify what electrical plant devices could be turned down on an ongoing basis with no impact on the building's performance or comfort for the occupants. This resulted in a significant reduction in overall electricity consumption and provided important financial savings.

The ADR value to connecting the BFC to the network was critical to gaining its participation. And it was important to convey different value propositions for different types of organization, such as

Those organizations that buy into:

- Contributing to energy security in Bracknell: "Be a good neighbor"
- Avoiding local disruption for their customers and staff
- Becoming large facilities with many employees
- Preventing disruption that could affect the business (e.g., retail, hotels)
- Tending to be large national/multinational organizations

Those organizations that buy into the CSR and green aspects of the project

Those organizations that see their participation in the project helping reinforce their reputation by:

- Leveraging the project's publicity
- Being part of a groundbreaking, first-in-Europe Ofgem-sponsored project
- Tends to be the larger, national/multinational organizations

Trial of Promotion Success

Early and ongoing engagement and communication with organizations and communities were vital to garner and maintain interest in their participation in the ADR program. A range of tactics was used to provide this:

- A customer engagement plan with target organizations located on the Bracknell network was developed by SSEPD. This included deciding whom in those organizations to invite and to which events and which value proposition to present.
- Early on in the TVV project, town hall breakfast meetings were held to introduce the project and how organizations could contribute to and participate in it. The involvement of the Thames Valley Chamber of Commerce, which was able to formulate the target marketing and facilitate the events, was vital.
- SSEPD and Honeywell held a number of small focus groups of individuals from various organizations that SSEPD wanted to join the ADR project. The purpose of these was primarily to test the effectiveness of the different value propositions developed in the customer engagement plan.
- High St Advisory Center in Bracknell acted as a central focal point for the project Your Energy Matters, which was used in addition to many other elements of the TVV project to inform organizations about the ADR project, its current participants, the benefits of participation, and how to participate. The center was also used for project update events and as a central meeting point for the ADR user group.
- The TVV Project website provided additional information and ongoing news about the ADR project. To inform the project sponsors and interested parties, DECC, Ofgem, and relevant trade associations were invited to Bracknell where they were introduced to the ADR project and its progress to date and were given a demonstration of ADR in operation, which was received positively by all attendees.
- Short case studies were developed for the various organizations participating in the ADR project, explaining the various aspects of their involvement. These were used to promote the project and act as references to encourage other organizations to join.
- Press releases about milestone achievements in the ADR project were distributed to the trade press and on the Internet to promote the ADR project and build awareness.
- SSEPD and Honeywell helped BFC to become involved in the TVV project (i.e., awards in 2012, where the Council won an award, further raising awareness of the project and ADR).
- Whereas building awareness of the ADR project was an important upfront activity, face-to-face meetings with the principal decision makers from organizations were key to gaining participation. The project could be presented in detail and its value proposition explained. No organization joined the project without having such an enrolment meeting.
- SSEPD will lead and facilitate an ADR user group of all organizations participating in the ADR project, The group meets regularly to discuss load-shed events and quarterly to discuss the organizations' experiences—good and bad—and to obtain tips and advice from others on how they are using ADR to obtain benefits in the operation of their buildings. Such events strengthen ownership in the project and the feeling of being in a unique group leading and proving the application of a major new technology that benefits the Bracknell area and UK plc as a whole.

Case Study 2: US Utility-Driven ADR Programs

The United States has many electrical DNOs using DR programs to manage demand growth on their networks and address peak demands. Three American DNOs have invested significantly and successfully in DR programs. The main reasons these programs are successful are:

- Demand peaks on their networks occur only for certain hours of certain days of the year. DR enables the DNOs to implement a far more flexible and cost-effective solution than traditional upgrade reinforcement organizations can.
- Regulator-driven financial incentives encourage DNOs to make these demand-response programs successful.
- All the DNOs provide attractive payment structures to encourage organizations to participate in their demand-response programs. No programs offer lower tariff rates.

Three successful American DNOs are discussed next.

CenterPoint Energy, Houston, Texas, United States

CenterPoint Energy is an electric transmission and distribution utility serving the Houston metropolitan area.

- It delivers power to more than 2.2 million homes and businesses in the 5000 square mile service territory.
- It owns, operates, and maintains the poles, wires, and substations that on behalf of 85 retail electric providers deliver electricity from power plants to customers over 3700 miles of transmission lines and nearly 50,000 miles of distribution lines.
- The Public Utility Regulatory Act requires each investor-owned electric utility to achieve savings goals through market-based programs: 20% reduction of demand in 2011 (288 MW), 25% in 2012 (271 MW), and 30% (258 MW) in 2013.
- It has implemented 15 energy reduction programs targeted to large and small commercial organizations, schools, residential areas, new construction, and hard-to-reach citizens. Its 2011 target reduction was 39 MW; actual achieved reductions far exceeded this.
- CenterPoint Energy's distribution network infrastructure suffers from severe peak demand congestion in the summer when air-conditioning loads are at their highest. As a result, CenterPoint Energy has launched its commercial load management program and industrial energy-efficiency program.
- Program participants are required to have a normal aggregate peak demand of 750 kW or more; each participating site must have at least 250 kW of normal peak demand and the capability to curtail at least 100 kW.

- Participants must be nonresidential customers taking service at the distribution level and/or be a nonprofit customer or government entity, including educational facilities. They must also be equipped with an interval data recorder meter.
- Participating customers agree to a one-year commitment, but there are no penalties if customers opt out of the program earlier. By its participation, the commercial/institutional customer commits to a maximum of five curtailments each year as follows: one scheduled curtailment of 1-3 h duration during each year of participation plus a maximum of four unscheduled curtailments of up to 4 h each during each year of participation. The availability period each year is limited to June through September, 1-7 pm weekdays, excluding federal holidays.
- CenterPoint Energy will pay a participating customer (or the project sponsor, if different) up to $35 per kW of verified curtailed load each year of participation. For example, a project sponsor who contracts for 1000 kW and consistently curtails 1000 kW or more when asked would earn up to $35,000 each year.
- The Public Utilities Commission, the energy regulator, allows energy-efficiency program costs to be recovered if they meet savings targets and offers an energy-efficiency cost recovery factor (EECRF) financial incentive. The EECRF is calculated as a performance bonus that is the lower of the percentage of net benefits or 20% of program costs.
- In 2011, CenterPoint's total spending on the implementation of energy-efficiency programs was $31,665,698. Because it exceeded the 2011 goal by 159%, it reached the maximum allowable performance bonus by rule of 20% of program costs, or $6,333,140.

Consolidated Edison, New York, United States

The DR programs of Consolidated Edison (Con Ed) are strictly focused on easing distribution congestion within its distribution system. For more than a decade, the distribution utility that serves about 3 million customers in New York City and Westchester, Connecticut, has offered contingency-based DR focused on the condition of distribution equipment across its more than 80 networks. More recently, it developed a peak DR program.

Currently, Con Ed must size its distribution system to meet its 13,000 MW peak even though those conditions may exist for only 10 h of an entire year. The alternative is to build bigger, costly network assets for these few hours.

Con Ed's original DR program had participants, but the economy caused it to struggle to attract organizations to enroll. When Con Ed added a peak shaving program that also emphasized improving the use of its distribution network, it needed to address the value versus incentives question. The utility commissioned Freeman, Sullivan & Co. to do a study on its existing DR programs and to identify how best to move forward. The study found the total resource cost (TRC) test, which is used more widely to assess the full incremental cost of the DR resource, was the better measure for Con Ed and that the utility could significantly increase its payments based on the benefits to the system.

Con Ed recently recalculated the value of DR and adopted a single metric for both of its programs. The result is that payments have doubled what they were a year ago, and there are new incentives designed to entice commercial customers of all sizes to take part in multiyear agreements.

- Customers in the Distribution Load Relief Program can receive up to $15 per month for each kilowatt they pledge to reduce (an availability payment) depending on where their building is located. They receive these payments even if Con Edison never asks them to reduce their use.
- These customers can also receive $1 for each kilowatt-hour they save during an event. Customers in this program receive notice to start reducing electrical use 2 h or less in advance.
- Con Edison has also increased the incentives in its Commercial System Relief Program. Customers in that program receive 21-h notice to start reducing electrical use and receive a second notice 2 h in advance.
- These customers can get up to $15 per month for each kilowatt they pledge to reduce, depending on how often events are called. They receive these payments even if Con Edison never asks them to reduce use. They receive an additional $1 for each kilowatt-hour they save when they reduce use at Con Edison's request.
- Both programs have other flexible options. A new feature starting this year is that customers in both programs can receive additional incentive payments of $5 per kW per month for participating for three straight years.

Con Ed now has about 200 MW in its DR programs. With a sizable stable of better-compensated, longer-term DR partners to call on, Con Ed can curb the amount of investment needed to maintain its distribution grid fit for purpose no matter what the demand.

CPS Energy, San Antonio, Texas, United States

CPS Energy is the largest US municipally owned energy utility providing both natural gas and electric service. Acquired by the city of San Antonio in 1942, it serves more than 741,000 electric customers in a 1514 square mile area.

- CPS Energy's Save for Tomorrow Energy Plan (STEP) seeks to reduce growth in electrical demand by 771 MW by 2020.
- CPS Energy's ADR program (provided by Honeywell) conserves electricity during periods of peak use, such as on hot summer days. By lowering demand during peak times, program participants help CPS manage its network resources better, which improves the way

electricity is delivered and used, making the grid cleaner, safer, and more reliable and efficient for all customers.

- Following a successful trial, Honeywell and CPS Energy intend to enroll 60 additional sites, bringing the potential reduction to nearly 6 MW, enough electricity to meet the typical load of more than 2200 homes.
- CPS Energy's customer receives two incentive payments: $20 per kW for the nonsummer months (October-May) and $50 per kW for the summer months (June-September).
- Customers receive a free DR audit, a free ADR gateway and access to an online portal to view building load data in real time in order to better troubleshoot any equipment issues.
- Maximum hour numbers of participation are 20 in nonsummer and 50 in the summer.

CPS Energy paid $4.4 million in DR rebates to 210 customers in 2013: $53,373 for average DR payments and $956,023 for the largest payment.

CHAPTER 12

Critical Issues for Data Center Energy Efficiency

Colin Pattinson, Ah-Lian Kor*, Roland Cross
Leeds Beckett University, Leeds, UK
**Corresponding author. a.kor@leedsbeckett.ac.uk*

Introduction

Brown et al. (2007) reported that for the year 2000-2006, the estimated energy consumption for US data centers was about 61 billion kWh (approximately 1.5% total US electricity consumption) or about $4.5 billion in electricity costs. From 2005 to 2010, electricity used by US data centers increased by approximately 36%; globally, it increased by about 56% (Koomey, 2011). In 2010, the energy consumption by data centers accounted for 1.7-2.2% total electricity use; globally, it was 1.1-1.5% (Koomey). Rising data center-related energy consumption will remain the overarching trend because of an Internet use boom (Troianovski, 2011), which sees a widespread demand for Internet services (e.g., video on-demand, social networks, telephony, music downloads, and VOIP; Koomey, 2008) and extended corporate data center services resulting from developments in mobile-, cloud-, and business-integrated analytics (IDC, 2014a).

Gantz and Reinsel (2012) define a "digital universe" as a world that encompasses digital data (e.g., images, videos, movies, communication data, monitoring data for the health/transport/building sectors, security data, banking data) that can be created, replicated, and consumed. The digital universe was expected to grow from 2013 to 2020 by a factor of 10 (i.e., from 4.4 trillion gigabytes to 44 trillion gigabytes), and data from embedded systems (the signals form a major component of the internet of things) from 2% of the digital universe in 2013 to 10% in 2020 (IDC, 2014b). Gartner (2014) estimates that IoT will include 26 billion units installed by 2020 and that its deployment will incur the generation, processing, and analysis of large quantities of data that will exert an increased workload (and energy consumption) on data centers. Consequently, it is imperative that data center energy efficiency is prioritized in order to help reduce its impact on the environment and climate change. Microsoft Corporation (2012) views the data center as an ecosystem comprising IT systems (e.g., servers, storage, network), mechanical systems (e.g., chillers, cooling systems, pumps), and electrical systems (internal transformers and UPS) that facilitate environmental controls, monitoring, and networking. The only effective means of measuring the efficiency of data center operations is to take a holistic approach (IBM Corporation, 2012a) that considers the measurements (and trade-offs) of all components within the ecosystem.

This chapter is based on a project funded under the JISC Greening ICT initiative (Pattinson and Cross, 2013). It was an experimental analysis providing real data relating to how the energy consumption in different parts of data centers could be accurately measured. This involved comparing the situation in which all the different elements of data center energy consumption were measured on a single meter together with other parts of the university built estates to a situation in which a high level of detailed measurement was obtained. This allowed the calculation of the power usage effectiveness (PUE) of the data center and the comparative measurement of different parts of the data center.

Aim and Objectives

The aim of the JISC Measuring Data Center Efficiency project was to monitor the energy consumption in different parts of its data center, varying the metering methodology during the project, so that metering was done at high granularity (measuring individual devices and locations); in other cases, the metering was done at lower granularity (giving the combined energy use of a number of devices and locations from a single meter). The primary focus was the temperature parameter and an investigation of its effect on data center energy consumption improvement using PUE.

The broad objectives included in the project were establishing baseline energy measurement methods, evaluating the effects of different granularities of metering on the baseline, and investigating the effect of consumption reduction methods. The subgoals and corresponding specific objectives of the project were as follows:

1. Identify appropriate candidate technologies for test purposes, undertake desk research to categorize possible power metering technologies and the criteria against which an informed choice can be made, and identify the appropriate power consumption reduction measures that can best be implement in the data center.
2. Identify appropriate candidate locations for installation including the locations and applications that are used as test beds, focusing particularly on the main university data center including server room equipment, UPS, and environmental control systems.
3. Identify user needs by consulting data center support staff before deployment to allow their work practices to be taken into account in the decision-making process and to ensure that there was no significant impact on the effectiveness of the IT systems and services offered from the data center.
4. Determine suitable metrics and measurement methods by reviewing the range of possible measurement metrics and metering approaches, selecting those that best allow

the evaluation of the relative performance of selected systems against the project requirements.

5. Install required power metering and monitoring devices by ensuring that the measurement and monitoring facilities required were in place, which involved the assistance of colleagues from estates and members of the technical staff. The work included integrating new metering devices into existing software systems used for collecting the data and producing meaningful management reports.
6. Record data center power consumption, calculate PUE; only campus-/building-level power consumption data are currently available and will be compared to the more granular data available after metering installation. These data were gathered over a period of at least two months of typical of business activity.
7. Implement power reduction measures identified earlier in the data center that could then be implemented in a staged way.
8. Record data over a period of the academic cycle of activities (to gather data at high/medium/low use levels); part of the experiment referred to earlier was to gather data at different use levels in order to create a full overview of the performance of the systems to be compared.
9. Record feedback from support staff, customers, and end users through helpdesk call logging as well as questionnaires and interviews during and after the project.
10. Report on outcomes by delivering a detailed description of the results and analyzes of outcomes, including guidelines for future deployment of thin and thick client systems.

Literature Survey

Green ICT

Terms used synonymously with *green IT* are *green computing, green ICT, sustainable computing, sustainable IT, environmental sustainable IT, environmentally friendly IT,* and *computing*. The strategy-focused definition of green IT given by the IEEE Computer Society[1] is rather narrow and refers to strategies for reducing energy consumption and the environmental impact of products, equipment, services, and systems. However, a broader definition of green IT addresses strategies related to environmental and social issues, government policies, and consideration of responsible ecological and innovative ways to exploit computing resources. The green IT definition provided by Murugesan and Gangadharan (2012) is environment sustainability-focused. It encompasses environmentally friendly computer, information systems, applications, and practices that aim to enhance energy efficiency, reduce GHG emissions, and reduce e-waste. The two categories of green ICT are typically:

[1] http://www.computer.org/portal/web/buildyourcareer/JT28.

1. *Greening of IT:* This encompasses energy-efficient and environmental sustainable designs, operations, use, and disposal ICT equipment, infrastructure, and systems to mitigate the environmental impact of ICT itself.
2. *Greening by IT:* This harnesses IT (via ICT-enabled solutions) to mitigate the environmental impact of other sectors: power, transportation, manufacturing, agricultural, building, service, and consumer (listed in SMARTer2020[2]). It involves the application of ICT to create energy-efficient and environmental sustainable operations, processes, and practices.

Data Centers

Data Center Efficiency

The core function of a data center is to provide computational capacity, storage, applications, and telecommunications to support an array of enterprise businesses and individual consumers (Cisco, 2007; Microsoft Corporation, 2012). This is made possible with the use of a range of servers (Web servers, application servers, and database servers) and network infrastructures. In order to operate efficiently and reliably, the servers require the support and interaction of multiple dependable systems: electrical (e.g., appropriate voltage level, uninterruptible power supply, backup power generators in the event of electrical outages), mechanical (e.g., cooling systems to provide appropriate temperature and humidity control), and control (Gao, n.d.; Microsoft Corporation, 2012). Figure 12.1 shows an overview of the key elements in a green data center (adapted from IBM Corporation, 2012b).

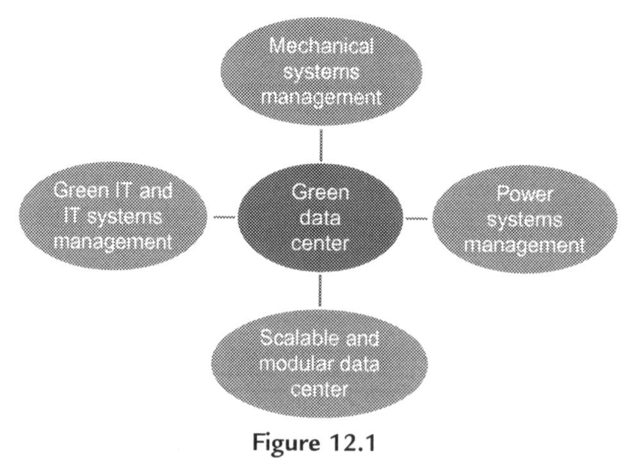

Figure 12.1
An overview of the key elements in a green data center.

[2] http://gesi.org/SMARTer2020.

Energy efficiency techniques based on green technologies (including greening of ICT and greening by ICT as discussed in section "Green ICT") form the cornerstones for a green data center. They are coded into three groups: IT-related techniques, facility-related techniques, and integration techniques (Figure 12.2).

Virtualization of the IT infrastructure has yielded unprecedented productivity accompanied by increased operational and energy efficiency (IBM Corporation, 2008a). Consolidation, which involves the physical consolidation of data centers, servers, and storage and network devices, helps lower energy costs (IBM Corporation, 2013). Guidelines on virtualization and consolidation can be found in NCB (n.d.); details on energy efficiency in mechanical and power systems are shown in Figure 12.3 and in the following publications: US Department of Energy (2011), IBM Corporation (2012b), and Energy Star (n.d.). It is possible for scalable and modular data centers to be deployed in a wide range of configurations to optimize the use of available physical space, lower operating costs, provide an efficient means to consolidate IT infrastructure, and increase energy efficiencies (IBM Corporation, 2008b). Data center infrastructure management (DCIM) integrates the IT and facility systems. Some DCIM solutions (CDW, 2013) are:

Schneider Electric's StruxureWare Data Center Operation: provides interface virtualization systems
IBM Maximo Data Center Infrastructure Management: provides real-time monitoring of the IT and facilities within the data center
CA ecoMeter: monitors UPS systems, air-conditioning units, power distribution units, IT equipment, and generators in real time
Raritan dcTrack: offers capability as an asset manager, visualization tool, and the means for capacity and change management
Emerson Network power's Trelli: monitors everything in the data center to provide an insight into the systems' interdependencies

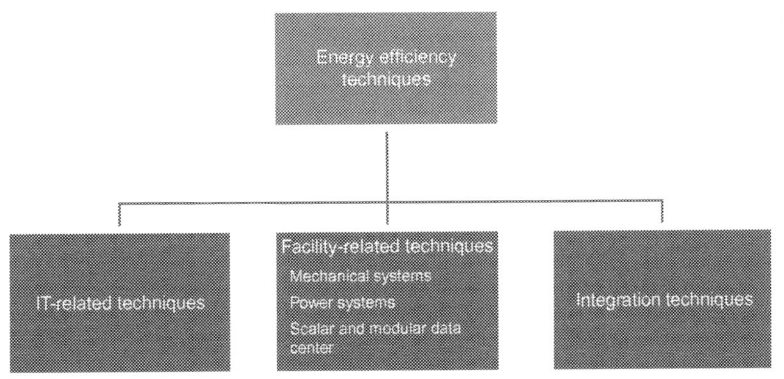

Figure 12.2
Energy efficiency techniques within a data center (adapted from IBM Corporation, 2012b).

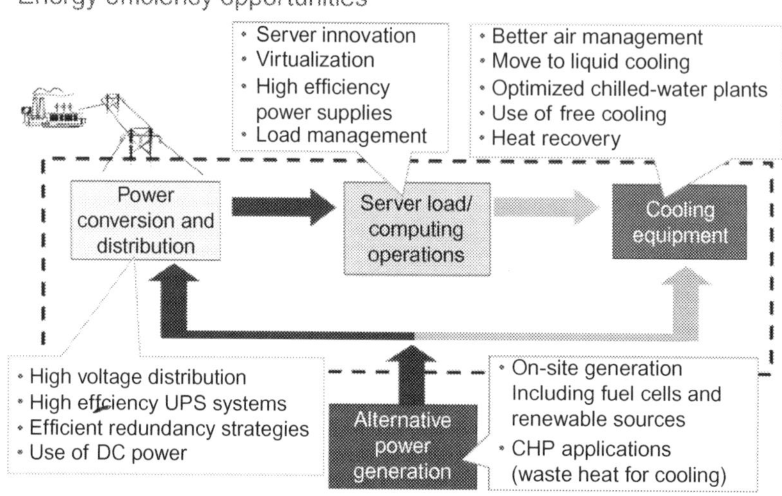

Figure 12.3

Energy efficiency opportunities for a data center (http://www1.eere.energy.gov/manufacturing/tech_assistance/pdfs/doe_data_centers_presentation.pdf).

The European Union Code of Conduct (European Commission, 2008) is a voluntary initiative, developed in collaboration with the industry to promote energy efficiency performance standards for data centers. Its aim is to help improve understanding of energy demand within the data center, raise awareness of the need to reduce energy consumption, and recommend energy efficiency best practice for the data center[3] and targets. The RAL-UZ 161[4] certification is the new Blue Angel[5] eco-label for energy-conscious data centers that helps data centers reduce their carbon and ecological footprints through the provision of clear guidance accompanied by an appropriate set of ecological criteria. It certifies the energy efficiency of data centers and evaluates hardware resources handling. The US Environmental Protection Agency has launched its Energy Star for green data centers.[6] It offers certification for data center products[7] (e.g., servers, storage, UPS, networks) and buildings.[8] In the United Kingdom, the (BCS) has specifically developed the Certified Energy Efficient Data Center Award (CEEDA)[9] to enable data center operators to enhance energy efficiency and reduce their environmental impact. The

[3] http://iet.jrc.ec.europa.eu/energyefficiency/sites/energyefficiency/files/files/documents/ICT_CoC/2014_best_practice_guidelines_v5_1_1r.pdf.

[4] http://www.blauer-engel.de/de/fuer-unternehmenwie-bekommen-sie-den-blauen-engel/uebersicht-vergabegrundlagen.

[5] http://www.blauer-engel.de/de/unser-zeichen-fuer-die-umwelt.

[6] http://www.greenbiz.com/blog/2012/02/16/how-energy-star-working-green-data-center-future.

[7] https://cio.gov/wp-content/uploads/downloads/2012/09/EPA-Energy-Stat-for-Data-Cneter-Products-UPS.pdf.

[8] https://portfoliomanager.energystar.gov/pm/glossary#DataCenter.

[9] http://ceeda.bcs.org/category/15880.

data center assessment includes the entire data center operations, processes, technologies, personnel, and buildings. The three awards given are gold, silver, and bronze. Finally, the ASHRAE (2011) has formulated a revised set of thermal guidelines for the data center (this is the responsibility of ASHRAE Technical Committee 9.9).

Data Center Efficiency Measurements and Metrics

According to IBM Corporation (2008a), data center efficiency measurement will lead to management, control, and enhancement of energy efficiency. This involves tracking as well as analyzing energy usage history and service levels which could be exploited for the monitoring and control of the previously mentioned systems within a data center (i.e., IT, power, and mechanical).

The PUE metric (The Green Grid, 2007, 2008a, 2012) is the *de facto* global standard for measuring infrastructure energy efficiency in ta data center. It enables data center operators to easily estimate the data center energy efficiency, to benchmark it against other data centers and to explore ways to enhance energy efficiency. PUE is a relative measure of the efficiency of an energy consuming resource: It expresses the *proportion* of total power supplied to a unit that is used in the actual work of the unit. The total power includes the energy consumed in doing useful work—data storage and processing *and* the energy used in "other" activities (lighting, air conditioning, etc.); actual work involves the data processing and storage activities of the data center. A data center in which there is no "other" activity (i.e., *all* supplied energy is used for data processing) would have a PUE of **1.0**. This is probably impossible: Leakage and other losses would preclude it, even if it were possible to remove all other energy consumption. Theoretically, a PUE of less than 1.0 is achievable: It would require the data center to generate energy that could be used elsewhere using the waste heat created by the processor. However, in practice, PUE values approaching 1.2 are seen as "very good." Major suppliers' data centers are typically at or around this 1.2 value, indicating that the most likely way of delivering these PUE values is by new build, purpose-designed units of a very large scale. A 2011 survey[10] by the Uptime Institute indicated that the current average PUE is 1.8 (based on a survey of 500 data centers, assumed to be US based), whereas in the United Kingdom, a BREEAM report from 2010[11] suggested the "typical" value is 2.2.

Undeniably, PUE is quick and easy to use. However, it fails to consider the many factors (which could offset each other) that affect power usage in a data center environment and the distribution of power or trend of electricity consumption with respect to time.[12] PUE merely

[10] https://www.datacenterknowledge.com/archives/2011/05/10/uptime-institute-the-average-pue-is-1-8/.

[11] https://www.bsria.co.uk/news/breeam-data-centres/.

[12] http://www.itwatchdogs.com/environmental-monitoring-news/data-center/getting-at-the-heart-of-data-center-monitoring-and-energy-usage-609950.

provides a one-time, instantaneous measurement of energy consumption as discussed by Stanford University researchers Yuventi and Mehdizadeh (2013), who suggested that a more accurate practice is the calculation of the average PUE over a period of time (at least a year). Another question raised about the accuracy of PUE is whether the measurement is taken during a maximum or minimum load for one or more of the previously listed systems within the data center.[13] The JISC Measuring Data Center Efficiency Project has addressed most of these concerns. First and foremost, it has two years of real-time monitoring of energy consumption and has collated half-hourly energy consumption for its relevant data center facility (power and mechanical), and IT systems are accompanied by corresponding calculated PUE and their daily or monthly averages (details are in section "Methodology"). Another critique of PUE is that its calculation has not factored in the use of renewable energy[14] (this is not included in the main electricity measurement). However, the latest formula for PUE (The Green Grid, 2014) has addressed this issue. Additionally, a new metric, green power usage effectiveness[15] has also considered this concern as proposed by GreenCloud. In order to refine the effectiveness of PUE, the Green Grid has developed more productivity metrics[16] (for all major power-consuming subsystems in a data center) and a summary of tabulated data center efficiency metrics is in Table 12.1.

Methodology

The details of the methodology are found in Appendices A and B of Pattinson and Cross (2013). At the onset of the JISC Data Center Efficiency project, new metering installation was carried out in the Woodhouse building LV switch panel (see Figure 12.4) and server room (see Figure 12.5). The system schematic diagram of the metering topology is shown in Figure 12.6. All metered data for the Woodhouse building was collated via the EGX300 gateway, which was configured to work with the Web query function of Microsoft Office 2010. A spreadsheet was created using the Web query command to auto update with real-time information from the meter's logging interval. When the spreadsheet was opened, it automatically connected to the EGX server and updated with logged data from the meters. A LTHW pipe joint burst in the Portland building resulted in flooding the main LV switch room (shown in Figure 12.7). Consequently, this caused a data outage for the period October 31, 2012 to December 17, 2012; this was the period without any collected data.

[13] http://www.datacenterjournal.com/it/is-pue-still-useful/.

[14] http://www.prismpower.co.uk/pp-downloads-08/whitepapers/pp-PuE%20Limitations.pdf.

[15] https://www.greenqloud.com/greenpowerusageeffectiveness-gpue/.

[16] http://www.greendatacenternews.org/articles/share/705283/.

Table 12.1 Green Data Center Efficiency Metrics

Metric Acronym	Metric Name	Formula and Description	Unit	Reference
PUE	Power usage effectiveness	The Green Grid has provided a series of iterations for the enhancement of the PUE formula. The formula below is extracted from The Green Grid (2014) $\text{PUE} = \frac{\text{Total facility source energy}}{\text{IT equipment source energy}}$ where Total facility source energy (kWh) is defined as the total energy measured at the utility meter with consideration of weighting factors for the type of energy source (details are found in The Green Grid, 2014) IT equipment source energy (kWh) is defined as the total energy sourced to all IT equipment within the data center	No unit, form: ratio	The Green Grid (2007, 2008a,b, 2009, 2010a, 2011a, 2012, 2014)
DCiE	Data center infrastructure efficiency	It is the reciprocal of PUE $\text{DCiE} = \frac{1}{\text{PUE}}$ $\text{DCiE} = \left(\frac{\text{IT equipment energy}}{\text{Total facility energy}}\right) \times 100\%$	No unit, form: ratio or %	The Green Grid (2007, 2008a,b, 2009, 2010a, 2011a, 2012)
PUE_x		There are four categories of PUE: $\text{PUE}_0 = \text{PUE category 0}$ It is a demand (unit is kW) based calculation based on a peak load during a 12-month measurement period $\text{PUE}_1 = \text{PUE category 1}$ It is a consumption-based (unit is kWh) calculation during a 12-month total kWh reading of the following: total IT load (sum of the reading of all relevant installed Uninterruptible Power Supply (UPS) system/s output); total facility source energy is obtained by adding up the 12 consecutive monthly kWh readings obtained from the company utility bills (note: the consumption of other sources such as fuels, natural gas, etc. will have to be converted to kWh) $\text{PUE}_2 = \text{PUE category 2}$ It is a consumption-based (unit is kWh) calculation during a 12-month total kWh reading of the following: total IT load (sum of the reading of all relevant installed Power Distribution Units' (PDUs) output); total facility source energy is the same as above $\text{PUE}_3 = \text{PUE category 3}$	No unit, form: ratio	The Green Grid (2010a, 2012, 2014)

Continued

Table 12.1 Green Data Center Efficiency Metrics—cont'd

Metric Acronym	Metric Name	Formula and Description	Unit	Reference
		It is a consumption-based (unit is kWh) calculation during a 12-month total kWh reading of the following: total IT load (sum of the reading taken at the point of connection of the IT devices to the electrical system); total facility source energy is the same as above Further details are found in the Green Grid references		
pPUE	Partial power usage effectiveness	Partial PUE (pPUE) takes into consideration all of the infrastructure components and it is a metric for energy use within specified boundaries which partitions the data center into zones. The boundaries could either be physical or logical (e.g. all cooling components, all power components, and all departmental components) pPUE = Total Energy within a boundary divided by the IT Equipment Energy within that boundary (note: pPUE can only be calculated for zones with IT) $\text{pPUE (of zone a)} = \frac{\text{NonIT power (of zone a)} + \text{IT power (of zone a)}}{\text{IT power (of zone a)}}$ For example, there are zones 0 to n where zone 0 does not have any IT equipment while the rest of the zones have IT equipment. Each appropriate zone will hold a portion (r) of the IT load Zones 1 to n: Assume No id the non-IT power for zone 0; $pPUE1$ is the partial PUE for zone 1, $pPUE2$ for zone 2,…, $pPUEn$ for zone n Assume $I1$ is the IT power for zone 1, $I2$ for zone 2,… In for zone n $r1 = \frac{I1}{I1 + I2 + \ldots + In} \quad r2 = \frac{I2}{I1 + I2 + \ldots + In} \quad rn = \frac{In}{I1 + I2 + \ldots + In}$ $\text{PUE} = \frac{N0}{I1 + I2 + \ldots + In} + (r1 * pPUE1) + (r2 * pPUE2) + \ldots + (rn * pPUEn)$ (note: $r1 + r2 + \ldots + rn = 100\%$)	No unit, form: ratio or %	The Green Grid (2011a)

DCeP	Data center productivity V_i	It is an equation that quantifies useful work that a data center produces based on the amount of energy it consumes $$\text{DCeP} = \frac{\text{Useful work}}{\text{Total facility source energy}}$$ where Useful work is a sum of tasks that are completed in a specified period of time, where each task is weighted based on a specified relative value $$\text{Useful work} = \sum_{i=1}^{m} Vi \times Ui(t, T) \times Ti$$ m is the number of tasks initiated during the assessment window V_i is a normalization factor that allows the tasks to be summed numerically $T_i = 1$ if task i completes during the assessment window, and $T_i = 0$ otherwise $U_i(t,T)$ is a time-based utility function for each task t is elapsed time from initiation to completion of the task T is the absolute time of completion of the task	No unit, form: ratio	The Green Grid (2014)
ScE	Server compute efficiency	This metric is time-based and it measures the proportion of time a server spent, providing active primary services (a list of criteria is found in The Green Grid, 2010b). The ScE for an individual server (in percentage) over a specific time period is calculated by summing the number of samples where the server is found to be providing primary services (p) and dividing this by the total number of samples (n) taken over that time period and multiplying by 100. (note: $p_i = 1$ when the observation is true and 0 if it is otherwise) $$\text{ScE} = \left(\frac{\sum_{i=1}^{n} pi}{n}\right) \times 100$$	No unit, form: %	The Green Grid (2010b, 2011a)
DCcE	Data center compute efficiency	The average ScE of individual servers provides a Data centre compute efficiency (DCcE). For a given data center with a total number of m servers, DCcE is calculated by simply averaging the ScE values of all servers for the same time period $$\text{DCcE} = \left(\frac{\sum_{j=1}^{m} \text{ScE}j}{m}\right)$$	No unit, form: %	The Green Grid (2010b, 2011a)

Continued

Table 12.1 Green Data Center Efficiency Metrics—cont'd

Metric Acronym	Metric Name	Formula and Description	Unit	Reference
GEC	Green energy coefficient	It is a metric that quantifies the portion of a facility's energy that comes from green sources. GEC is computed as the green energy consumed by the data center (measured in kilowatt-hours or kWh) divided by the total energy consumed by the data center (kWh) $GEC = \frac{\text{Green energy used by the data center}}{\text{Total data center source energy}}$	No unit, form: ratio or %	The Green Grid (2014)
ERF	Energy reuse factor	It is a metric that identifies the portion of energy that is exported for reuse outside of the data center to another area within a mixed-use building or to another facility. ERF is computed as reuse energy divided by the total energy consumed by the data center $ERF = \frac{\text{Reuse energy outside of the data center}}{\text{Total data center source energy}}$	No unit, form: ratio	The Green Grid (2010c, 2011a, 2012, 2014)
ERE	Energy reuse effectiveness	Both PUE and ERE are useful metrics in the analysis of data centers that reuse energy $ERE = (1 - ERF) \times PUE$	No unit, form: ratio	The Green Grid (2010c, 2011a)
SER_{DC}	Data centre source energy ratio	This is a data center energy reuse metric. It tracks the source energy consumed by the utility grid (or onsite cogeneration) to deliver each kWh of electricity to the IT equipment $SER_{DC} = \frac{\text{Primary energy from power station}}{\text{IT energy}}$	No unit, form: ratio	BCS (2011)
CUE	Carbon usage effectiveness	CUE is a metric that enables an assessment of the total GHG emissions of a data center, relative to its IT energy consumption. CUE is computed as the total carbon dioxide emission equivalents (CO2eq) from the energy consumption of the facility divided by the total IT energy consumption $CUE = \frac{\text{Total CO2 emissions caused by the total data center source energy}}{\text{IT equipment source energy}}$	$kgCO_2eq/kWh$	The Green Grid (2010a, 2014) BCS (2011)

		Mathematically, it is equivalent to multiplying the PUE by the data center's carbon emission factor (CEF) which can be obtained here.[1] $CUE = CEF \times PUE$ Alternative representation is: $CUE = \left(\frac{CO2\ \text{emitted in kgCO2eq}}{\text{Unit of energy in kWh}}\right) \times \left(\frac{\text{Total data center source energy}}{\text{IT equipment source energy}}\right)$ This is also known as the IT carbon intensity which is essentially the carbon intensity of the electricity delivered to the IT equipment including data center infrastructure losses but less any benefit from the heat re-use system		
WUE	Water usage effectiveness (site)	A site-based metric that is an assessment of the water used on-site for operation of the data center. This includes water used for humidification and water evaporated on-site for energy production or cooling of the data center and its support systems $WUE = \frac{\text{Annual site water usage}}{\text{IT equipment energy}}$	Liters/kilowatt-hour (L/kWh)	The Green Grid (2011b)
WUE_{source}	Water usage effectiveness (site + source)	WUE_{source}, a source-based metric that includes water used on-site and water used off-site in the production of the energy used on-site. Typically this adds the water used at the power-generation source to the water used on-site $WUE_{source} = \frac{\text{Annual source energy water usage} + \text{Annual site water usage}}{\text{IT equipment energy}} = [EWIF + PUE] + \frac{\text{Annual site water usage}}{\text{IT equipment energy}}$ where EWIF is an energy water intensity factor and it is a value which is a measurement of the volume of water used for generating energy. A list of EWIF values can be found in The Green Grid (2011b)	Liters/kilowatt-hour (L/kWh)	The Green Grid (2011b)
CADE	Corporate average data center efficiency	$CADE = \text{Facility efficiency} \times \text{IT asset efficiency}$ where $\text{Facility efficiency} = \text{Facility energy efficiency}\ (\%) \times \text{Facility utilization}\ (\%)$ $\text{IT asset efficiency} = \text{IT energy efficiency}\ (\%) \times \text{IT utilization}\ (\%)$ Further explanation can be found in Neudorfer and Ohlhorst (2010)	No unit, form: ratio or %	McKinsey & Company and Uptime Institute (McKinsey and Co., 2008)

[1] http://www.epa.gov/climateleadership/documents/emission-factors.pdf.

Figure 12.4
Woodhouse building LV switch panel.

Figure 12.5
Meter installation for servers in the server room.

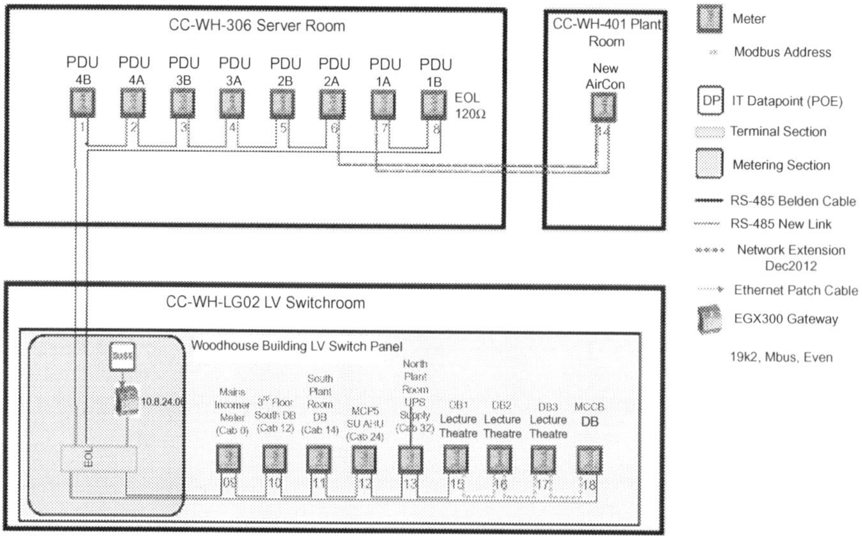

Figure 12.6
System schematic diagram of the metering topology.

Figure 12.7
Flood at Portland causes outage.

Implementation

The project focused on two major changes to the data center provision:

1. *The target temperature "set point":* Long-established processes call for data centers to be maintained at low temperatures (the ASHRAE[17] recommendation is currently 27 °C[18]); however, most data centers have traditionally operated at somewhat lower temperature, hence the well-known "chilling" effect experienced when entering a data center. This project had historically used a value of 21 °C. The justification for this was the need to operate devices within their "safe" operating range; it was driven by a view that there is a direct relationship between operating temperature and reliability. Historically, there was some justification for this view, but it had been overtaken by the increased reliability and temperature tolerance of contemporary components. One contributory factor to this increased reliability has been the reduction in electro-mechanical devices and the increased use of solid-state systems. Therefore, the project sought to explore the relationship between temperature and energy use.
2. *The use of alternative cooling mechanisms:* During the course of this project, an upgrade of cooling systems was carried out. The existing refrigeration-based cooling unit was replaced by an evaporative cooling unit. The former technology is known to be energy

[17] http://www.ashrae.org/.

[18] http://www.techweekeurope.co.uk/news/ashrae-says-data-centres-can-get-hotter-21896—February 2011 report.

consuming, so the experiment sought to provide some quantitative evidence on the change in energy consumption brought about by such a change.

Operation of the Experiment

Following commissioning of the equipment, data logging was undertaken throughout the project as described in Appendix A of Pattinson and Cross (2013) with key events being recorded in the project log, reproduced in Appendix B (Pattinson and Cross, 2013). This allowed researchers to investigate some questions and draw conclusions to them regarding the data center efficiency.

Assumptions

The main assumption made was that load for the servers remained approximately constant throughout the entire experiment.

Results and Discussion

Results

The changes implemented include varying the AHU set point temperature between 21 and 26 °C in 2 °C steps and transferring the air-cooling function to a more efficient free cooling system.

PUE Analysis

An analysis of the average PUE for Leeds Met data center had been conducted for data collected from mid-July 2012 until December 2012. As previously mentioned, there was a data outage from October 31, 2012 to December 17, 2012. Consequently, there was no analysis for November 2012 and the first half of December 2012. However, analysis for other time periods are found in Figures 12.8–12.13 and Table 12.2.

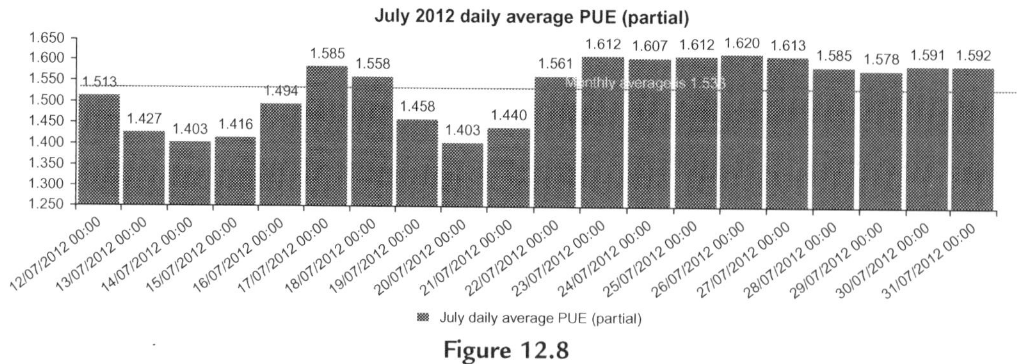

Figure 12.8
July 2012 daily average PUE (partial).

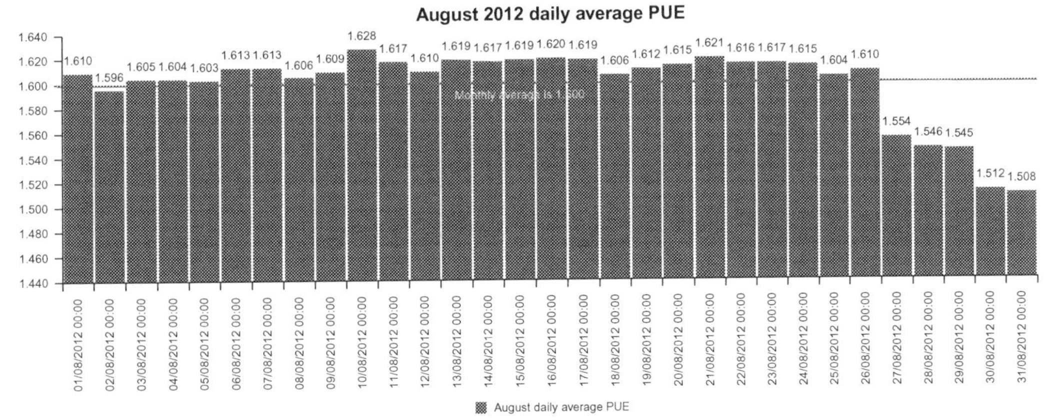

Figure 12.9
August 2012 daily average PUE.

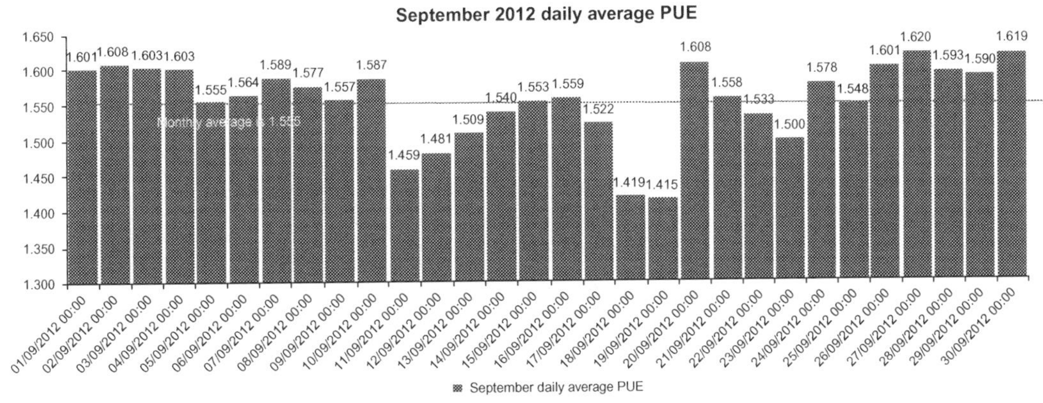

Figure 12.10
September 2012 daily average PUE.

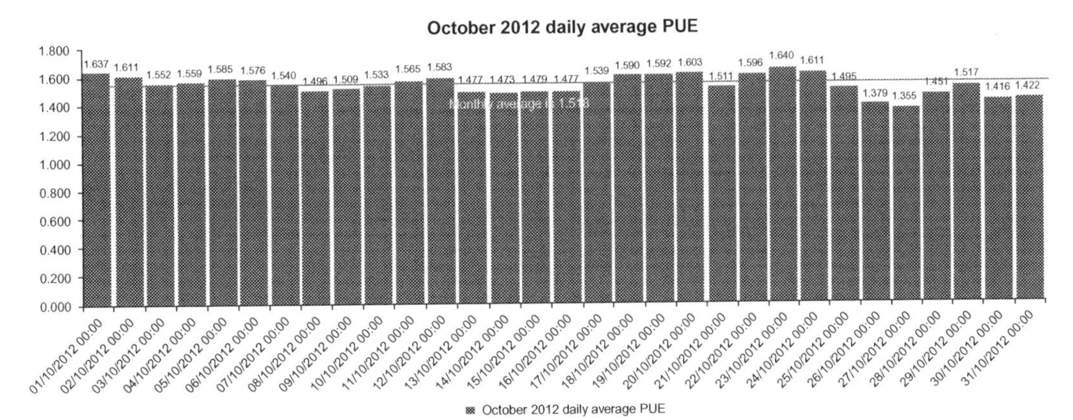

Figure 12.11
October 2012 daily average PUE.

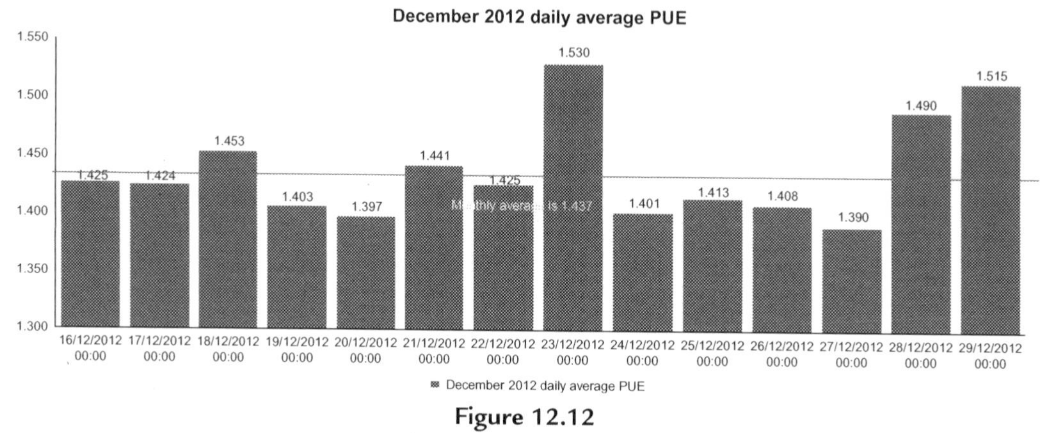

Figure 12.12
December 2012 daily average PUE.

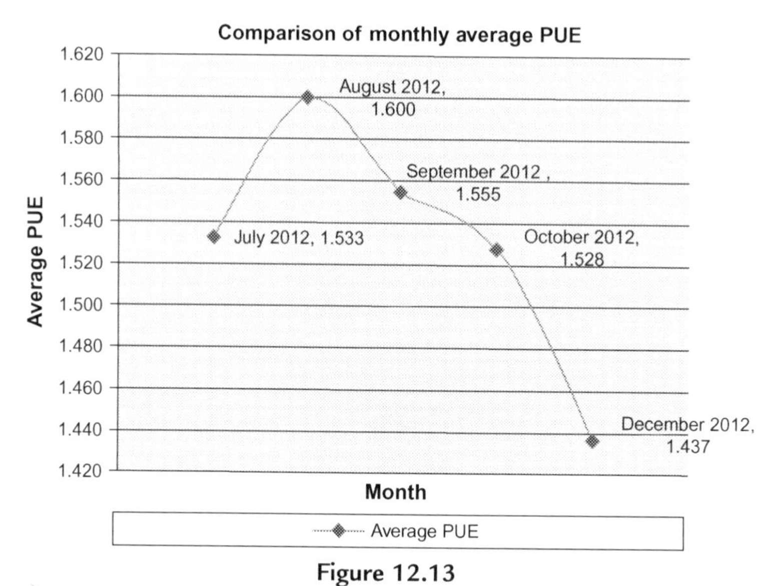

Figure 12.13
Comparison of the average PUE for July 2012-December 2012.

Table 12.2 Average PUE for July 2012 to December 2012

Month	Average PUE
July 2012	1.533
August 2012	1.600
September 2012	1.555
October 2012	1.528
December 2012	1.437

During the period shown in Figure 12.13, the cooling system and set temperature (at 26 °C) had remained constant. The graph shows that the PUE peaked in the month of August 2012 but was at its minimum in the month of December 2012.

Effect of Set Point Temperature

Figure 12.14 depicts the effect of a set point temperature change (from 24 to 26 °C on February 6, 2012) for the data center cooling system on the computer room air handlers (CRAH) energy consumption. The graph in Figure 12.14 shows that during this snapshot of the timeline, the average CRAH energy consumption seemed to increase when the set point temperature was increased from 24 to 26 °C. Figure 12.15 shows that the trends for the CRAH energy consumption and the total input energy looked similar. Thus, it could be inferred that the rest of the components contributing to the total input energy seemed to remain approximately constant. Figure 12.16 shows that the total ICT energy consumption was rather constant. Consequently, it could be concluded that for this time period, when the set point temperature increased from 24 to 26 °C, the total input energy increased while total ICT energy consumption remained quite constant, thus, increasing the PUE.

Effect of a Change in the Cooling System

In order to investigate this, a comparison of the total energy consumption for the CRAH systems had been made for December 2011 and 2012 (see Table 12.3). The rationale for choosing the month December was that the other factors were assumed to be constant. In December 2011, the CRAH units were Airedale AH DF (80 kW cooling duty) with Trend OEM controls. As for December 2012, a new evaporative cooling system was used. A snapshot of the latter half of December was taken because of the partial data outage for December 2012.

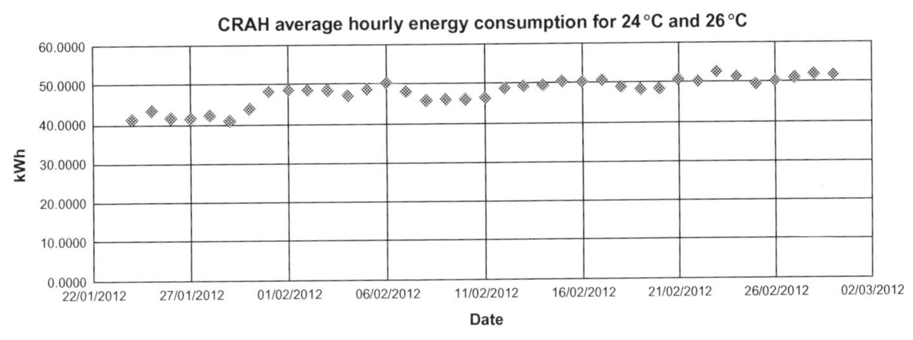

Figure 12.14

Graph for the CRAH at South Plant Room average half-hourly energy consumption for two set temperatures 24 and 26 °C. *Note*: Points before 06/02/2012 are for 24 °C while other points are for 26 °C.

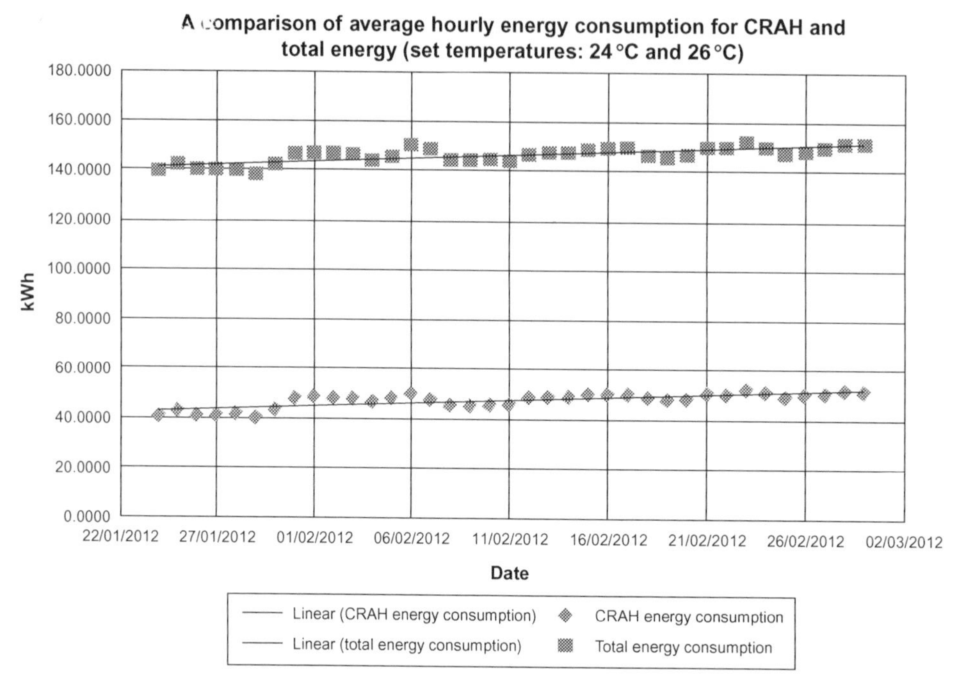

Figure 12.15

A comparison of the average half-hourly energy consumption for CRAH at South Plant Room and the average total input energy. *Note*: Before 06/02/2012 the set temperature was 24 °C and after was 26 °C.

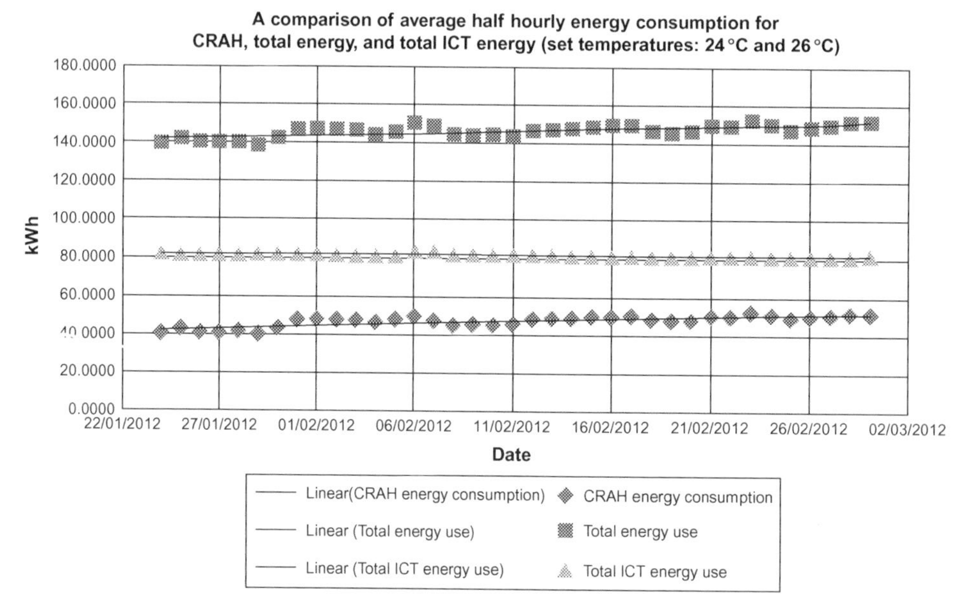

Figure 12.16

A comparison of the average half-hourly energy consumption for CRAH at South Plant Room, average total energy, and average total ICT Energy. *Note*: Before 06/02/2012 the set temperature was 24 °C and after was 26 °C.

Table 12.3 A Comparison of CRAH Energy Consumption for December 2011 and 2012 Using Matched Pair of Dates

December	Average Hourly CRAH Energy Consumption (2011) kWh	Average Hourly CRAH Energy Consumption (2012) kWh	Increase in Efficiency (Matched Pair) (%)
16			
17		18.0399	
18		18.9042	
19	46.2655	15.1068	67.35
20	44.0311	13.9004	68.43
21	45.6396	17.0127	62.72
22	47.099	16.997	63.91
23	46.5635	25.3643	45.53
24	45.9577	15.3362	66.63
25	48.2612	16.2882	66.25
26	49.3558	15.9623	67.66
27	47.6522	14.6726	69.21
28	46.2446	22.3359	51.70
29	45.4318	24.1982	46.74
30	45.1723	12.8627	71.53
31	48.2956		

Note: Set temperature for the following critical dates: 19/12/2011–24 °C; 20/12/2011–26 °C; 23/12/2011–24 °C December, 2012–26 °C; empty slots are due to missing data.

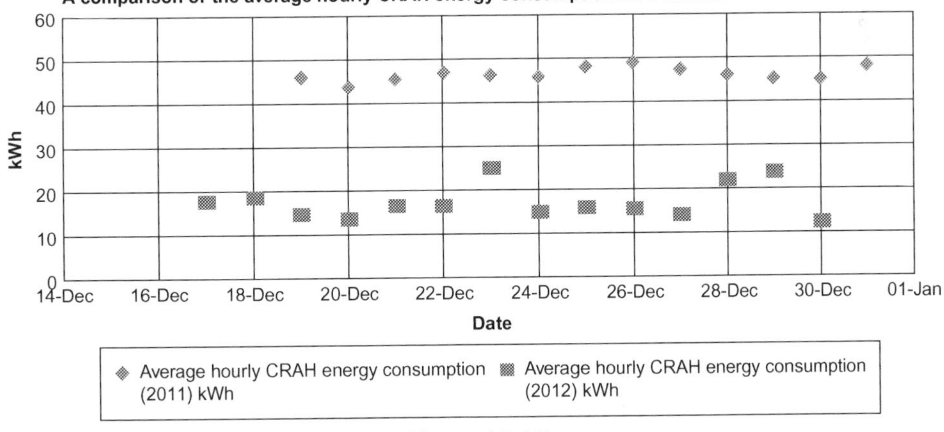

Figure 12.17
A comparison of CRAH energy consumption for December 2011 and 2012 using matched pair of dates.

The graph in Figure 12.17 shows that the new evaporative cooling system was more efficient and thus decreased energy consumption. The percentage of increase in energy efficiency is shown in Figure 12.18.

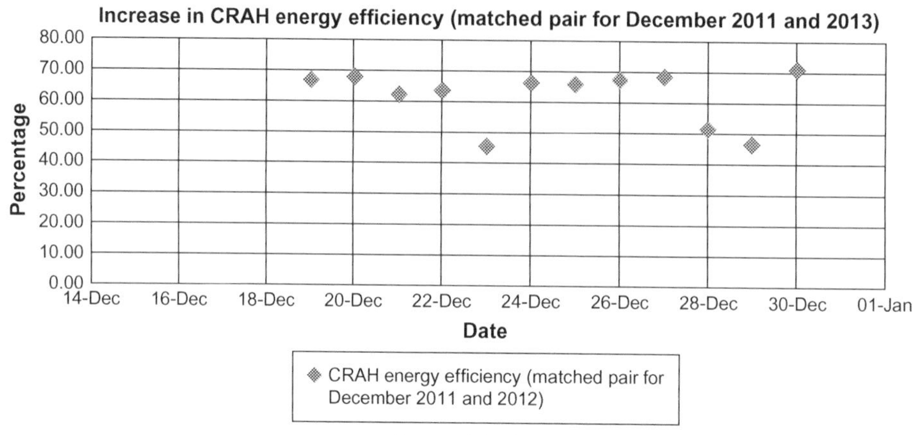

Figure 12.18
Increase in energy efficiency for the new evaporative cooling system (December 2012) compared with the Airedale AH DF units (December 2011).

The average PUE for December 2011 was 1.7696 and for December 2012 was 1.3262. Table 12.4 demonstrates that there was a decrease in PUE for the December 2011 and 2012 matched-paired dates. This analysis shows the new evaporative cooling system seemed to result in energy efficiency.

Immediate Impact

Indications in the project showed that the change in AHU set point value and the type of cooling system affected the PUE. This had created a gradual progression of average PUE in December 2011 of between 1.74-1.80 to an average PUE in December 2012 of 1.29-1.37.

Table 12.4 A Comparison of PUE for December 2011 and 2012 Using Matched Pair of Dates

December	Average Half Hourly PUE (2011)	Average Half Hourly PUE (2012)	Decrease in PUE (Matched Pair) (%)
20	1.7405	1.3148	24.46
21	1.7690	1.3378	24.37
22	1.7826	1.3217	25.85
23	1.7758	1.3745	22.60
24	1.7583	1.3083	25.59
25	1.7908	1.3141	26.62
26	1.8060	1.3112	27.40
27	1.7832	1.3013	27.02
28	1.7619	1.3531	23.20
29	1.7495	1.3619	22.16
30	1.7483	1.2892	26.26

Future Impact

The results from this experiment provided pointers to most effective changes in data center configuration. However, the effects of changes in the set point temperatures must be explored further for a longer time span and in the different seasons of the year. It also provided lessons to be learned relating to possible problems encountered during the experiments. This project also provided valuable insights into how a real data center behaved in practice. Additionally, it showed the use of a range of benchmark techniques for comparison (e.g., Figures 12.19 and 12.20).

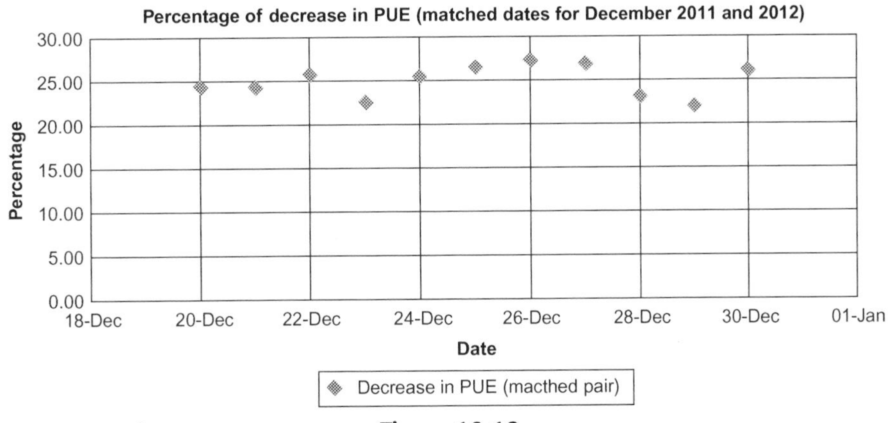

Figure 12.19

A comparison of PUE for December 2011 and 2012 using matched pair of dates.

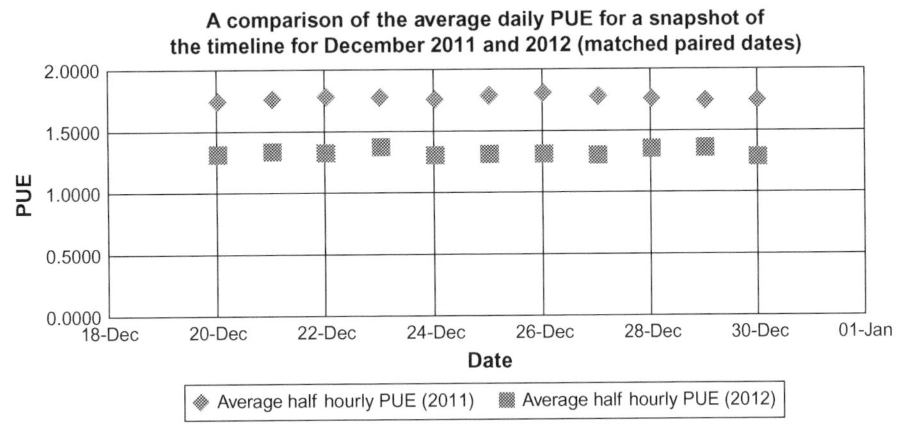

Figure 12.20

Change in PUE for the new evaporative cooling system (December 2012) when compared with the Airedale AH DF units (December 2011).

Conclusions

Many problems have been encountered in this project (refer to Appendices A and B of Pattinson and Cross, 2013). Thus, this brought about only a very short time span (toward the end of the project) when complete and correct meter readings had been collated. Consequently, the analysis conducted was rather limited. A more rigorous analysis at different granularities could have been carried out if sufficient and correct data (corresponding to each change to the data center configuration) had been collected. The results suggest that set point temperatures and the technology for the cooling system affected the PUE of the data center. However, the following are recommended for future work:

- Extend the timeline for data collection
- Investigate the effects of virtualization of the servers on PUE
- Include a wider range of parameter changes to the data center configuration

Implications for the Future

This project indicated that some changes were more energy effective than others and it would be worthwhile to explore further before implementing them. Based on the results of the experiments, radically different technologies (e.g., chip manufacturing or cooling systems) would override any effects of changes that had been made to the energy efficiency-related parameters. However, given the current technology and its likely life span, these conclusions should be valid for the short- to near-term future.

Acknowledgments

The authors record their thanks to the JISC Greening ICT project for funding this work and for ongoing support and advice. We also record our appreciation of technical colleagues at Leeds Metropolitan University for technical support throughout the project.

References

American Society of Heating, Refrigerating and Air-Conditioning Engineers, Inc. (ASHRAE), 2011. ASHRAE TC 9.9: Thermal guidelines for data processing environments—expanded data center classes and usage guidance. http://ecoinfo.cnrs.fr/IMG/pdf/ashrae_2011_thermal_guidelines_data_center.pdf (accessed 15.06.14).

British Computing Society (BCS), 2011. Data Centre Energy Re-UseMetrics. http://dcsg.bcs.org/sites/default/files/protected/Data%20Centre%20Energy%20ReUse%20Metrics%20v0.1.2.pdf (accessed 15.06.14).

Brown, R., et al., 2007. Lawrence Berkeley national laboratory: technical report to congress on server and data center energy efficiency: public law 109-431. http://escholarship.org/uc/item/74g2r0vg, (accessed 15.06.14).

CDW, 2013. DCIM: total management insight. http://www.edtechmagazine.com/higher/sites/edtechmagazine.com.higher/files/dcim-total-management-insight.pdf (accessed 15.06.14).

Cisco, 2007. Cisco data center infrastructure 2.5 design guide, technical report no: OL-11565-01. http://www.cisco.com/application/pdf/en/us/guest/netsol/ns107/c649/ccmigration_09186a008073377d.pdf (accessed 15.06.14).

Energy Star, n.d. Top twelve ways to decrease the energy consumption of your data center. http://www.energystar.gov/index.cfm?c=power_mgt.datacenter_efficiency (accessed 15.06.14).

European Commission, 2008. Code of conduct on data centres energy efficiency, Version 1.0. http://ec.europa.eu/information_society/activities/sustainable_growth/docs/datacenter_code-conduct.pdf (accessed 15.06.14).

Gantz, J., Reinsel, D., 2012. IDC analyze the future: IDC iView—the digital universe in 2020: big data, bigger digital shadows, and biggest growth in the far east. http://idcdocserv.com/1414 (accessed 15.06.14).

Gao, J., n.d. Machine learning applications for data center optimization. http://static.googleusercontent.com/media/www.google.com/en/us/about/datacenters/efficiency/internal/assets/machine-learning-applicationsfor-datacenter-optimization-finalv2.pdf (accessed 15.06.14).

Gartner, 2014. News release: Gartner says the internet of things will transform the data center. http://www.gartner.com/newsroom/id/2684616 (accessed 15.06.14).

IBM Corporation, 2008a. Creating a green data centre to help reduce energy costs and gain a competitive advantage. http://www-935.ibm.com/services/uk/igs/pdf/greenit_pov_final_0608.pdf (accessed 15.06.14).

IBM Corporation, 2008b. IBM scalable modular data center: the art of the possible. rapidly deploying cost-effective, energy-efficient data centers. http://www-935.ibm.com/services/us/its/pdf/smdc-eb-sfe03001-usen-00-022708.pdf (accessed 15.06.14).

IBM Corporation, 2012a. Data center operational efficiency best practices: enabling increased new project spending by improving data center efficiency. A research report, no: RLW03007USEN.

IBM Corporation, 2012b. A green IT approach to data center efficiency, Redbooks Document No: REDP-4946-00. http://www.redbooks.ibm.com/redpapers/pdfs/redp4946.pdf (accessed 15.06.14).

IBM Corporation, 2013. IBM and Cisco: together for a World Class Data Center, IBM Redbooks. http://www.redbooks.ibm.com/redbooks/pdfs/sg248105.pdf (accessed 15.06.14).

IDC, 2014a. Datacenter trends and strategies. http://www.idc.com/getdoc.jsp?containerId=IDC_P13027 (accessed 15.06.14).

IDC, 2014b. The digital universe of opportunities: rich data and the increasing value of the internet of things. http://www.emc.com/leadership/digital-universe/2014iview/executive-summary.htm (accessed 15.06.14).

Koomey, J.G., 2008. Worldwide electricity used in data centers. Environ. Res. Lett. 3, 034008.http://iopscience.iop.org/1748-9326/3/3/034008/pdf/1748-9326_3_3_034008.pdf.

Koomey, J.G., 2011. Growth in data center electricity use 2005 to 2010. Analytics Press, Oakland, CA. Technical report for The New York Times, August 1. http://www.analyticspress.com/datacenters.html (accessed 15.06.14).

McKinsey & Co., 2008. McKinsey & Company and the Uptime Institute. http://www.ecobaun.com/images/Revolutionizing_Data_Center_Efficiency.pdf (accessed 15.06.14).

Microsoft Corporation, 2012. Data Center 101. http://cdn.globalfoundationservices.com/documents/Strategy_Brief_Data_Center_101.pdf (accessed 15.06.14).

Murugesan, S., Gangadharan, G.R., 2012. Harnessing Green IT: Principles and Practices. Wiley, IEEE., ISBN 978-1-119-97005-7.

National Computer Board, NCB, n.d. Guideline on server consolidation and virtualisation. http://www.ncb.mu/English/Documents/Downloads/Reports%20and%20Guidelines/Guideline%20on%20Server%20Consolidation%20and%20Virtualisation.pdf (accessed 15.06.14).

Neudorfer, J., Ohlhorst, F.J., 2010. Data center efficiency metrics and methods. http://viewer.media.bitpipe.com/979246117_954/1279665297_327/Handbook_SearchDataCenter_efficiency-metrics_final.pdf (accessed 15.06.14).

Pattinson, C., Cross, R., 2013. Measuring data centre efficiency, JISC Green IT technical report January 2013. ISBN: 978-1-907240-33-1. http://repository.leedsmet.ac.uk/main/view_record.php?identifier=8250&SearchGroup=research (accessed 15.06.14).

The Green Grid, 2007. Green grid metrics: describing datacenter power efficiency. http://www.thegreengrid.org/~/media/WhitePapers/Green_Grid_Metrics_WP.ashx?lang=en (accessed 15.06.14).

The Green Grid, 2008a. Green grid data center power efficiency metrics: PUE and DCiE. http://www.eni.com/green-data-center/it_IT/static/pdf/Green_Grid_DC.pdf (accessed 15.06.14).

The Green Grid, 2008b. A framework for data center energy productivity. http://www.thegreengrid.org/~/media/WhitePapers/WhitePaper13FrameworkforDataCenterEnergyProductivity5908.pdf?lang=en (accessed 15.06.14).

The Green Grid, 2009. Usage and public reporting guidelines for the green grid's infrastructure metrics (PUE/DCiE). http://www.thegreengrid.org/~/media/WhitePapers/White%20Paper%2022%20%20PUE%20DCiE%20Usage%20Guidelinesfinalv21.ashx?lang=en (accessed 15.06.14).

The Green Grid, 2010a. Carbon usage effectiveness (CUE): a green grid data center sustainability metric. http://www.thegreengrid.org/~/media/WhitePapers/Carbon%20Usage%20Effectiveness%20White%20Paper_v3.pdf?lang=en (accessed 15.06.14).

The Green Grid, 2010b. The green grid data center compute efficiency metric: DccE. http://www.thegreengrid.org/~/media/WhitePapers/DCcE_White_Paper_Final.pdf?lang=en (accessed 15.06.14).

The Green Grid, 2010c. Ere: a metric for measuring the benefit of reuse energy from a data center. http://www.thegreengrid.org/~/media/WhitePapers/ERE_WP_101510_v2.ashx?lang=en (accessed 15.06.14).

The Green Grid, 2011a. Data center efficiency metrics: mPUE, partial PUE, ERE, DCcE, Green Grid Technical Forum, 2011. http://www.thegreengrid.org/~/media/TechForumPresentations2011/Data_Center_Efficiency_Metrics_2011.pdf (accessed 15.06.14).

The Green Grid, 2011b. Water usage effectiveness (WUE™): a green grid data center sustainability metric, http://www.thegreengrid.org/~/media/WhitePapers/WUE (accessed 15.06.14).

The Green Grid, 2012. PUE™: a comprehensive examination of the metric, http://www.thegreengrid.org/~/media/WhitePapers/WP49-PUE%20A%20Comprehensive%20Examination%20of%20the%20Metric_v6.pdf?lang=en, (accessed 15.06.14).

The Green Grid, 2014. Harmonizing global metrics for data center energy efficiency: global taskforce reaches agreement regarding data center productivity, http://www.thegreengrid.org/~/media/Regulatory/HarmonizingGlobalMetricsforDataCenterEnergyEfficiency.pdf?lang=en, (accessed 15.06.14).

Troianovski, A., 2011. Storage Wars: Web Growth Sparks Data-Center Boom. The Wall Street Journal. http://online.wsj.com/news/articles/SB10001424052702303763404576417531646400002 (accessed 15.06.14).

US Department of Energy, 2011. Best practices guide for energy-efficient data center design, technical report. http://www1.eere.energy.gov/femp/pdfs/eedatacenterbestpractices.pdf (accessed 15.06.14).

Yuventi, J., Mehdizadeh, R., 2013. A critical analysis of power usage effectiveness and its use in communicating data center energy consumption, Stanford University technical report, no: CIFE Working Paper #WP131. http://cife.stanford.edu/sites/default/files/WP131_0.pdf (accessed 15.06.14).

CHAPTER 13

Communitywide Area Network and Mobile ISP

Colin Pattinson, Ah-Lian Kor*, Richard Braddock
Leeds Beckett University, Leeds, UK
**Corresponding author: a.kor@leedsbeckett.ac.uk*

Introduction

The main aim of the project discussed here was to develop a framework for delivering an e-learning environment using free and open source software (FOSS) and creating a system that was transportable. This was made possible by the use of a vehicle that had very low-power (and thus "green") IT equipment while maintaining an acceptable level of functionality. The server machine that was employed contained a core operating system (OS) of all its connected machines. Thus, the user machines did not need traditional hard drives, which often fail in a mobile environment. This was considered a viable green solution because of its low power consumption. One of the challenges faced was the use of a single machine serving various roles. This was made possible by virtualization using VMWare ESXi to deliver the core functionality of a server within a modular, appliance-led software stack. The ESXi was chosen for its 32 MB disk footprint. Additionally, a mobile Internet Service Provider (mISP) platform provided a wire-free and gateway connectivity. The work carried out in this student project was the basis for a proposed contribution to the JANET Wireless LAN project, exploring the use of this technology to support students on field-trip types of activities.

Context of Application

A mobile learning environment is different from a typical bus environment. Such an environment needs to provide a more advanced method of communication (e.g., wireless local area network (WLAN)) and must address several environmental issues relating to the atmospheric conditions that could affect its performance quality. In the context of the project, the following issues were taken into consideration: geographical areas with limited power and high-speed Internet connection access; technical issues relating to the cost of bandwidth, content filtering, authentication, legal (or accounting) factors, and security. Services that were made accessible to the clients were the local content management system (CMS) or Web services.

Prototype Mobile Learning Environment

Architecture of Prototype

The deliverable project called for a transportable IT solution suitable for operation in a mobile vehicle. It was primarily designed to serve an optimized e-learning environment and to satisfy other requirements (e.g., to serve as a thin-client terminal server and provide periphery services to neighboring portable devices). Figure 13.1 shows a wide variety of OS-dependent applications that could contribute to an inefficient use of multiple hardware platforms. Consequently, a consolidated solution was necessary. A VMWare ESXi-based virtualization was selected because of its 32 MB disk footprint and the fact that it is an OS independent hypervisor (VMware, 2007) that can deliver the core functionality within a modular appliance-led software stack.

This virtualized approach required a networked hardware gateway (Squid caching proxy[1] used as a gateway) and more to establish a connection to the Internet because this approach was unable to enumerate modems in the same way that physical machines can. However, VMware ESXi virtual machines do not have USB controllers associated by default (Ncomputing, 2010). A VMkernel is the underlying OS that controls access to all hardware devices (e.g., network or storage) on the server (VMware, 2007) using modules via the vmklinux driver (currently, Vmware's ESXi 5.5 has two drivers: vmklinux, the legacy driver and native mode, the new driver (Emulex, 2013)). The implemented solution using a discrete efficient hardware platform is depicted in Figure 13.2.

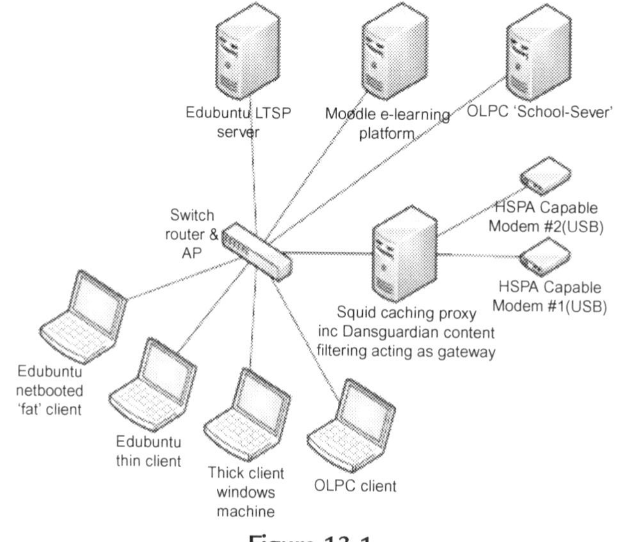

Figure 13.1
Multiple hardware platforms.

[1] http://www.squid-cache.org/.

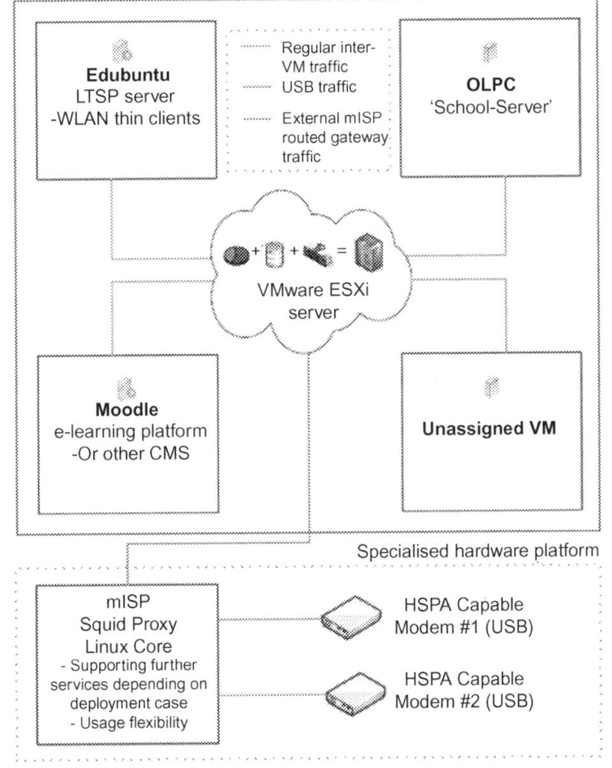

Figure 13.2
Discrete efficient hardware platform for the implemented solution.

General Requirements

To reiterate, the main goal of the project was to provide gateway services for mobile virtualized environments that could be repurposed as a mobile community that was not reliant on wired backhaul. This implemented solution was made possible with minor software-level changes and the addition of WLAN functionality. Including this level of flexibility allowed the service to be able to completely power down the central virtualization server and operate solely as an mISP (Braddock and Pattinson, 2009; Pattinson et al., 2010). However, the mISP infrastructure needed to comply with the electrical as well as thermal constraints within the target locale. Additionally, this infrastructure comprised appropriate hardware components and software topology that fostered a convergence of a range of services that would have been traditionally abstracted across multiple physical or virtual servers. The sample mISP deployment with caching considerations is shown in Figure 13.3. A list of technical capabilities of the specialized platform follows:

- Aggregating multiple cellular-based Internet connections to provide a redundant high-speed backhaul link

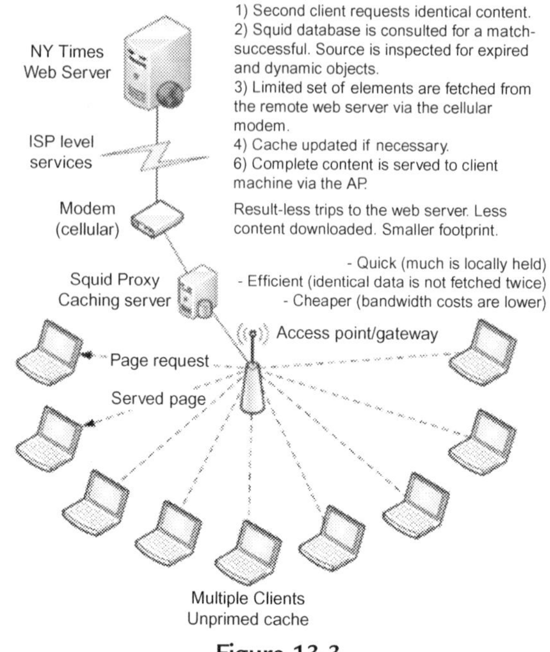

Figure 13.3
A sample mISP deployment with caching considerations.

- Incorporating a wired backhaul link when within range of such a service
- Optimizing transparent link/bandwidth
- Acting as a wireless gateway to authenticated or trusted nodes as well as performing this authentication via a Web-based interface
- Encapsulating session-level accounting and reporting, thus nullifying legal concerns that have plagued the wISP industry in developing nations (Mitta, 2009)
- Incorporating a high-powered 802.11 g radio (details are found in Cisco, n.d.-b) interface when acting as a mobile learning environment
- Provisioning a secure host OS on which to house the software payload
- Providing adequate processing power to allow further server-side applications to be integrated as necessary
- Providing local storage for server-side applications, possibly with precautions for further data defense and/or security
- Allowing remote diagnosis and management services to interact at all stages of the design, whether they are traditional network metrics or more environmental aspects such as current climate conditions or internal state of the energy source(s)
- Demonstrating a clear methodology for powering all services on and off the grid consistently in an autonomous fashion, including providing renewable energy collection
- Using FOSS at every stage to meet user needs while maintaining zero software expenditure

Infrastructure

The infrastructure comprised (WLAN)[2] and worldwide interoperability for microwave access (WiMAX, IEEE 802.16).[3] The former is directed at short-range applications using radio frequency to transmit data over a short range while the latter supports long-distance wireless broadband. Point-to-point link[4] adhered to site-specific requirements (e.g., geographic information in terms of latitude and longitude coordinate provisions) and site survey requirements (e.g., determination of the number and placement of access points (or mesh nodes) that would provide adequate signal coverage throughout a facility). Satellite broadband technology provided high-speed, reliable, and cost-effective Internet access for the project. A range of cellular services was also surveyed so as to select the best of the available options for generic deployment.

Hardware Stack

Several of the product requirements were low cost so that it would be accessible to a wider market segment and low power and compatibility with the latest electrical architecture. The end product was a small portable Web server and mobile teaching environment with incorporated graphical output capabilities, audio facility, input devices connection (such as USB), touch-screens/interactive whiteboards (Pro-zone), presentation aids, traditional keyboard/mouse combo, and interface capability with popular WAN adapters (e.g., USB cellular adapters).

To reiterate, the primary use of a VMware ESXi server was to create an educational environment through a Linux terminal server project (LTSP)[5] centric distribution. A transparent Squid caching proxy[6] with a content filter (e.g., Dansguardian,[7] OpenDNS,[8] Squid guard,[9] or SafeSquid[10]) acting as a gateway. The gateway services were accompanied by the efficient use of WAN resources via incorporated channel aggregation techniques. In so doing, several channels were combined to create another channel of greater bandwidth.

The hardware specifications for the implemented solution have been tabulated in Table 13.1.

---

2 http://www.cisco.com/c/en/us/tech/wireless-2f-mobility/wireless-lan-wlan/index.html.

3 http://computer.howstuffworks.com/wimax.htm.

4 https://publib.boulder.ibm.com/infocenter/iseries/v5r3/index.jsp?topic=%2Frzajw%2Frzajwptop.htm.

5 http://en.opensuse.org/LTSP.

6 http://wiki.squid-cache.org/SquidFaq/InterceptionProxy.

7 http://dansguardian.org/.

8 https://support.opendns.com/entries/26514730-Web-Content-Filtering-and-Security.

9 http://www.squidguard.org/.

10 http://www.safesquid.com/.

Table 13.1 Hardware Specifications

Hardware	Discussion
Server	The Intel® Atom™ processor[a] for the server was developed specifically for targeted performance and low power while maintaining Intel architecture instruction set compatibility. Intel Atom processors also featured Intel Hyper-Threading Technology[b] for better performance and increased system responsiveness. It was accompanied by smaller compact designs with a thermal design power specification ranging from less than 1 to 2.5 W for mobile devices. The battery life in select devices was extended with low idle and average power consumption, allowing the device to stay powered on while conserving energy
Network Routers	La Fonera 2.0[c] wireless router was employed because when upgraded with the powerful 802.11n (Cisco, n.d.-a) WiFi routing capability, it could foster the use of multiple antennas to increase data rates. Its benefits are as follows: 1. Download torrents and files directly to a USB device even when the computer is shut down 2. Upload videos to YouTube and multiple photos automatically to Facebook, Flickr, and Picasa 3. Connect a USB hub and add multiple USB devices (Webcams, printers, etc.) 4. Convert a 3G signal into Wi-Fi 5. Add new applications with a simple click 6. Enjoy secure Wi-Fi at home 7. Connect for free at any FON Spot[d] in the world Meraki mini Wi-Fi router[e] not only provided the vanilla 802.11b/g access but also acted as a node in a wireless mesh network capable of providing a viable wireless backbone
Cellular access	The General Packet Radio Service (GPRS, also known as 2.5G; details can be found in Ghribi and Logrippo (2000)) technology was used to transmit data at speeds up to 60 kbits per second. It was chosen over the High-Speed Downlink Packet Access (HSDPA, also known as 3G; details can be found in Holma and Toskala (2006)) technology because it was considered a battery friendly way to send and receive e-mails as well as to browse the Internet. Additionally, the Linux technology was leveraged to deliver seamless mobile services and applications
Front end access point	The front-end access point was the device that the users initially "saw" and connected to or connected via (in the case of the ESXi gateway device). It consisted of: 1. One or more high-powered 802.11 radios for maximum signal propagation, multispectrum support, future standard support, and application firewalling. 2. At least two switchports operating at Fast Ethernet[f] standard (i.e., maximum speed of 100 MBps). Two ports were needed because one was designed to be a WAN port, and the other acted as an internal LAN port to provide mainstream connectivity to the back end server. A third switchport was necessary when specifying a gateway device to allow that unit to be given connectivity. 3. Fast Microprocessor without Interlocked Pipe Stages (MIPS; details are found in Hennessy et al. (1981)) processor provided switching/routing at line speed. 4. Integrated voltage regulation to guard against surges or discrepancies in the DC input. 5. 8 MB + flash providing ample storage for device firmware and relevant packages.

Continued

Table 13.1 Hardware Specifications—cont'd

Hardware	Discussion
	6. 32 MB + RAM allowing sufficient application execution space and scope for the deployment to scale well when faced with a high number of users or greedy routing protocols. 7. USB connectivity to permit "breakout" mass storage space for administrative data. 8. External antennae connection to allow replacement with high gain alternatives.
Back end server node with USB modems	The back-end server node was largely transparent to the end user unless he or she was exchanging data with a service known to be hosted on the box. It comprised: 1. Dual 7.2 MBps HSPA modems to provide redundant WAN access. Huawei E220 HSDPA USB modem[g] was selected because it has a very small footprint design and high-speed performance. Its USB cable can be used by both laptops and desktop computers. A list of E220 features are as follows: • Works well on Linux systems[h] with kernel 2.6.20 and higher. • Has unlock software.[i] • Huawei 220 firmware upgrade allows 7.2 MBps access and downloads. It is much faster compared to the standard 3.6 MBps. a. Works with OpenDNS servers that provide reliability and redundancy of the Internet bandwidth (OpenDNS Security Labs, 2013) 2. Low power mini ITX[j] (mITX) motherboard capable of operating in a 15 W envelope with onboard power supply unit (PSU) circuitry and scope for multiple Network Interface (NIC) modules. The Intel D945JSEJT Johnstown[k] platform was ideal. 3. Small second-generation solid-state disks (SSD) to carry OS, applications, and user data
Switch regulator	Switching regulators to manage an efficient power topology (6-18 V input)
OpenWRT	OpenWRT[l] is a Linux-based distribution for embedded systems and provides facility for integration of network components
Integration of hardware	Accton technology (Accton VG2211i[m]) was used to combine gateway, wireless, and voice over Internet protocol (VoIP) functions. It is a versatile network device that provided high-speed broadband Internet connection and plain ordinary telephone service (POTS)-quality telephone service for wired and wireless users everywhere

[a]http://download.intel.com/pressroom/archive/reference/Next_Generation_Atom_briefing.pdf?iid=pr_smrelease_Pinetrail_rellinks1.
[b]http://www.intel.com/content/www/us/en/architecture-and-technology/hyper-threading/hyper-threading-technology.html.
[c]https://www.fon.com/en/product/fonera2nFeatures.
[d]http://maps.fon.com/.
[e]http://www.engadget.com/2006/08/03/meraki-mini-wifi-router-also-does-mesh/.
[f]https://www.princeton.edu/~achaney/tmve/wiki100k/docs/Fast_Ethernet.html.
[g]http://www.huaweie220.com/.
[h]https://wiki.archlinux.org/index.php/Huawei_E220.
[i]http://www.gsmliberty.net/shop/huawei-e220-unlock-software-p-2317.html.
[j]http://www.mini-itx.com/.
[k]http://www.logicsupply.com/d945gsejt/.
[l]https://dev.openwrt.org/.
[m]http://www.devicemanuals.com/support/pdf-manual-8324.html.

Software Stack

The following is a list of software requirements for the mobile learning environment:

1. CMS or server-side services (e.g., geophysical data processing).
2. Platform with security and account management facility.
3. Wi-Fi Protected Access II (WPA2) that is a certification program to indicate compliance with security protocol created by the Wi-Fi Alliance[11] to secure wireless computer networks; a popular Radius server with WPA2 security is the FreeRADIUS.[12]
4. Authentication plans such as Open Lightweight Directory Access Protocol[13] (OpenLDAP) suite[14] that includes the following: stand-alone LDAP daemon (server); libraries implementing the LDAP protocol; utilities, tools, and sample clients.
5. Connections to neighboring institutions via virtual private network (VPN). An example is OpenVPN.[15] Some of the OpenVPNs provide remote and secure access to networks and application resources. Site-to-site VPNs[16] are secured and scalable.
6. Mesh-based routing protocols (details can be found in Murthy and Manoj, 2004) are for optimizing the use of the network resources and reducing the power consumption of the network. Some relevant protocol examples are OpenWRT Better Approach To Mobile Ad-hoc Networking (B.A.T.M.A.N)[17] mesh that is a routing protocol for multihop ad hoc mesh networks and optimized link state routing protocol for mobile ad hoc networks that is an "optimization of the classical link state algorithm tailored to the requirements of a mobile wireless LAN" (Clausen and Jacquet, 2003).

Most of these software requirements are more compatible with the Intel X86 family of central processing units that encompass Intel's Core and Pentium lines and AMD's Athlon, Opteron, and Sempron. Linux is an open source computer OS designed primarily for the PC but also available for a wide range of other systems. Debian-based[18] distributions are free OSs. Ubuntu[19] is an OS built by a worldwide team of expert developers and is well documented. It contains a Web browser, office suite, media apps, instant messaging, and so on and is an open source alternative to Windows and Macintosh. A wealth of open source community projects exists, many of which overlap or perform similar functions. To ensure a quality end product for

[11] http://www.wi-fi.org/.

[12] http://freeradius.org/.

[13] http://www.openldap.org/software/download/.

[14] https://access.redhat.com/site/documentation/en-US/Red_Hat_Enterprise_Linux/3/html/Reference_Guide/s1-ldap-daemonsutils.html.

[15] http://openvpn.net/.

[16] http://www.cisco.com/c/en/us/solutions/enterprise-networks/site-to-site-vpn-solution/index.html.

[17] http://www.open-mesh.org/projects/open-mesh/wiki.

[18] http://www.debian.org/.

[19] http://www.ubuntu.com/.

the project, extensive testing was performed before a final payload was defined. However, with that proviso, a number of services that should prove their worth in the mISP mobile learning environment deployment had been identified. Processes were separated by handing off computationally expensive tasks to the back-end server demonstrating maximum network and processing efficiency. See Table 13.2 for examples of software that were considered for the project. As a matter of fact, plain gateway router functionality was simpler to achieve as the result of a significantly smaller package payload. A single mITX x86 solution was tried for this usage scenario, and documentation (Pattinson et al., 2010) was made available as appropriate.

Electrical Topology

All components within the system were powered from a 12v DC source. The TIER solar charge controller (a University of California, Berkeley product) (Ramos and Brewer, n.d.) met the project electrical requirements well. It combined a sealed lead acid battery, 60 W solar panel, Power over Ethernet (PoE), redundant 12 V output, voltage monitoring over Ethernet, peak power tracker (University of California, Berkeley, 2010). The board itself was 7.6 cm × 7.6 cm, which meant that it did not have to be implemented outside the solution enclosure, and in theory, the unit could cater for loads up to 120 W. The initial plan was to power the mITX back-end server node using the TIER controller and the 12v rail on the onboard molex connector to provide voltage to the front-end access point. This approach had been tested with a M1-ATX[20] PSU and WRT54GL[21] router, and it was ensured that the onboard PSU could be integrated onto the Intel D945JSEJT[22] board. Also, the vanilla RouterStation[23] had power consumption of approximately 7 W when passing traffic whereas the WRT54GL was lower at approximately 6 W, so the estimate for the unreleased pro edition was at an approximately 9 W footprint. When such situation arose, the TIER controller would require slight modification with a split charge module, thus creating dual 12 V outputs with some loss of efficiency. A PSU took an input of 6-24 V DC and regulated it into an ATX-style connector. The voltage could be stepped down to the 5 V rail, but because it focused on car-based solutions, it would encounter problems with booting back up. M1-ATX operated in dumb-PSU mode with arming single-pole single-throw[24] switch. When paired with a PicoUPS 100, it not only allowed seamless switchover from AC source to DC battery but also served to charge a lead acid source from 16 V input. This 16 V input could either be an AC brick or a solar panel. Based on a critical specification survey, the following power-related decision was made: (1) standard regulated "power brick' providing 15-18 V, (2) PicoUPS-100, and (3) sealed lead acid

[20] http://www.mini-itx.com/store/information/m1-atx-manual.pdf.

[21] http://support.linksys.com/en-eu/support/routers/WRT54GL.

[22] http://www.intel.com/content/www/us/en/motherboards/desktop-motherboards/desktop-board-d945gsejt.html.

[23] http://tier.cs.berkeley.edu/drupal/node/128.

[24] http://www.gaugemaster.com/instructions/switches_explained.pdf.

Table 13.2 Cross-Hardware Software Alignment

Service/App	Function	Front-End App	Back-End Server
CoovaAP[a]	CoovaAP is an OpenWRT-based firmware designed especially for HotSpots	X	
PHPMyPrepaid[b]	An interface for the creation and management of prepaid cards and accounts designed to work with a MySQL database and FreeRADIUS	X	
Dynamic DNS[c]	It is used to update the client to keep track of a dynamic IP address	X	
Syslog Server[d]	A server logs event messages sent by network devices		X
SNMP Client[e,f]	The manager's host runs the simple network management protocol (SNMP) client (consisting of manager and management information base (MIB)) (McGraw-Hill Companies, 2000) that is in charge of data collection and display. A manager sends a request for information about managed resources to an agent[g]	X	
SNMP server[g]	An agent runs the SNMP server (consisting of MIB and data of managed devices) (McGraw-Hill Companies, 2000). The agent gathers the requested data for the manager and returns a response[g]		X
SNMP trap	"SNMP traps enable an agent to notify the management station of significant events by way of an unsolicited SNMP message" (Cisco, 2006b). Traps are network packets that contain data relating to a component of the system sending them[h]	X	X
OpenSSH[i]	OpenSSH provides secure tunneling capabilities and several authentication methods and supports all SSH protocol versions	X	
OpenVPN[j]	"Open VPN listens on TUN/TAP devices, takes the traffic, encrypts it, and sends it to other VPN partners where another openVPN process receives the data, decrypts it, and hands it over to the virtual network device, where an application might already waiting for the data" (Feilner, 2006, p. 33)	X	X
Debian OS[k]	This is a free open operating system for a computer		X
Web management interface	A example is Webmin[l] that is a Web management interface for Unix system administrators	X	
Iptables[m]	Iptables is a firewall installed by default on all official Ubuntu distributions (Ubuntu, Kubuntu, Xubuntu)	X	X
Ifenslave[n]	Ifenslave is a tool to attach and detach slave network devices to a bonding device		X

Continued

Table 13.2 Cross-Hardware Software Alignment—cont'd

Service/App	Function	Front-End App	Back-End Server
DansGuardian[o]	DansGuardian is an open source Web content filter that currently runs on Linux, FreeBSD, OpenBSD, NetBSD, Mac OS X, HP-UX, and Solaris		X
FreeRADIUS[p]	This refers to a Remote Authentication Dial-In User Service (RADIUS) server that is usually a daemon process running on a UNIX or Windows NT machine. It can support a variety of methods to authenticate a user, and when provided with the user name and original password given by the user, it can support PPP, PAP or CHAP, UNIX login, and other authentication mechanisms (Cisco, 2006a)		X
Squid[q]	It is a caching proxy for the Web supporting HTTP, HTTPS, FTP, and more		X
mOnOwall[r]	This is a free embedded firewall software package that provides all the important features of commercial firewall boxes		X
VNC server[s]	This is a program that shares a desktop with other computers over the Internet[t]		X
SAMBA[u]	SAMBA provides secure, stable, and fast-file and print services for all clients using the SMB/CIFS protocol, such as all versions of DOS and Windows, OS/2, Linux, etc.		X
OLDAP[v]	This is an open source implementation of the Lightweight Directory Access Protocol (LDAP). The suite includes SLAPD (server); libraries implementing the LDAP protocol; and utilities, tools, and sample clients		X

[a]http://coova.org/CoovaAP.
[b]http://phpmyprepaid.soft112.com/.
[c]http://www.noip.com/support/knowledgebase/getting-started-with-no-ip-com/.
[d]http://www.techiecorner.com/1479/how-to-setup-syslog-server-in-ubuntu/.
[e]http://www.cyberciti.biz/faq/debain-ubuntu-install-net-snmpd-server/.
[f]http://www.debianhelp.co.uk/snmp.htm.
[g]http://docs.oracle.com/cd/E23943_01/web.1111/e13743/snmpagent.htm.
[h]http://support.microsoft.com/kb/172879.
[i]http://www.openssh.com/.
[j]http://www.serverubuntu.it/openvpn-bridge-configuration.
[k]http://www.debian.org/.
[l]http://www.webmin.com/.
[m]https://help.ubuntu.com/community/IptablesHowTo.
[n]http://linux.die.net/man/8/ifenslave.
[o]http://dansguardian.org/.
[p]http://freeradius.org/.
[q]http://www.squid-cache.org/.
[r]http://m0n0.ch/wall/.
[s]http://www.realvnc.com/products/vnc/documentation/5.0/guides/upgrade/ad1058811.html.
[t]https://help.ubuntu.com/community/VNC/Servers.
[u]https://www.samba.org/.
[v]http://www.openldap.org/.

battery because it was widely available with 13.5 float voltage, a nominal voltage of 12.6 V (6×2.105 (Series)), most robust accompanied by deep cycle batteries.

Prototype

Prototype Specifications

A summary of the prototype specifications (depicted in Figure 13.4) follows:

1. Hardware built on Intel Atom or VIA Nano platforms
2. Multiple implementations to cater for distinct user needs in terms of portability, runtime, processing power, general-purpose computation on Graphics Processing Units,[25] acceleration, data protection using redundant array of independent disks technology, and information security using Advanced Encryption Standard
3. Open source software (OpenLDAP, FreeRADIUS, phpMyAdmin, MYSQL, PHP5, Joomla, Moodle (or other CMS), OpenVPN, OpenWRT/DDWRT/HyperWRT/xWRT, Linux Distribution, interface management built from AJAX (or other appropriate attractive coding front end)
4. Low power approach encompassing a power architecture that facilitates the use of renewable energy sources
5. Low cost because it is built from pre-existing components. Established and proven components are being applied to a different usage schema
6. Uses lead acid-based power approach because associated circuits are readily available (PicoUPS)

The platform provided an efficient and effective 802.11-based WLAN, but the x86 based "server" was repurposed to provide additional functionality besides simply caching content and authenticating credentials. This took the form of a centrally served collaboration tool, project management environment, CMS, or file store.

Configuration

The configuration achieved in the project consisted of the following (the step-by-step configuration guide is found in Pattinson et al. (2010)): (1) Access Point, (2) Ubuntu, (3) HSDPA Adapter, (4) IPTables configured through IPKungfu,[26] (5) set up a transparent proxy with Squid, and (6) LAMP (Linux, Apache, MySQL and PHP). How to build a LAMP server is found here.[27]

[25] http://gpgpu.org/.

[26] http://freecode.com/projects/ipkungfu.

[27] http://lamphowto.com/.

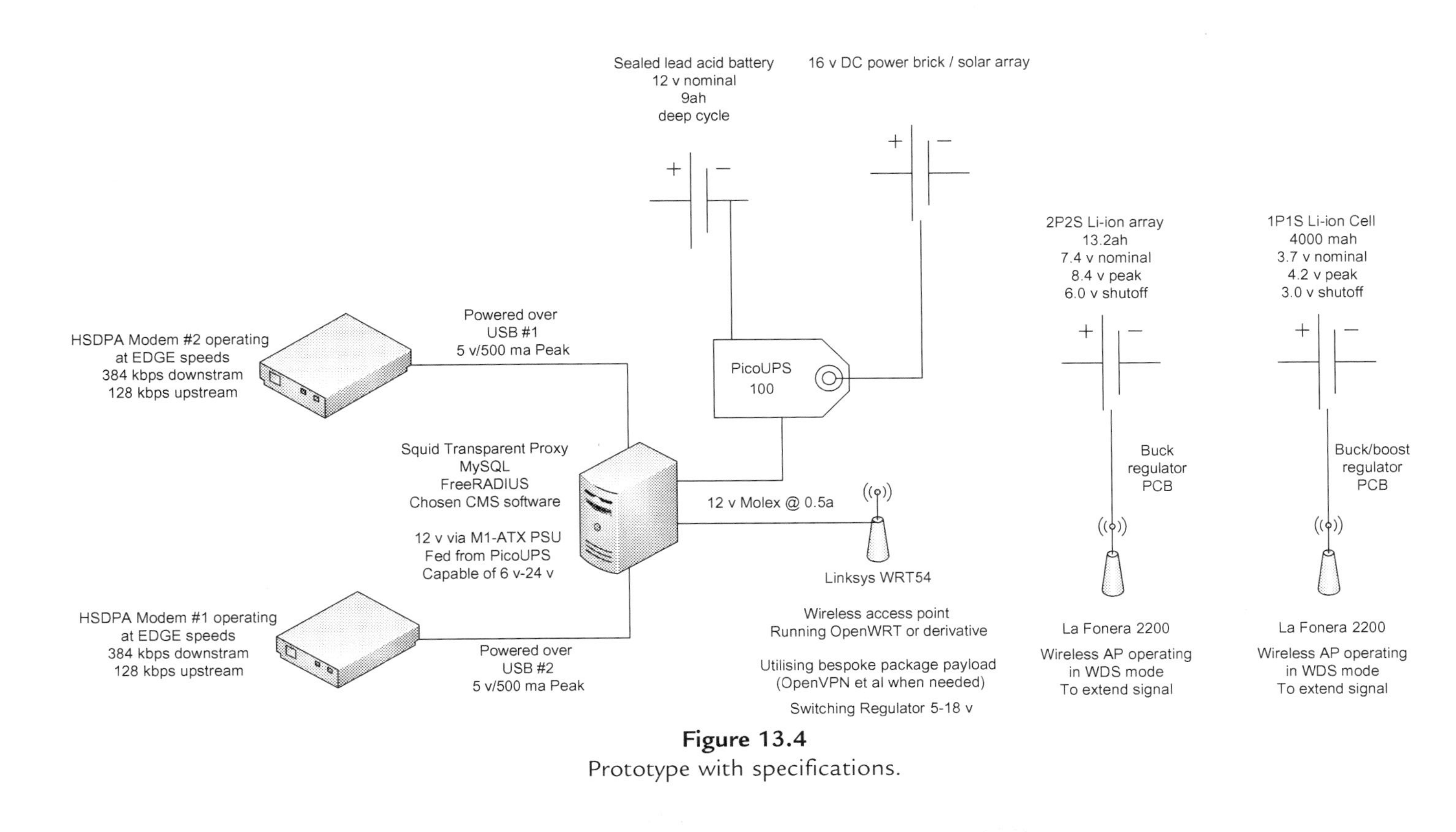

Figure 13.4
Prototype with specifications.

Langdale Pilot Study

Three pilot studies were conducted for this project (see Pattinson et al., 2010). However, only the pilot study for the Langdale Outdoor Activity Adventure (held on October 12, 2009) will be discussed here. The trial took place from September 28 to October 2, 2009 and involved the Carnegie Mellon faculty and their Outdoor Adventure activity undergraduate module, which was voted module of the year 2008 by the University. The trial location was High Close, Loughrigg, Ambleside, Cumbria LA22 9HJ (latitude = 54 26′18.53 N; longitude = 3 01′17.07 W; elevation = 583 ft.).

Requirements

There was a need to support ~60 students on a week-long residential course in Langdale (a complete module). Facilities were provided to allow students to access the module learning materials and skills for learning materials that helped guide students for their presentation preparation that took place at the end of the module. Students were permitted to bring their own IT equipment (e.g., laptops) of use those provided by the University. In the previous year of delivery (2008), students requested technology support and online assistance. Thus, students were allowed access to the materials throughout their studies. Additional information about the actual location of the residential course and its layout was obtained in order to ascertain a smooth deployment. However, it was noted that factors affecting the X-Stream's performance were X-Stream's access traffic (in its "normal" mode), the location, and the layout.

Core Equipment

1. Kit used for the Langdale trial start date (September 28, 2009): the Option Icon 505,[28] HSPA adapter from Easy Devices,[29] and the antenna/pigtail assemblies from MS (Distribution).[30] The trial was carried out with existing testbed hardware supplemented by the Icon 505 adapter.
2. The majority of testbed hardware was privately owned: (1) 3UK WAN access and Huawei E160G dongle,[31] (2) Vodafone WAN access, (3) 1 × Advent 4211 netbook (identical architectural specs to proposed equipment), (4) 1 × WRT54G v2.0 running DD-WRT v24, (5) 1 × RJ45 Cat 5E Cable, (6) 1 × Huawei E160G HSPA adapter (locked to 3), and (7) 1 × Option Icon 505 HSPA adapter.

---

[28] http://www.option.com/product/icon-505/.

[29] http://easydevices.co.uk/.

[30] http://www.msdist.co.uk/.

[31] http://www.panorama-antennas.com/shop/3g_networks/3/E160G/.

3. Testbed equipment used:
   a. In order to maintain consistency, netbooks with the same technical specs were used throughout the entire experiment. The only difference is the electrical level implementation which is a concern only for outdoor/off-the-grid deployments.
   b. The WLAN AP was lower powered and so was not able to support as many concurrent users or greedy routing applications. The radio had a lower effective range as a result of the inferior receipt sensitivity. The project plan originally conveyed a requirement of 60 concurrent users that did not pose a problem. The 60 students were split into nine groups of 6-7, and together with staff laptops, gave a maximum of 12 concurrent clients.
   c. Net access and intranet functionality did not require additional running processes or demanding protocol routing.
   d. The low gain (2.2 dBi) antennas[32] included with the WRT54G had been well suited to a multifloor deployment.

Client Equipment

1. The client hardware was provided by the Carnegie Mellon University Media Loans Department. Staff in the faculty were briefed that to use the service, students would require a device fitted with an 802.11g capable WLAN card as well as a modern Web browser.
2. Students were told that they were free to bring their own laptops at their own risk. In previous years, they tended to avoid supplying their own terminals, and this trend continued with only two users bringing their own kit. These two laptops were a 2009 Macbook and a Dell Inspiron 1545. Note that users attempted to use the service with their smartphones, which represented an interesting shift in the expected access pattern.
3. The Carnegie Mellon faculty provided four Toshiba P100-222 laptops.
4. The Media Loans Department provided three Toshiba Satellite Pro A60 laptops and several digital cameras.

Measuring Cellular Backhaul Signal Strength

1. A number of measurements had been taken with various packages for both Windows 7 and Ubuntu 9.04. They were (1) Option GlobeTrotter Connect,[33,34] (Win7-Locked to Option dongles), visual and received signal strength indicator (RSSI),[35] (2) UMTSmon[36] (Ubuntu,

[32] http://martybugs.net/wireless/rubberducky.cgi.

[33] http://www.softpedia.com/get/Network-Tools/Misc-Networking-Tools/GlobeTrotter-Connect.shtml.

[34] http://www.comparecellular.com/images/phones/userguide1007.pdf.

[35] http://networkengineering.stackexchange.com/questions/1673/what-is-the-meaning-of-rssi.

[36] http://umtsmon.sourceforge.net/.

FOSS)—Visual Only, (3) NetworkManager,[37] (Ubuntu, FOSS)—Visual Only, and (4) Minicom[38] (Ubuntu, FOSS).
2. MWconn[39] had been a prime candidate for inclusion, but unfortunately configuration could not be finalized in time. It has since been established that this piece of software probes signal strength using AT+CSQ.[40]
3. HSOConnect[41] was not used because it would have negated NetworkManager and, based on its documentation, because it was clear that it used AT+CSQ to probe signal strength. Please note there was no standard visual representation of the signal strength metric.
4. Option GlobeTrotter Connect was provided by Option Wireless Technology, which worked only with Windows machines.
5. Telco agnostic was locked to Option device so it could be tested only with the Icon 505.
6. Display's visual cue and RSSI reading had not been captured because they were supported by Minicom, which is also provided by Telia, Sweden.

UMTSmon

UMTSmon[42] is a basic Linux client for connecting to cellular networks. It also provides for profiles and so on. It is really useful before enhancing the level of integration under Linux. Arguably, it still offers great features such as SMS management, although this was irrelevant to the project. Its technical data output is not in alphanumeric form. However, it was a problem getting UMTSmon to work correctly with Vodafone. In reflection, this could have been as simple as the Access Point Name, which could be automatically detected under UMTS but required input for Global System for Mobile (GSM) Communications.[43]

NetworkManager

The NetworkManager[44] sponsored by Red Hat is a cornerstone of Linux on the desktop. However, there has been a conflict between this and WICD[45] (an open source wired and wireless network manager for Linux) and the latter is expected to replace the former. A huge database of compatible adapters, telcos, plans, and their associated configuration details have been built. Both Vodafone and 3 can be set up via a wizard without requiring specialist knowledge and information about their active connection is shown in Figure 13.5.

37 https://wiki.gnome.org/action/show/Projects/NetworkManager?action=show&redirect=NetworkManager.
38 http://linux.die.net/man/1/minicom.
39 http://www.mwconn.net/MWconn_Manual.pdf.
40 http://m2msupport.net/m2msupport/atcsq-signal-quality/.
41 https://wiki.archlinux.org/index.php/Hsoconnect.
42 http://umtsmon.sourceforge.net/.
43 http://www.gsma.com/aboutus/gsm-technology/gsm.
44 https://wiki.gnome.org/Projects/NetworkManager.
45 http://wicd.sourceforge.net/.

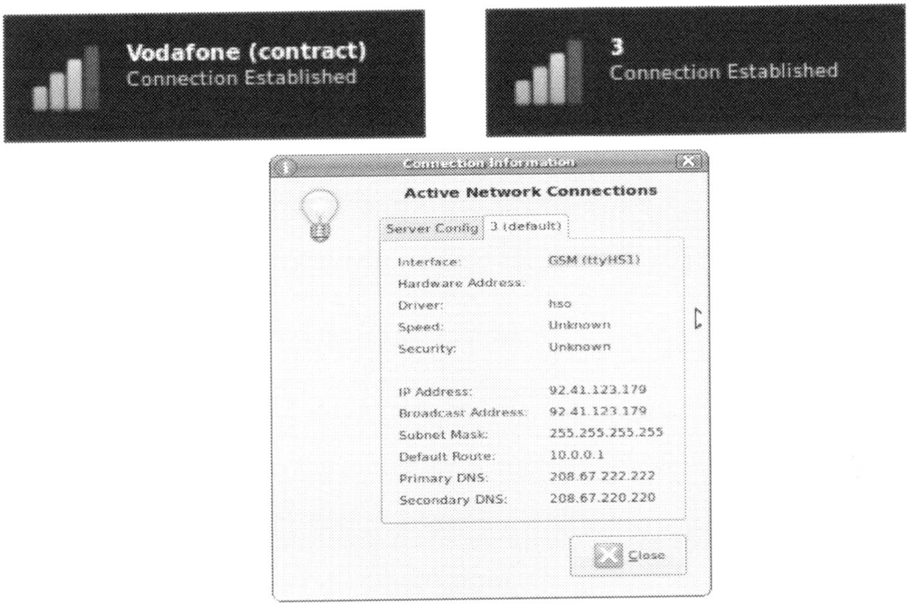

Figure 13.5
Connection information.

The lead developer's blog[46] makes an excellent read regarding Linux WWAN development. The native Ubuntu 9.04 RC with NetworkManager Applet 0.7.0.100 was the network management system chosen for this project. However, a problem that incurred was that signal strength was always shown as ¾ bars regardless of actual strength and actually provided this datum only at connect time.

Minicom[47]

1. Minicom is a communication program that somewhat resembles the shareware program TELIX[48] but is free with source code and runs under most unices. It allowed us to open a text-based communication channel directly through a serial port that was enumerated by the modem.
2. Most operator front ends simply interpreted the output of extended commands that could be issued through a Minicom session.
3. Hyperterminal, which came with Windows XP, could be used as a substitute for Minicom. It is important to note that Windows uses different port-naming conventions (i.e., COM instead of tty). The command that the project focused on was AT+CSQ, which provided

[46] http://blogs.gnome.org/dcbw/.

[47] http://linux.die.net/man/1/minicom.

[48] http://web.teipir.gr/STDN/configur.html.

output for the RSSI) and bit error rate (BER), respectively. It provided a more granular view than all the visual interpretations, which merely interpreted this data. The first number represented signal strength above −113 dB. Numbers of 10 or higher (−93 dBm) were considered "acceptable." The second number represented the BER, 0: less than 0.2%, 1: 0.2-0.4%, 3: 0.8-1.6%, 4: 1.6-3.2%, 5: 3.2-6.4%, 6: 6.4-12.8%, and 7: more than 12.8%. Some modems provided output "99" to BER if they did not support the command or if the value was null. This seemed to be true for Huawei.

Configuring Minicom (Figure 13.6)

Information for configuration follows:

1. We had to instruct Minicom to communicate with our modem on its communications/ diagnostic port, as opposed to the data port.
2. Ports were enumerated as/**dev/tty*** and could vary from manufacturer to manufacturer. On the Huawei E169, two (ttyUSB0 and ttyUSB1) were created. However, four on the Option Icon 505 were created: ttyHS0 (control port), ttyHS1 (application port), ttyHS2 (diagnostics port), and ttyHS3 (modem port). Once the modem was plugged in, the ports could be listed by issuing **ls/dev/tty***. If the ports were not visible, the device was stalling at the mass storage or CD detection stage.
3. Minicom ought to be launched with the –s switch so as not to load any presets.
4. Minicom required root privileges. Once launched, the user was presented with a configuration menu.
5. It was necessary to enter the serial port setup area and amend appropriately. Note the use of the diagnostic port with the Option icon.

Using Minicom[49]

The process is as follows:

1. Hit Exit and the modem will initialize.

```
+-------------------------+
|                         |
|    Initialising Modem   |
|                         |
+-------------------------+
```

2. AT + GMI outputs the manufacturer.
3. AT + GMM outputs the model.
4. AT + CSQ provided outputs for the RSSI.
5. Many other commands could be inputted but were irrelevant for this review.

[49] http://www.cyberciti.biz/tips/connect-soekris-single-board-computer-using-minicom.html.

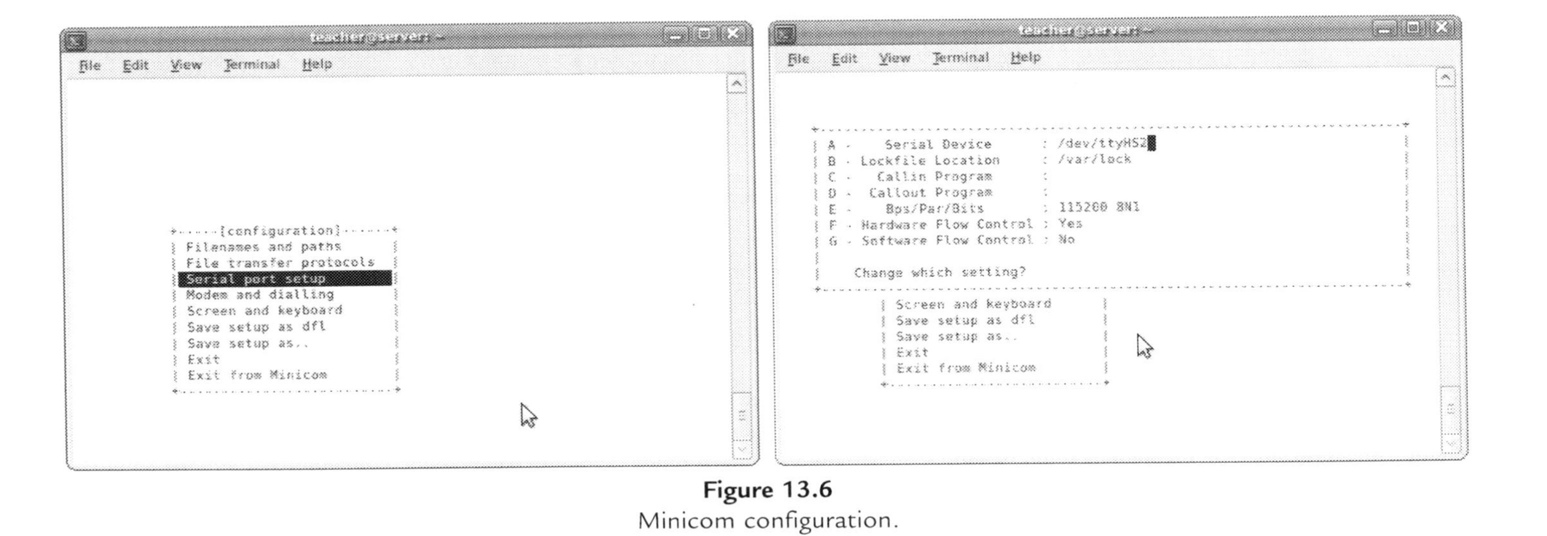

Figure 13.6
Minicom configuration.

AT+CSQ/RSSI Lookup (Table 13.3)

Table 13.3 Signal Strength[a]

Mode	Description
Marginal	Levels of −95 dBm or lower. At these levels, it is very likely that the throughput is low and that disconnection occurs because of cell loading/breathing even with an outdoor antenna
Workable under most conditions	Levels of −85 to −95 dBm. Probably it is worth considering an outdoor gain type antenna. Could suffer poor throughput and disconnections because of cell loading/breathing
Good	Levels between −75 and −85 dBm. Typically, there is no problem holding a connection at this level (even with cell breathing) without the use of an external antenna
Excellent	Levels above −75 dBm. These should not be affected by cell breathing/loading and should not require an external antenna

[a]http://mybroadband.co.za/vb/showthread.php/61703-How-to-check-Signal-Strength.

Measuring WAN Bandwidth/Latency

1. Speedtest.net[50] (Manchester server) handled bandwidth measurements.
2. Pings[51] of three key servers handled latency.
3. The key difference was the access method when Vodafone used GPRS whereas 3UK used HSPA.
4. All tests were performed directly from the local server. Previous testing had confirmed that clients connected via the server experienced neither speed loss nor a latency increase other than that of the wireless traversal.

Measuring WAN Bandwidth

Three tests were performed between 11.00 PM and 11.30 PM on September 9, 2009 (this time slot was selected because the network was not expected to be loaded). Results of the WAN bandwidth measurement are shown in Figure 13.7.

[50] http://www.speedtest.net/.

[51] http://www.pingtest.net/.

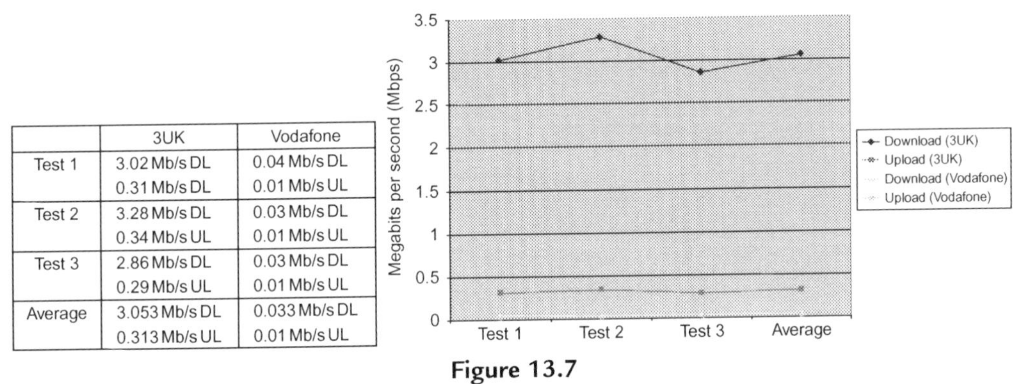

	3UK	Vodafone
Test 1	3.02 Mb/s DL 0.31 Mb/s DL	0.04 Mb/s DL 0.01 Mb/s UL
Test 2	3.28 Mb/s DL 0.34 Mb/s UL	0.03 Mb/s DL 0.01 Mb/s UL
Test 3	2.86 Mb/s DL 0.29 Mb/s UL	0.03 Mb/s DL 0.01 Mb/s UL
Average	3.053 Mb/s DL 0.313 Mb/s UL	0.033 Mb/s DL 0.01 Mb/s UL

Figure 13.7
WAN bandwidth results.

Measuring WAN Latency

1. Three tests were performed between 12 PM and 12.30 PM on September 9, 2009 (this time slot was selected because the network was not expected to be loaded). Results are depicted in Figure 13.8.
2. OpenDNS servers (secondary) were used for original Google lookup. DNS Cache was flushed before lookup.

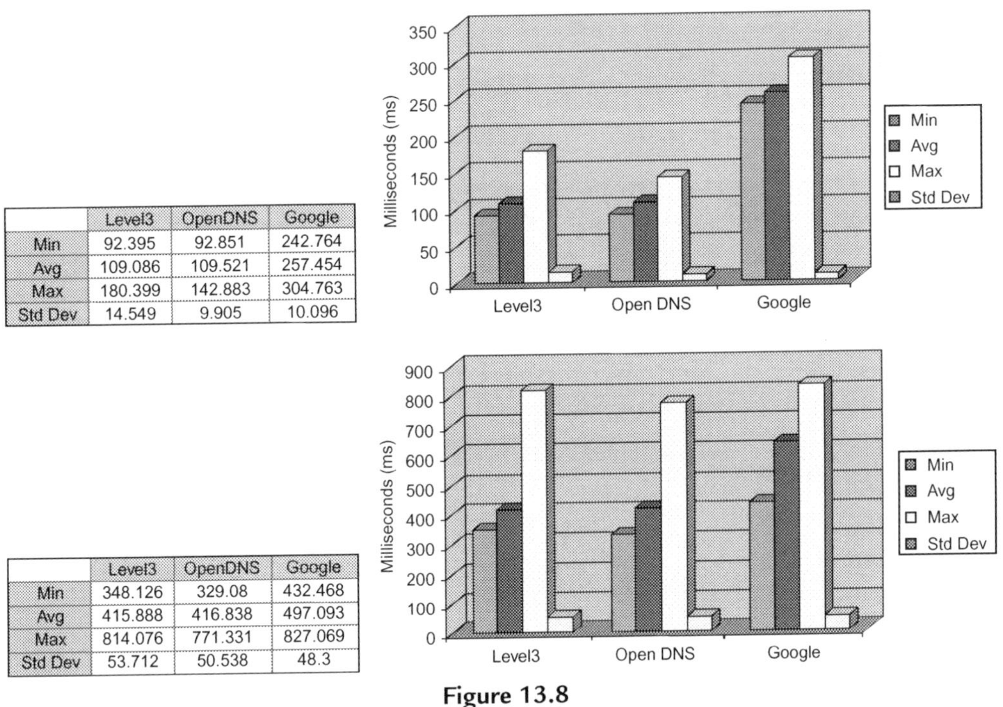

	Level3	OpenDNS	Google
Min	92.395	92.851	242.764
Avg	109.086	109.521	257.454
Max	180.399	142.883	304.763
Std Dev	14.549	9.905	10.096

	Level3	OpenDNS	Google
Min	348.126	329.08	432.468
Avg	415.888	416.838	497.093
Max	814.076	771.331	827.069
Std Dev	53.712	50.538	48.3

Figure 13.8
WAN latency results.

3. OpenDNS server and Level3 DNS server had been chosen for direct ping-to-IP tests.
4. Both used the anycast[52] protocol, so a geographically closest parent was selected.

WAN Backhaul Conclusion

1. The quality of HSPA connection though 3UK was not expected:
   a. Existing research (during the project life cycle) had no evidence of this phenomenon.
   b. During the study, it was hard to determine which cell we were connected to.
   c. The mast sharing agreement with T-Mobile has possibly helped in the study.
   d. Ofcom Sitefinder was updated only quarterly; thus, the coverage maps were possibly not up to date.
   e. Surprisingly, 3 had not done any promotion through its coverage checker.
   f. There was a possibility that better HSPA coverage in the Loughrigg Fell area depended on the radio propagation through the line of sight down to that area.
2. The Vodafone GPRS service was not acceptable:
   a. The received signal strength was respectable by virtue of being near its mast.
   b. The service was not Enhanced Data rate for GSM Evolution[53] (EDGE).
   c. Basically, it had been far too slow for general Web access, although all was well when cached sites were accessed.
3. The hostel operator had been contacted to feed back the preceding findings of the study.

WLAN Planning

1. The construction of YHA Langdale (see Figure 13.9) was unfavorable for installation of a WLAN for the following reasons: (1) walls are mostly 2 ft thick, (2) heavy stone was used for construction; (3) doors closed automatically, and (5) AP was placed at the far end of the building to accommodate the cellular link. Results of the site survey are shown in Figure 13.10.
2. Ekahau heatmapper[54] had been used to create an approximate ground floor coverage map.
3. inSSIDer Wi-Fi network scanner[55] is a free wireless troubleshooting tool for both XP and Vista; Vistumbler[56] is a wireless network scanner for Vista. Both facilities had been used for additional measurements. Both produced reports that no background WLANs were operating in a/b/g modes when client hardware was used.

[52] http://searchnetworking.techtarget.com/definition/anycast.

[53] http://www.gsmarena.com/glossary.php3?term=edge.

[54] http://www.ekahau.com/wifidesign/ekahau-heatmapper.

[55] http://inssider.en.lo4d.com/.

[56] http://www.vistumbler.net/.

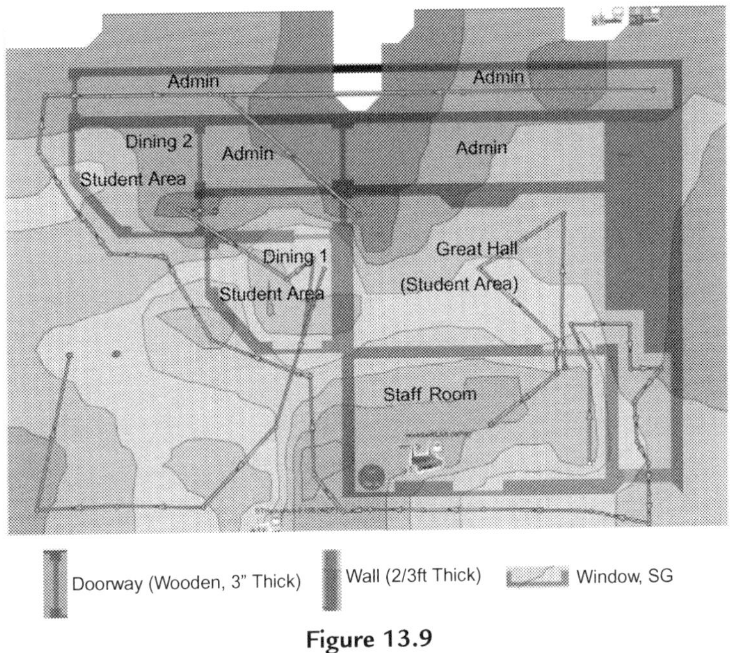

Figure 13.9
Layout of the Langdale YHA.

4. 70 mW output is widely regarded as the safe limit for WRT boxes, and 65 mW had actually been used to meet the UK equivalent isotropic radiated power (EIRP) limit.[57]
   a. A 2.2 dBi antenna is standard, and because RP-TNC was used instead of R-SMA, the standard antenna could not be swapped for the 5 dBi ones that had been purchased for the project.
   b. Total net EIRP (including 0.5 db connector/cable loss) = 96.142 mW.

WLAN Propagation Losses

To reiterate, the thick walls dampened the signal, and the measurements had to be taken with additional tools inside the staff room right next to the access point and in the great hall just on the other side of the staff room. The inSSIDer and Vistumbler were wireless network scanners for XP/Vista/7. The latter was akin to Netstumbler but with updated detection strategies to remain compatible with newer OSs. The results of the study were that inSSIDer reported a 10 dB loss, and Vistumbler reported a 14 dB loss.

Squid Proxy—Control Measures

Squid worked very well, although when a WAN connection had been disconnected and subsequently reconnected, it required restarting. Squid had also been used to track client use

[57] http://stakeholders.ofcom.org.uk/binaries/consultations/powerlimits/summary/powerlimits.pdf.

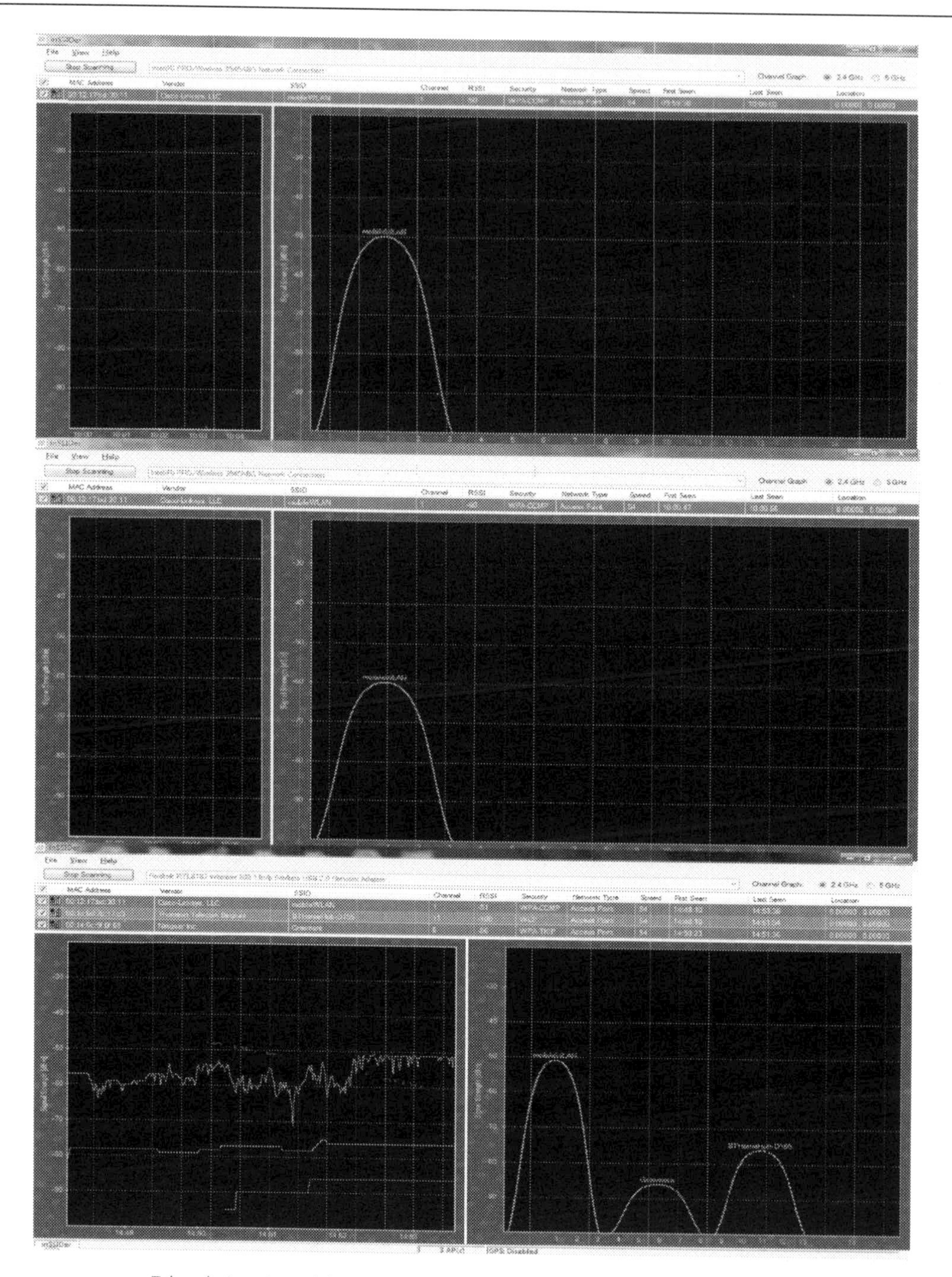

Taken during a tour of the premises, both internal and external over a 10 min period
Realtel RTL8187 in mixed g/b mode — 500 mW & 9 dBi omnidirectional antenna — 2422 mW EIRP(!)

Figure 13.10
Site survey.

and ensure that it did not exceed acceptable levels: (1) a catchall rule of 300 MB/day has been used to prevent clients from exceeding the 5 GB cap assigned to the 3 contract and (2) as a secondary measure, 3UK has a relatively accurate (not real-time) Web-based usage checker. Calamaris had been the intended package to retrospectively create usage reports. However, Squid, or a command issued to it, seemed to have destroyed the relevant log files. This was an issue that was further analyzed to determine future courses of action. However, evidence of the proxy usage is shown in Figure 13.11.

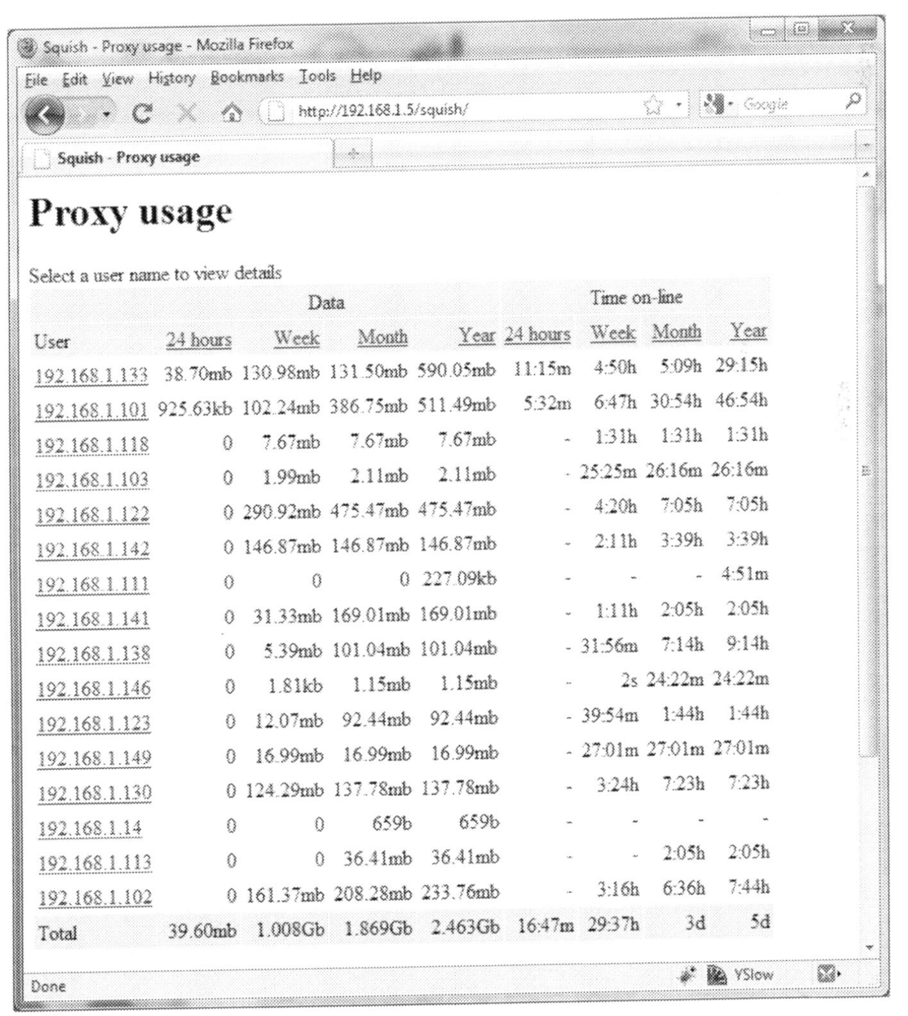

Proxy usage

Select a user name to view details

	Data				Time on-line			
User	24 hours	Week	Month	Year	24 hours	Week	Month	Year
192.168.1.133	38.70mb	130.98mb	131.50mb	590.05mb	11:15m	4:50h	5:09h	29:15h
192.168.1.101	925.63kb	102.24mb	386.75mb	511.49mb	5:32m	6:47h	30:54h	46:54h
192.168.1.118	0	7.67mb	7.67mb	7.67mb	-	1:31h	1:31h	1:31h
192.168.1.103	0	1.99mb	2.11mb	2.11mb	-	25:25m	26:16m	26:16m
192.168.1.122	0	290.92mb	475.47mb	475.47mb	-	4:20h	7:05h	7:05h
192.168.1.142	0	146.87mb	146.87mb	146.87mb	-	2:11h	3:39h	3:39h
192.168.1.111	0	0	0	227.09kb	-	-	-	4:51m
192.168.1.141	0	31.33mb	169.01mb	169.01mb	-	1:11h	2:05h	2:05h
192.168.1.138	0	5.39mb	101.04mb	101.04mb	-	31:56m	7:14h	9:14h
192.168.1.146	0	1.81kb	1.15mb	1.15mb	-	2s	24:22m	24:22m
192.168.1.123	0	12.07mb	92.44mb	92.44mb	-	39:54m	1:44h	1:44h
192.168.1.149	0	16.99mb	16.99mb	16.99mb	-	27:01m	27:01m	27:01m
192.168.1.130	0	124.29mb	137.78mb	137.78mb	-	3:24h	7:23h	7:23h
192.168.1.14	0	0	659b	659b	-	-	-	-
192.168.1.113	0	0	36.41mb	36.41mb	-	-	2:05h	2:05h
192.168.1.102	0	161.37mb	208.28mb	233.76mb	-	3:16h	6:36h	7:44h
Total	39.60mb	1.008Gb	1.869Gb	2.463Gb	16:47m	29:37h	3d	5d

Figure 13.11
Evidence of proxy usage.

Squid Proxy—Policy

The service was set up to form part of an adult learning environment. Thus, there had been no block or filter imposed by the project at access level. However, it became apparent that users were utilizing the increased bandwidth of the 3 service to browse social networking sites. Blocks had been integrated through Squid access controls and seemed to be sufficiently robust. Other more "parental" filters could be integrated through OpenDNS.

Feedback on the Mobile Learning System

Feedback (by Module Leader)

The feedback given by module leader was as follows:

1. The hardware was thought to be good although he would like to see it developed so it really would be easy enough for a staff member to use after a short staff development session.
2. The module had been enhanced through the provision of a rich intra and inter Web content that otherwise would not have been easily accessible.
3. In the future, it would be good to allow the staff and students to start using the system earlier to familiarize themselves with it. In doing so, they could be persuaded to add content to the system, thus efficiently exploiting their knowledge. Additionally, an additional step could be taken by considering the integration of the use of IT into the marking scheme.
4. There were certainly more opportunities available to use the system for other modules.

Feedback (Tutors)

The feedback given by two tutors were as follows:

1. The system had been utilized, and it has been great that the whole preloaded "bank of information" had been made accessible. It has also enhanced the delivery of the module.
2. Staff would like to rely on the accessibility of this material and thus be able to simply direct pupils to the relevant part of the site during other forms of feedback.
3. This module in particular was probably too busy in terms of activities and other demands on student time and deterred them from making the most of the collaborative facilities that had been offered. It would perhaps benefit students more if they had been exposed to this system before the module began.
4. A one-page "crib sheet" would be useful because nontechnical staff would like to use the hardware in the future.
5. A huge cost benefit was bringing data locally. In this module, there were many rich teaching resources that use imagery, video, and audio. Serving these remotely would have cost a lot financially.

6. Students probably needed more precise advice regarding bringing their own equipment.
7. It would be great, but probably quite technically difficult, to implement some kind of transparent backup system. Many users had lost or corrupted their work through bad practice. When a member of IT staff was not available on-site, these users would simply have to start from the beginning again.
8. Coverage in mobile phone signal differed from what they had been told. Inconsistencies in coverage maps were inconvenient.

Conclusions

The discussion for this section addresses three strands: student experience, methodology evaluation, and future work. First, the residential week had been very useful and fruitful. However, it had been rather difficult to engage with the students and find the time to perform analysis on the technical work for the following reasons: (1) the student activities were staggered, so the building was empty for only a short period every day, (2) students had not made full use of the interactive elements of the blog because of the lack of motivation, (3) the repository had been used by students, (4) it was suggested that the module delivery pattern be changed in future years and that students should be able to access their module marks directly, (5) although the backhaul could be considered a success, valuable lessons had been learned during the planning process, (6) conducting a follow-up study when the specialist project materials had been sourced, configured, and finalized would be worth considering, and (7) the building itself (in the project) was a case study of a challenging Radio Frequency (RF) environment.

As for methodology evaluation, gathering IPs would have allowed the researchers to perform Geo lookups. Looking back at the speed test results, the Vodafone connection had routed over a much larger distance than 3. Traceroutes[58] should have been performed both locally and internationally to ascertain route selection. When data provision and scripting were available, it would have been interesting to see how the networks perform at different times of day. The performance of GPRS was so bad that a new speed testing service should have been be selected to offer further granularity. An alternative was to establish a Transmission Control Protocol (TCP) or User Datagram Protocol(UDP) throughput[59] using iPerf,[60] although this would require a dedicated server. It would have been helpful to track cellular RSSI at various times of the day to analyze the effects of cell breathing. Since the trial, methods had been developed internally to estimate the values of RSCP and Ec/Io. Netstumbler[61] should have been used in preference to both inSSIDer and Vistumbler because it probes at a lower level in the hardware

---

[58] http://navigators.com/traceroute.html.

[59] http://www.laynetworks.com/Comparative%20analysis_TCP%20Vs%20UDP.htm.

[60] http://sourceforge.net/projects/iperf/.

[61] http://www.netstumbler.com/downloads/.

stack and gives more precise results. Alternatively, Kismet[62] could have been used, but this would have required a modification of the default configuration to suit the available hardware and preferences. Although Ekahau[63] was a valuable tool, more time should have been allotted for a more thorough through analysis with it on the ground floor and other floors as well. This could be supplemented by signal measurements at each point to give meaning to colors.

Future development should incorporate more robust dial-up or watchdog system for the WAN and consider possible migration to HSOConnect[64] or Command Line Interface (CLI)[65] for NetworkManager (e.g., cnetworkmanager[66]). A better filtering method should be integrated (e.g., Dansguardian[67]) but subject to overheads. Automated scripting for restarting core processes should be collated, and more investigation into RF mapping procedures/approaches should be conducted, both for cell and WLAN models. A comparison should be made for multiple client equipment types (i.e., different WLAN cards or antenna setup). A data cap or limiting method should be applied to each individual user. However, another alternative is simply to control individual nodes using a combination of Squid and Dynamic Host Configuration Protocol (DHCP).[68] Calamaris[69] could be employed to conduct detailed Squid log analysis. Finally, it would be useful to compare back-end Wired Ethernet and 802.11 g.[70]

References

Braddock, R., Pattinson, C., 2009. Bridging the community network gap with FOSS and mobile ISPs. In: IEEE Proceeding of the Third International Conference on Next Generation Applications, Services and Technologies, pp. 509–514. http://dx.doi.org/10.1109/NGMAST.2009.69.

Cisco, 2006a. How Does the Radius Work? http://www.cisco.com/c/en/us/support/docs/security-vpn/remote-authentication-dial-user-service-radius/12433-32.html, (accessed 15.05.14.).

Cisco, 2006b. Understanding Simple Network Management Protocol (SNMP) Traps. http://www.cisco.com/c/en/us/support/docs/ip/simple-network-management-protocol-snmp/7244-snmp-trap.html, (accessed 15.05.14.).

Cisco, n.d.-a. http://www.cisco.com/c/en/us/solutions/enterprise-networks/802-11n/index.html, (accessed 15.05.14.).

Cisco, n.d.-b. http://www.cisco.com/c/en/us/td/docs/wireless/access_point/12-4_10b_JA/configuration/guide/scg12410b/scg12410b-chap6-radio.html, (accessed 15.05.14.).

Clausen, T., Jacquet, B., 2003. Optimized Link State Routing Protocol (OLSR). http://www.ietf.org/rfc/rfc3626.txt, (accessed 15.05.14.).

[62] http://www.kismetwireless.net/.

[63] http://www.ekahau.com/wifidesign.

[64] http://peck.org.uk/Downloads.html?relPath=Debian_Packages.

[65] http://docs.fedoraproject.org/en-US/Fedora/13/html/User_Guide/sect-User_Guide-Connecting_to_the_Internet-NM_CLI.html.

[66] http://manpages.ubuntu.com/manpages/maverick/en/man1/cnetworkmanager.1.html.

[67] http://dansguardian.org/.

[68] http://www.isc.org/downloads/dhcp/.

[69] http://cord.de/calamaris-home-page.

[70] http://www.pcbuyerbeware.co.uk/Networking-Wireless-Networks.htm.

Emulex, 2013. Emulex® Drivers for VMware ESXi 5.5 Migration Guide. http://www-dl.emulex.com/support/elx/rt960/b122a/esxi5.5_guide_elx.pdf, (accessed 15.05.14.).

Feilner, M., 2006. Openvpn: Building Virtual Private Networks. PACKT Publishing, Birmingham. ISBN: 1-904811-85-X.

Ghribi, B., Logrippo, L., 2000. Understanding GPRS: the GSM packet radio service. Comput. Netw. 34, 763–779.

Hennessy, H., Jouppi, N., Baskett, F., Gill, J., 1981. MIPS: a VLSI processor architecture. In: VLSI Systems and Computations, pp. 337–346.

Holma, H., Toskala, A., 2006. HSDPA/HSUPA for UMTS: High Speed Radio Access for Mobile Communications. John Wiley and Sons, Chichester, ISBN: 978-0-470-01884-2.

McGraw-Hill Companies, 2000. Chapter 23: Simple Network Management Protocol. http://medusa.sdsu.edu/network/CS596/Lectures/ch23_SNMP.pdf, (accessed 15.05.14.).

Mitta, M., 2009. Hosts liable only for own content on Net. Times of India New Delhi Edition. 26th February, 2009, p. 16, http://timesofindia.indiatimes.com/india/Hosts-liable-only-for-own-content-on-Net/articleshow/4192635.cms, (accessed 15.05.14.).

Murthy, C.S.R., Manoj, B.S., 2004. Ad Hoc Wireless Networks Architectures and Protocols. Prentice Hall, NJ, USA, ISBN: 978-0-13-147023-1.

Ncomputing, 2010. Server Virtualization Infrastructure Deployment Guide: Deploying NComputing L-series and vSpace on Virtual Machines. http://www.ncomputing.com/kb/afile/168/188/, (accessed 15.05.14.).

OpenDNS Security Labs, 2013. How OpenDNS Achieves High Availability with Anycast Routing. http://labs.opendns.com/2013/01/10/high-availability-with-anycast-routing/, (accessed 15.05.14.).

Pattinson, C., Braddock, R., Kor, A.L., 2010. JANET UK wireless LAN project. Technical report June 2010, disseminated to JANET UK as a user manual (for managers and engineers). ISBN:978-1-907240-32-4.

Ramos, M., Brewer, E., n.d. The tier solar controller, http://openarchitecturenetwork.org/system/files/TIER%20solar%20controller%20research.pdf, (accessed 15.05.14.).

University of California, Berkeley, 2010. Power. http://tier.cs.berkeley.edu/drupal/power(accessed 15.05.14.).

VMWare, 2007. The Architecture of VMware ESXi. http://www.vmware.com/files/pdf/ESXi_architecture.pdf, (accessed 15.05.14.).

CHAPTER 14

Thin-Client and Energy Efficiency

Colin Pattinson, Roland Cross, Ah-Lian Kor*
Leeds Beckett University, Leeds, UK
**Corresponding author: a.kor@leedsbeckett.ac.uk*

Introduction

A *thick client* is a personal computer with integral disk storage and local processing capability. It also has access to data and other resources via a network connection and is accepted as *the* model for providing computing resources in most office environments. The Further and Higher Education sector is no exception to that, and therefore most academic and administrative offices are equipped with such desktop computers to support users in their daily tasks. This system structure has a number of advantages: There is a reduced reliance on network resources, users access a system appropriate to their needs, and they may customize "their" system to meet their own personal requirements and work patterns. However, the system also has disadvantages: Some are beyond the scope of our JISC project, but of most relevance to the green IT agenda is the underuse (in terms of processing power) of relatively complex and expensive (in first cost and in running cost) desktop systems and servers. Whereas some savings are achieved through use of power management mechanisms, in most configurations, only a small portion of the overall total available processor resource is used. As a matter of fact, enterprises are seen to repeatedly buy computing resources that are beyond their requirements, thus increasing their carbon profiles (Hayes, 2010). Consequently, this has led to the emergence of an alternative paradigm, the thin client. In a *thin-client* system, the desktop is stripped of most of its local processing and data storage capability and essentially acts as a terminal to the server, which now takes responsibility for data storage and processing. The energy benefit is derived through resource sharing when a shared server process can service multiple user processes. In summary, the advantages offered by this design paradigm are lower hardware and administration costs, lower energy consumption, and a more efficient use of resources (Staehle et al., 2008). However, according to Hayes (2010), even though thin clients require fewer complex components and could possibly affect 25% or more power savings, their use could result in the need for additional resources at the server end.

Currently, there is limited primary research that involves measured data comparing thin- and thick-client systems in operation in similar real situations that allow direct and fair comparisons to be drawn. The main goal of our JISC-funded project was to address this issue by investigating the relative performance of thin- vs. thick-client systems within a higher education (HE) environment. The project team consisted of multiple stakeholders across Leeds Beckett University (academic researchers, technical services, and estates division), and the project was an extension of work undertaken for the joint development and implementation of the university's green IT strategy[1] (presented in a Salix technical workshop in 2010). This strategy is part of the university's overall commitment to sustainability, which has consistently placed it in the top 10 in the annual People and Planet University League's table of green universities[2]. The school was named the greenest university in 2008 by the *Sunday Times*[3] for being the first English university to achieve the International Organization Standard (ISO) 14001 for environmental management. Additionally, the JISC project was built on previous work undertaken at Leeds Metropolitan University (Pattinson and Siddiqui, 2008) that explored the real costs and benefits of alternative technology implementations. For the purpose of our research, variations of thin and thick clients have been considered. First, it is not uncommon for thin-client deployments to continue to use their existing PCs as thin-client workstations with or without modification. Also, attempts by PC makers to reduce the power requirements of their products have given rise to a further variation: the incorporation of low power features in otherwise standard PC technology, working as thick clients.

Aims and Objectives

The principal aim of the JISC thin-client project was to carry out direct measurements of energy consumption (electrical power) on a sample set of users whose patterns of work were representative of a typical university environment. Each user was allocated one of the combinations of client systems with application support to permit them to carry out their normal workloads. We set out to monitor the *total* energy draw of this operation. Therefore, we identified and configured a server specifically to support the thin-client components of our service with their energy use monitored. During the project, we also investigated users' experience with the various systems. Although we aimed to develop a thin-client desktop identical to that of the thick client, we anticipated some possible change in user experience, whether real or perceived, and we were aware of the potential for resistance arising from a number of causes ranging from genuine performance issues through lack of opportunity for personalization of the computing environment to general resistance to change. We therefore liaised with users to gather their overall experiences of the project.

[1] http://www.eventlink.org.uk/uploads/DOCS2/53-LMU_Salix_Presentation.ppt.

[2] http://peopleandplanet.org/gogreen/greenleague2007.

[3] https://www.leedsbeckett.ac.uk/strategicplan/Leeds-Metropolitan_Strategic-Plan_2010-2015.pdf.

Literature Review

Green IT

Undeniably, energy efficiency, and what is often referred to as the "green agenda," has become a very significant part of the business and education landscapes internationally, nationally, and locally.[4] Organizations may adopt "green-ness" for a variety of reasons: to save money, to meet regulatory requirements, as part of their commitment to (corporate) social responsibility, in pursuit of a real desire to save the environment, or for the perceived marketing benefits; one outcome has been an upsurge of interest in ways of monitoring and reducing energy use.

In most organizations, the last 20 years have seen a domination of office IT provision by the PC o or under a desk, running applications in the device's processor and memory with data being stored either locally or in a networked server (a configuration termed *thick client*). Higher education and further education (HE/FE) teaching, research, and administrative functions are no different from many others in this respect. The typical office desk has a monitor and keyboard connected to a separate base unit comprising processor, hard disk, and associated hardware and software providing network connectivity, printing (remote and/or local), and the facility to attach universal serial bus (USB)-compliant devices, thus allowing easy connection of peripherals (input-output devices, video cameras, off-line storage, etc.). In almost all cases, these facilities are attached to a network, allowing data to be retrieved from, and stored on, servers as well as for data to be exchanged between/among users. The most typical means of operation in this setting is for a document (e.g., a report) to be retrieved from a network server and stored temporarily on the local machine while it is processed by a program (a locally resident word processor in this example). When the task is completed, the revised document is transferred back to the server for storage until it is required again. Some users opt to use their local disk for long-term data storage, moving the data across the network only when required by the need to share files with colleagues.

In essence, each user has a fully functional, stand-alone computer system with the added ability to access remote file storage (and possibly printing facilities) provided by a separate server device located elsewhere in the organization. Some combination of organizational policies and personal preference governs the mix of local and remote storage use by each user. This is the classic "workstation-server" model of computing, which has the advantages of being robust and reliable. However, this model also has disadvantages: most significantly, the computing resource at the desktop is underused, and the energy requirements (mainly for the operation of the hardware but also the cooling required) are significant. The tendency for most office PCs to

[4] http://www.naturaledgeproject.net/SustainableIT.aspx; Gabriel (2008); https://www.gov.uk/government/uploads/system/uploads/attachment_data/file/85968/uk-government-government-ict-strategy_0.pdf; http://oro.open.ac.uk/10677/1/paper5.RoyPotter%26Yarrow_IJSHEPaperJuly07.pdf.

remain switched "on" throughout the working day, albeit increasingly often in standby or sleep mode, means that the proportion of useful work to power drawn is coming under increasing scrutiny. Several power-saving initiatives have been developed, including sleep mode, where power is removed (or significantly reduced) on all components in the computer except for the memory and processor activity needed to detect a wake up signal (internal from the keyboard or externally via the network) and respond to it. Sleep mode is also referred to as standby or hibernate mode by some manufacturers. Whereas sleep mode is very effective in switching unused equipment away from a high-power "active" state, it obviously does not address the use of power while the system is awake.

More radical approaches involve a realignment of the processing activity so that more work (and/or data storage) takes place at a central location with users sharing the processing facilities, offering economies of scale available from more intensive resource use and delivering lower overall energy consumption for equivalent activities. The ultimate extension of this model is the cloud computing paradigm[5] by which very large data and server centers store and process data remotely and are accessed by the user via an Internet connection.

In the type of office computing environment widely employed in HE/FE, as well as in other business settings, the thin-client is the one shared processor technology that is attracting much interest. The goal of these systems is the transfer of the processing workload from often underused desktop computers (thick clients) to a configuration in which that work is carried out at a server whose processing capacity can be shared among a number of users who could access this resource via a simpler (and hence lower-powered) device. The availability of network bandwidth and the enhanced reliability of contemporary networks make it possible to rely on networked equipment in this way. The superficial appearance is of an updated version of the time-sharing, multiuser minicomputers[6] of an earlier age with much improved user interface and many additional capabilities.

Proponents (Greenberg et al., 2001) suggest that the overall benefits from this distribution of workload (in thin-client paradigm) include an overall energy saving and efficiency (Davis, 2008; Vereecken et al., 2010a,b). The majority of recent research work has focused on the potential for thin-client systems in mobile and wireless scenarios (Satyanarayanan et al., 2009; Miettinen and Nurminen, 2010; Simoens et al., 2010; Lange et al., 2011; Tang et al., 2011) in which energy efficiency is an important consideration. This would seem to support claims for the efficiency of thin-client methods. However, there are few hard data to confirm this, and one recent paper has suggested that the number of users needed to deliver any worthwhile improvement is very large (Abaza and Allenby, 2009). In particular, there is a lack of reliable data describing the specific energy requirements of such systems (either

[5] http://www-304.ibm.com/businesscenter/cpe/download0/158304/IBM_ForwardView_Cloud_computing.pdf.

[6] http://web.mit.edu/smadnick/www/MITtheses/02285292.pdf.

at the data center or the desktop) in the HE/FE environment. Furthermore, the impact of other factors, such as staff training, the consequences of a move to a more controlled desktop configuration, deployment costs, and the need to replace existing technology, is not fully understood (Doyle et al., 2009). Finally, the existence of gradations of "thin-ness" and the tendency to reuse erstwhile "thick" PCs as thin-client terminals mean that there is a need to look beyond the simple "thick vs. thin" dichotomy. It is necessary to identify the various characteristics that contribute to thin-ness and to determine whether some parts of thin-client technology are more or less effective in particular circumstances.

The classic study comparing thin and thick clients by the Fraunhofer Institute (UMSICHT 2006, 2008a,b, 2011) focused on the "environmental effects" of the whole life cycle of thin and thick implementations of a single system (for their full lives: manufacture, deployment, and disposal) and defined a set of "typical usage scenarios" based on levels of the use of standard applications. It does not specifically address the type of deployment typically seen in HE/FE when between 20 and 50 users (whether students in a laboratory session, academics preparing teaching and research material, or administrative workers processing student records) use the same application program set in a broadly similar way. The question of which applications are more or less appropriate to deployment on a thin-or thick-client platform is also unclear. Some reports have suggested that the thin-client is best suited to kiosk mode (details of configuration of such mode can be found here,[7,8]) applications (Doyle et al., 2009), but here again, there is a shortage of real data to back this up and very little of such data exists for the kind of operational requirements seen in HE/FE.

Comparison of Thick and Thin Client Power Consumption

Many research projects have been conducted to demonstrate a difference in power consumption of thin-client and thick- (or fat-) client PCs. Murugesan (2008) recommends thin-client computers to organizations because they draw only about one-fifth (approximately 20%) of the power of a desktop PC. On the other hand, the experiments commissioned by Wyse Technology Inc. (n.d.) to compare the power consumption of thin-client devices and personal computers reveal that the former was more energy efficient with some models using 85% less power than their PC rivals in real-world environments. Vereecken and colleagues (2010a) compared the energy consumption for thick clients (PC desktops, Dell OptiPlex360 [Intel Core 2 Duo E7400]) and thin clients (Wyse S10 devices) in three contexts (customer premise, the network, and the data center). For the network environment, three types of network technologies (ADSL2, VDSL2, and PON) are used to compare thin-client settings and typical desktop scenarios with thick clients (Vereecken et al., 2010b). The result has been

[7] http://h10032.www1.hp.com/ctg/Manual/c00120884.pdf.

[8] http://www.vmware.com/files/pdf/VMware-View-KioskMode-WP-EN.pdf.

consistent with significantly lower power consumption for the thin-client setup (their experiments reveal a power saving of approximately 67% of the desktop power consumption). Additionally, using ADSL2 is found to be of the greatest advantage whereas the power savings for VDSL2 or PON are not significant.

Project Reports on the Performance of Thin-Client Systems

Some of the project reports on the performance of thin-client systems have already been noted. However, in this section, we discuss the underlying data used in earlier work.

1. Queen Margaret University
   A 2008 report[9] describes an installation then recently undertaken at Queen Margaret University at its campus in Musselburgh, near Edinburgh. This is an almost wholly thin-client site (a small number of PCs were available for certain applications that "are difficult to run over the thin-client network"). It is also a location where subject areas are delivered "such as applied health sciences, [which] do not need massive amounts of computing power or data analysis." The measured energy of the thin-client systems is quoted as 25 watts (W), compared to 120 W for an "average" PC. The maximum number of thin clients per server is suggested as 40.
2. SusteIT
   The SusteIT report provides the most complete and comprehensive review of work in this field, and it would be superfluous to repeat that here. The associated spreadsheet makes use of data (energy use etc) "... obtained from vendors and correct at the time [of writing]". These data inform the spreadsheet that is used to calculate comparative energy requirements and allows users to estimate the relative energy performance of thick vs. thin clients for a particular user base. The costs and carbon comparison tool for thick versus thin clients accompanied by a user guide can be found here.[10]
3. Fraunhofer Institute
   The Fraunhofer Institute (Fraunhofer UMSICHT, 2006) published a report that conducted a "whole-life" comparison of thin- and thick-client systems then available and assessed raw material requirements, energy in use, and disposal costs. The "use" phase is based on identifying three workload profiles (light, medium, and heavy)(ibid, Table 2, p. 4). The "medium user" category most closely resembles the profile of our users. It is suggested that 35 of this category of users in thin-client mode means "the system is already working to full capacity" (ibid, p. 8). Note that later in the document, where power use is assessed numerically, the "worst-case" measurement reported is based on 20 users per server

[9] http://www.jisc.ac.uk/media/documents/publications/greenict-queenmargaret.pdf.

[10] http://www.susteit.org.uk/files/category.php?catID=5.

(ibid, p. 21). The reported measurements[11] (for the use phase) are thin client: 26-31 W and thick client 68-96 W, depending on particular system specification.

Methodology

Overview

To reiterate, our JISC thin-client project (Pattinson and Cross, 2011; funded under the JISC Greening ICT initiative) aimed to provide some actual measurement data against which to test the received wisdom. We installed within Leeds Metropolitan University a thin-client deployment in an office environment that supported a mixture of academic and administrative colleagues in their daily work. This deployment was a direct replacement for the normal staff desktop (PC-based thick client); thus, some staff received the thin-client replacement with a purpose-built thin-client terminal; some continued to work with their normal PC in thick-client mode, and others were provided with low-energy but still desktop-based PC systems running either as thin or thick clients. The limited timescales and our concern not to affect the day-to-day work of our volunteer colleagues led us to endeavor to deliver thin-client systems as nearly identical in look and feel as possible to the thick-client system being replaced. Once installed and operational, power consumption was measured by socket-mounted meters at each desk and by recordings of the server system associated with the thin-client deployment.

We measured usage during the autumn/winter term of 2010 to capture what we believed to be a typical sequence of operation with academic staff preparing and delivering course material, gathering information in support of research, writing research papers, and dealing with administrative duties; administrative colleagues were handling issues of student enrollment, committee documentation, and data reporting. We were unaware of any particular circumstances that would make this particular term or this particular office significantly different from any other across the university.

Variations of Thick and Thin Clients

Our methodology required us to identify the candidate thin- and thinner-client systems to be used, the thick-client variation being that provided by the university's standard user desktop/application support. We also identified suitable candidate locations for identifying specific criteria and appropriate metering and data-gathering methods. Each is now discussed in turn:

1. *Thin-client systems*: Our primary criterion for selection of this technology was that it would allow us to deploy application support in a form that was approximately identical to that of

[11] https://www.igel.com/fileadmin/user/upload/documents/PDF_files/Green_IT/Green%20IT%20TCecology_en.pdf.

the thick-client systems they replaced. Undeniably, it was not appropriate to involve users in a major change of system environment. We recognized that this might not be the most effective use of thin-client systems in a longer term and wider scale deployment where the cost of user training might be recouped by the improvements and savings made from optimal use of thin-client devices. However, the scale and duration of our experiment meant that it was not appropriate to expect users to reorient themselves to a different user environment. We also sought to select a commonly used combination of hardware and software for our deployment and determined that we would use Wyse terminals[12] supported via a Citrix server platform on a Dell Poweredge server.[13] In addition, we configured a number of low-powered desktop PCs[14] to operate as thin-client terminals.

2. *Thinner client systems*: There is a spectrum of systems that *could* fit under the term "thinner client"; we defined it as being a desktop PC that is specifically promoted as being a *low-power* device but is otherwise comparable to our standard desktop platforms in form and operation.[15]
3. *Thick-client systems*: Our thick-client platforms, as noted previously, used a variety of the standard university deployment systems purchased over the last three years.[16]

We ensured that our chosen testbed contained a group of users whose daily workloads represented a cross-section of those expected across the institution. The location for the experiment was relatively self-contained, making it easier for us to install and maintain the test systems. Ideally, we selected a location that also had integral power monitoring, removing the need to manually read and record data from separate power meters. The final choice of our sample was our technology-enhanced learning team, not specifically because of their job roles (although their support and willingness to work with us was a benefit) but because they occupied a building (the Old School Board Building in central Leeds) that is physically separate from but close to our other locations. Other reasons for choosing this team were its size (office occupancy varying between 10 and 20 at any time), which allowed us to deploy our range of test systems, and the workload that offered a suitable mix of academic and administrative activities.[17] A disadvantage was that this location did not possess the built-in metering available

[12] Thin-client terminals: Wyse terminals C30LE, 1 GHz GOGX processor, 512 MB memory, 64 MB flash memory, Citrix ICA 10.17 (build 104).

[13] Thin-client server platform: "Dell Poweredge 2950, 2x quad core 3 GHz Intel Xeon C5450 CPU, 10 GB RAM, Server OS: Windows Server 2003 Enterprise SP" running Citrix Metaframe XP FR3.

[14] Thin-client low-power desktops: FX160, Atom 1.6 GHz processor, 4 GB memory, 64 GB solid state HD, Citrix 10.150.58643.

[15] Thick-client low power desktops: FX160, 4 GB memory, 64 GB solid state HD, running Windows XP and standard staff build.

[16] Thick-client desktops: models included 1x Dell Opitplex GX780 desktop, 1x Dell Opitplex GX760 desktop, 3x Dell Opitplex GX745 desktops, 4x Dell Opitplex GX620 desktops, 2x Dell Precision T1500 workstations.

[17] In reality, the user base became wider than this group and location because we relocated equipment to maintain levels of use throughout the project.

elsewhere in the university; hence, we needed discrete metering using plug-in meters at each desktop. However, the requirement for one of our team having to physically read the data meant that those involved in the experiment were able to make contact with a member of the project team on a regular basis. Power meter readings at the desktop were provided by using individual inline power meters.[18] At the server, we measured energy by using Dell OpenManage Server Administrator.

Implementation

Our goal to achieve a common user experience across all platforms meant that we had to develop, test, and deploy a thin-client system that matched, as nearly possible, the user interface offered by the standard desktop. The reasons for this were discussed earlier. We made no adjustment to our power-saving approaches used across the university and made no special effort to reinforce, modify, or remove the general advice and guidance given to all users about power use.[19] Therefore, control of the switch off was with the individual user, and we anticipated that the usage patterns in this respect were comparable to those across the institution. The only exception to this was that all participants were advised not to turn off power at the wall socket because that would cut off the power meters and lose historical data. When buildingwide interruptions to power supply occurred, the time of these was noted and meter readings adjusted accordingly.

Operation of the Experiment

Following the commissioning of the equipment, users followed normal working practices with no constraints or added requirements placed on them by the team. Readings of the energy meter attached to each desk-based system were recorded at frequencies ranging from daily to twice weekly, depending on access constraints together with any other information, such as reported issues and other experiences. We anticipated—correctly—that most operational problems would be reported and addressed at the time they arose through our normal maintenance procedures. We gathered data between October 22 and December 17, 2010, with measurements of the full thin-client deployment commencing on November 4, 2010. In total, 25 individuals took part in the trial. Additionally, we had three thin-client systems available for general use in a development laboratory.

[18] Inline desktop power meters: Pro Elec PL09564 see http://www.amazon.co.uk/Plug-Power-Energy-Monitor-Meter/dp/B000Q7PJGW.

[19] Users were requested to turn off devices at the end of the work day and before long periods of absence from their desks, but no specific measures were enforced beyond our standard installation of screen saving and configuration of sleep mode.

Assumptions

Some discussion was held within the project team about how to display results and what exactly should be measured. It was agreed that because we were interested in comparing the power consumption of different services rather than different technologies, we would measure power consumption regardless of whether the period involved normal business hours, weekends, bank holidays, part-time/full-time employee, or intensity with which an individual user operated the equipment. The considered choice of the experiment location described previously meant the following assumptions were made:

1. We had a mixture of staff broadly representative of the HE sector in terms of
   a. Part-time or full-time
   b. Administrative, support, academic, and managerial
2. Participants behaved in a similar fashion as they had before the experiment.
3. Participants used both services to perform the same business functions.

Results

Table 14.1 shows the results of power use by technology and individual devices. The power usage of a Toshiba laptop seems to be the lowest. Although it is an interesting finding, it is not relevant to the comparison of the thick and thin clients because of its advanced power management features. The Wyse thin-client terminal seemed to be very energy efficient compared to Dell thin and thick clients. However, the difference between the power consumption for both Dell 160 series thick and thin clients seemed to be insignificant. The results of the individual devices were combined to produce thick- and thin-client systems. The total power consumption for such systems is depicted in Tables 14.2 and 14.3. (The values in Table 14.3 are extracts from Table 14.2.) The findings shown in these two tables reveal that a thick-client system using a lower-power PC will have a significantly lower power use compared to thin clients using Wyse terminal and low-power PC. However, this finding is contrary to the thick-client system using a range of PCs. On the other hand, a thin-client system using Wyse terminal has a slight advantage over a thin-client system with low power PC.

Immediate Impact

For further comparison, the Queen Margaret University (QMU) and Fraunhofer Institute data quoted above (assuming 24-hour operation) give values of:

QMU (Queen Margaret University) Thin-Client Project[20]:
Thin client (with server): 22 W (0.022 kW); thick client: 100 W (0.1 kW)

[20] http://www.ucisa.ac.uk/~/media/Files/events/ucisa2010/unishowcasesucisa2010/UCISA_Harrogate_2010%20fraser%20pdf.ashx.

Table 14.1 Breakdown of Power Use by Type of Technology

Number of participants			25	
Number of participants discarded			3	
Total number of measurements			329	
Total number of measurements discarded			10	
Duration of experiment			54 days	
Average period of measurement of single device			23.6 days	
Device	**No.**	**Daily Power/kW**	**Error**	**Notes**
Dell 160 thin client	6	0.177	±0.002	Low-power PC
Dell 160 thick client	2	0.163	±0.005	Low-power PC Note: Power consumption varied little with type of client
Wyse terminal	4	0.052	±0.003	Thin-client terminal
A Range of thick-client PCs				
Dell GX620	4	0.749	±0.003	Thick-client desktops
Dell GX 745	3	1.144	±0.003	Thick-client desktops
Dell GX760	1	0.43	±0.01	Thick-client desktops
Dell GX780	1	0.26	±0.01	Thick-client desktops
Dell Precision Workstation T1500	1	1.59	±0.01	Thick-client workstation
Average thick-client PC	10	0.841	±0.01	
Toshiba laptop	1	0.04	±0.01	Interesting but of no real value to this study
Server	16 clients	6.1	±0.1	
Server per client		0.379	±0.005	

Table 14.2 Breakdown of Power Use by Device

A Comparison Per User (Factoring in the Per Client Value for the Thin-Client Server) Gives the Following Results (kWh): Service Comparison				
	Client (kW)	**Server (kW)**	**Total (kW)**	**Error**
Thin client using Wyse terminal	0.052	0.379	0.43	±0.01
Thin client using low-power PC (Dell 160)	0.177	0.379	0.56	±0.01
Thick client using low-power PC (Dell 160)	0.163	0.00	0.163	±0.005
Thick client using a range of PCs	0.841	0.00	0.841	±0.01

Table 14.3 Comparison of Power Use by Service

System	Total Power Usage (kW)
Thin client using Wyse terminal	0.43
Thin client using low-power PC (Dell 160)	0.56
Thick client using low-power PC (Dell 160)	0.16
Thick client using a range of PCs	0.84

Fraunhofer Institute (2006) Study on Environmental Comparison of PC and Thin-Client Desktop Equipment[21]:

Thin client (IGEL): 26-31 W (0.026-0031 kW); thick client: 68-96 W (0.068-0.096 kW)

Without considering the server power consumption, our results (shown in Tables 14.2 and 14.3) for the thin and thick clients are comparable to that of QMU and Fraunhofer Institute. However, in order to further improve the results, it would have been more beneficial if the power consumption of the clients had considered the various modes: active, idle, and sleep. Additionally, it would have been more useful if the experiment had covered a wider range of computer brands and series, as well as reviewing and updating the power consumption estimation tools employed for the purpose of the JISC project. There were several service issues that emerged in the project:

1. We observed no issues of reliability specifically affecting the performance of the thin-client equipment. Two power outages were experienced, both general losses of power to geographical locations that affected all users in equal measure.
2. Two of our thin-client users withdrew from the experiment, citing performance shortcomings when using particular graphics and resource-intensive applications. We were not aware that these users were predisposed to expect such problems. They had not made any prior detailed study of thin-client systems and therefore would not have been aware that this was often suggested as a problem for thin-client users. We therefore concluded that these perceptions of unacceptable performance were genuine and formed from actual experience. See Appendix A for examples of comments.

Future Impact

The comparison of our measured data with those of reports from up to five years earlier indicate that there have been significant improvements in the power requirements of many ICT devices and that the power requirements of a typical unit are lower than those of an older, comparable device. This indicates the need for estimation tools and reports based on them to be updated regularly to ensure that any comparisons that are made are carried out on a true "like-for-like" basis.

[21] https://www.igel.com/fileadmin/user/upload/documents/PDF_files/Green_IT/Green%20IT%20TCecology_en.pdf.

Conclusions

Although our measurements reinforce the widely held and published information that thin-client technology offers a significant per user energy saving under conditions of normal use when compared to standard thick-client configuration, the lowest energy consumption in our measurements was from the use of *low-energy PCs used in the thick-client mode*.

It should be repeated at this point that the nature of our experimental design, in particular the decision to replicate the thick-client desktop, might have meant that we were not fine-tuning the thin-client deployment to the utmost. However, we do believe that the difference is significant and would continue even with such tuning in place.

Also, it is possible that we were underusing the server in thin-client mode, but we assumed that we could double the number of clients without increasing power consumption or degrading the speed of the service (which is unlikely). The figures are:

Server Serving 32 Clients	Client	Server	Total	Error
Thin client using Wyse terminal	0.052	0.19	0.24	±0.01
Thin client using low-power PC (Dell 160)	0.177	0.19	0.36	±0.01
Thick client using low-power PC (Dell 160)	0.163	0.00	0.163	±0.005
Thick client using a range of PCs	0.871	0.00	0.87	±0.01

Recommendations

Our results suggested that the development of low- power PC systems meant that these devices were able to offer a solution that is better (in terms of energy use) than thin-client systems operating in the same environment. When added to the consideration that a change to these systems would have less impact on infrastructure and user experience, the advantages were clear. We had not been able to assess the claims of longevity for thin-client systems (it generally understood that these systems are viable for double [or more] of the life span of a standard PC). This does not seem unlikely in respect to the technical specification of thin-client terminals: Their relative simplicity suggests that they should indeed have a longer life than the more complex PC solution with consequent impact on costing. However, the technological life span may not be the limiting factor in user selection and choice or in decisions to upgrade.

Implications for the future

Clearly, the decision to adopt a particular technology (thick- or thin-client system) is not based solely on power use. As a matter of fact, other technical factors are also relevant: reliability, expandability, flexibility, and suitability for processing (and data) and hungry applications. User experience and resistance are significant nontechnical barriers to change, and matters such as software licensing are often as reasons that make thin-client "difficult,"[22] but perhaps the most significant reason for a reluctance to take up thin-client technology to any great extent is that thick-client technology has served for 20 years and users and technical staff are familiar with its operation. When this major intellectual investment is added to the significant financial commitment to thick-client systems, it seems likely that any widespread adoption of thin-client technology would require a combination of raising awareness, successful major implementations, and possibly stimuli such as pricing.

Our experiments have of necessity been limited in scale regarding the numbers of users and the actual devices used. We therefore recommend that further experimental measurements be undertaken to assess the use of thin-client systems in different operational areas (in particular in student laboratories, both computer science and generic) and in larger-scale office environments where the actual limitations of server capacity can be addressed. We also recommend that further tests similar to ours but with different platforms be carried out.

The intentional decision not to affect the user experience might have meant that we were not using our systems to their best effect, and we would recommend that work be undertaken to explore using alternative user interfaces and software applications that are tailored to take advantage of the thin-client paradigm. Finally, we should be aware of what could be seen as the next expression of client-server methods: cloud computing. Whereas doing so raises a number of legal, operational, and organizational questions, taking *all* data processing and storage equipment off-site (even off-shore) offers a large potential energy saving at an organizational level.

Acknowledgements

The authors record their thanks to the JISC Greening ICT project for funding this work and for ongoing support and advice. We also record our appreciation of technical colleagues at Leeds Metropolitan University for technical support throughout the project, and in particular to those colleagues whose willingness to allow their workstations to be measured and—in many cases—replaced by different technologies, allowed us to gather the data that informs this project.

Appendix A: User Feedback Comments

"Needs to have access to laptop if things need to be done quickly and ThinC is not performing."

"Not appropriate – has been told, by [participant's line manager], that she can be a control station."

[22] Note that these licensing matters can be and are addressed very effectively in thin-client deployments.

"Crucially I need the capacity to render videos which might require extra memory/processor. I currently have about 10 GB of files I am working on (TEL competition, ALT conference, TEL website) – these are time critical and I could not afford any disruption to my work on them."

"Using a USB to transfer my files could cause problems as I have 28 GB for CETL ALiC on the h drive (though as long as this is secure and at least read accessible I don't need to transfer it)."

"That's all I can think of for now but I would want some assurance that an immediate response to any problems could be guaranteed, i.e., a named person who could fix any problems quickly."

References

Abaza, M., Allenby, D., 2009. The effect of machine virtualization on the environmental impact of desktop environments. Online Journal on Electronics and Electrical Engineering 1 (1). http://www.infomesr.org/OJEEE-V1N1_files/W09-0010.pdf, accessed (15.06.2014.).

Davis, E., 2008. Green Benefits Put Thin-ClientComputing Back On The DesktopHardware Agenda. http://www.meritalk.com/uploads_legacy/whitepapers/ForrestorGreen_and_TCs.pdf, accessed (15.06.2014.).

Doyle, P., et al., 2009. Case studies in thin client acceptance. UbiCC J. 4 (3), 585–598. http://www.ubicc.org/files/pdf/6_377.pdf, accessed (15.06.2014.).

Fraunhofer UMSICHT, 2006. Environmental Comparison of PC and Thin Client Desktop Equipment. Fraunhofer UMSICHT, Oberhausen, Germany. https://www.igel.com/fileadmin/user/upload/documents/PDF_files/Green_IT/Green%20IT%20TCecology_en.pdf, accessed (15.06.2014.).

Fraunhofer UMSICHT, 2008a. Environmental Comparison of the Relevance of PC and Thin Client Desktop Equipment for the Climate. Fraunhofer UMSICHT, Oberhausen, Germany. https://www.igel.com/fileadmin/user/upload/documents/PDF_files/White_Paper_EN/TCecology2008_en.pdf, accessed (15.06.2014.).

Fraunhofer UMSICHT, 2008b. "PC vs Thin Client": Economic Evaluation. https://www.igel.com/fileadmin/user/upload/documents/PDF_files/White_Paper_EN/PCvsTC-en.pdf, accessed (15.06.2014.).

Fraunhofer UMSICHT, 2011. Thin Clients 2011—Ecological and Economical Aspects of Virtual Desktops. https://www.igel.com/fileadmin/user/upload/documents/PDF_files/White_Paper_EN/thinclients2011-en.pdf, accessed 15.06.2014.).

Gabriel, C., 2008. Why it's not naive to be green. Bus. Inf. Rev. 25, 230.

Greenberg, S., Anderson, C., Mitchell-Jackson, J., 2001. Power to the People: Comparing Power Usage for PCs and Thin Clients in an Office Network Environment. Thin Client Computing, Scottsdale, AZ. http://www.lamarheller.com/technology/thinclient/powerstudy.pdf, accessed (15.06.2014.).

Hayes, J., 2010. Thin client's fat challenge [IT desktop computing]. Eng. Technol. 4 (21), 52–53.

Lange, C., Kosiankowski, D., Weidmann, R., Gladisch, A., 2011. Energy consumption of telecommunication networks and related improvement options. IEEE J. Sel. Topics Quantum Electron. 17 (2), 285–295.

Miettinen, A.P., Nurminen, J.K., 2010. Energy efficiency of mobile clients in cloud computing. In: Proceedings of the 2nd USENIX Conference on Hot Topics in Cloud Computing. USENIX Association, Berkeley, CA, USA, pp. 4–10. https://www.usenix.org/legacy/events/hotcloud10/tech/full_papers/Miettinen.pdf, accessed (15.06.2014.).

Murugesan, S., 2008. Harnessing green IT: principles and practices. IT Professional 10 (1), 24–33. http://ieeexplore.ieee.org/stamp/stamp.jsp?tp=&arnumber=4446673, accessed (15.06.2014.).

Pattinson, C., and Cross, R. (2011). Does "thin client" mean "energy efficient"? JISC Green IT Technical Report JISC Technical Report March 2011. ISBN: 978-1-907240-34-8.

Pattinson, C., Siddiqui, T.M., 2008. A performance evaluation of an ultra-thin client system. In: Proceedings of International Conference on E-Business, Oporto, Portugal, July 2008, pp. 5–11.

Satyanarayanan, M., et al., 2009. The case for VM-based cloudlets in mobile computing. Pervasive Comput. 8 (4), 14–23. http://www-inf.telecom-sudparis.eu/COURS/MOPS-RM/Articles/Satyanarayanan09-VMBasedCloudlets.pdf, accessed (15.06.2014.).

Simoens, P., et al., 2010. Cross-layer optimization of radio sleep intervals to increase thin client energy efficiency. IEEE Commun. Lett. 14 (12), 1095–1097.

Staehle, B., et al., 2008. Quantifying the influence of network conditions on the service quality experienced by a thin client user. In: Proceedings of the 14th GI/ITG Conference Measuring, Modelling and Evaluation of Computer and Communication Systems (MMB), pp. 1–15.

Tang, W., Lee, J.H., Song, B., Islam, M., Na, S., Huh, E.N., 2011. Multi-platform mobile thin client architecture in cloud environment. Procedia Environ. Sci. 11, 499–504.

Vereecken, W., et al., 2010a. Energy Efficiency in Thin Client Solutions. Networks for Grid Applications. Lecture Notes of the Institute for Computer Sciences, Social Informatics and Telecommunications Engineering, 25, pp. 109–116.

Vereecken, W., et al., 2010b. Power efficiency of thin clients. Eur. Trans. Telecommun. 21 (6), 479–490. http://onlinelibrary.wiley.com/doi/10.1002/ett.1431/full, accessed (15.06.2014.).

Wyse Technology Inc. (nodate). Desktop Energy Consumption: A Comparison of Thin Clients and PCs, http://www.athena.dk/files/userdir/documents/energy_study.pdf accessed (15.06.2014.).

CHAPTER 15

Cloud Computing, Sustainability, and Risk

Case Study: A Quantitative Fuzzy Optimization Model for Determining Cloud Inexperienced Risks' Appetite

Babak Akhgar[1]*, Ashkan Tafaghodi[2], Konstantinos Domdouzis[1]
[1]*Sheffield Hallam University, Sheffield, UK*
[2]*University of Tehran, Tehran, Iran*
**Corresponding author: b.akhgar@shu.ac.uk*

Introduction

Today, almost all organizations use information technology in any way to enable their businesses to increase their competitive power, reduce cost, and so on. Of course, technological and business risks of using information technology (IT), especially regarding new trends, should not be ignored cursorily. In a number of enterprises, IT-related risk is part of operational risk. However, in cases when IT is the key factor of new business initiatives, it is better not to show IT risk with a hierarchic dependency on one of the other risk categories (ISACA, 2009). Cloud computing is the most influential IT-enabling trend in business today. Thus, many organizations adopt cloud computing and consider it the next major milestone in technology and business collaboration.

According to US National Institute of Standards and Technology (NIST), *cloud computing* is defined as a model for enabling flexible, on-demand network access to a shared pool of configurable computing resources with minimal management effort or service provider interaction (Mell and Grance, 2009).

Security and privacy have topped the list of concerns related to clouds. Many studies have been done on cloud computing risks and its business effects. (Bhardwaj and Kumar, 2011; Saripalli and Walters, 2010) have offered cloud risk assessment based on a quality indicator for rehabilitative care (QuIRC) framework. (Mounzer et al., 2010) has offered a model of risk assessment, control, and mitigation of the amount of damage based on cascades of failures.

(Grobauer et al., 2011; Heiser and Nicolett, 2008; Jansen, 2011) have noticed primary cloud vulnerabilities and risks, mainly from technical perspectives. (Marston et al., 2011; Chorafas, 2011) discussed business- and strategy-related issues surrounding cloud use.

According to ISO/IEC (13335-1:2005), *IT risk* is defined as "the potential that a given threat will exploit vulnerabilities of an asset or group of assets and thereby cause harm to the organization. It is measured in terms of a combination of the probability of an event and its consequence."

Risk and reward are two sides of a coin. Organizations must embrace risk and make a trade-off between the two in pursuing their goals. The more an organization can tolerate reasonable risk, the more benefits it can earn. Therefore, applying good IT risk management and aligning it with the organization's ERM[1] will provide tangible business benefits. ERM is a process affected by an entity's management board and other personnel, and it is implemented in a strategy across the enterprise. It is designed to identify potential events or factors that might affect the entity and to manage risk within its appetite level. ERM also provides reasonable assurance regarding the achievement of entity objectives (Rittenberg and Martens, 2012). ERM attempts to holistically view all risks that a firm undertakes; at its heart, it is what is defined as *risk appetite*. Only if an organization thinks clearly about its risk appetite can it balance risks and opportunities. (Xanthopoulos, 2010) defines risk-related terms as follows:

Risk appetite Willingness of an enterprise to take on risk in order to achieve the desired returns

Risk preferences Management choices regarding various risk-return trade-offs; a first indication of which risks are considered to be acceptable and which not

Risk tolerance Willingness to quantify the risk appetite at the enterprise level; appropriate risk metric for setting specific maximum limits for each risk

Risk target Expression of the optimal amount of risk that an organization is willing to assume in order to achieve it strategic goals and objectives

(Xanthopoulos, 2010) points out some considerable comments when dealing with risk appetite:

- It is strategic and related to the pursuit of organizational objectives.
- It forms an integral part of corporate governance.
- It guides resource allocation.
- It guides an organization's infrastructure that supports the activities related to risk recognition, assessment, response, and monitoring in pursuit of organizational objectives.
- It influences the organization's attitudes toward risk.
- It is multidimensional when applied to the pursuit of value in short-term and long-term strategic planning.
- It requires effective monitoring related to the organization.

[1] Enterprise risk management.

Furthermore, we discuss the previous concepts and terms related to taking a holistic and more accurate business view of the risks surrounding cloud architecture. We need a quantitative model to generalize a given organization's risk appetite of CIA security categories to make sure that it adheres to the main ERM plan intelligibly. It is especially useful for unprecedented IT risks for which sufficient information, records, and logs are not available to use for precisely calculating its impact and resulting risk appetite.

The section Cloud Architecture and Risk Preferences discusses cloud architecture layers and clarifies risk preferences. In the section Cloud Governance and Operations Model, we discuss risk criteria and a basic quantitative approach. The next section explains the development of an optimization model and how its results were calculated. A case study is examined to verify the model's functionality before the conclusion.

Cloud Architecture and Risk Preferences

The most common types of cloud computing deployment models, according to the National Institute of Standards of Technology, are *private*, *hybrid*, *community*, and *public*. Moving from a private deployment model to a public development continuum provides less direct control of IT and information assets and results in more inherent risks. From the architectural perspective of cloud computing (Marks and Lozano, 2010), we discuss four major models of holistic cloud computing that could be the main points of concern when dealing with cloud risk preferences.

Figure 15.1 depicts a cloud computing model composed of four supporting tiers of concern when dealing with its risk issues.

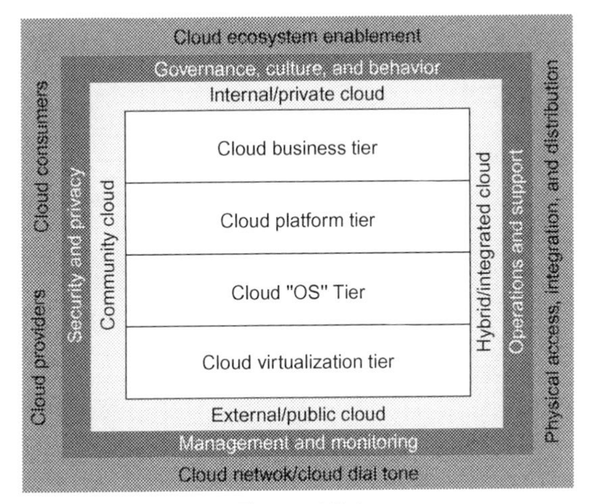

Figure 15.1
Cloud computing reference model.

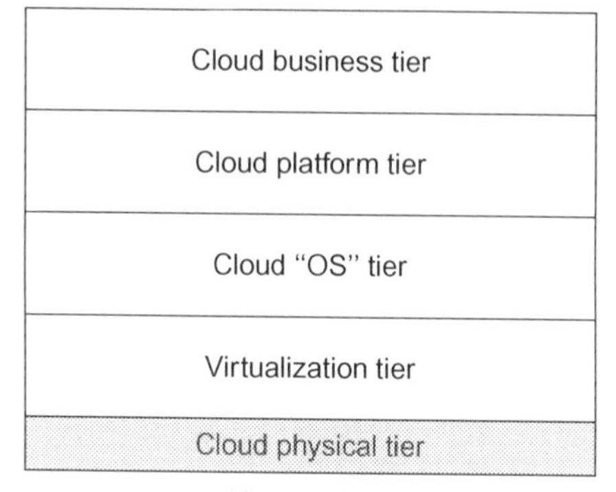

Figure 15.2
Cloud enablement model tiers.

I. The *cloud enablement model* is the core of the reference model. It describes the fundamental technology tiers of cloud computing capabilities that are provided by the cloud platform and Cloud Service Providers (CSPs) to potential consumers of cloud-enabled technology and business capabilities (Marks and Lozano, 2010) (Figure 15.2).

A cloud's *physical tier* provides the physical computing, storage, network, and security resources that support cloud requirements (Marks and Lozano, 2010).

A cloud's *virtualization tier* helps CSPs to share infrastructure resources with multiple clients in order to distribute the cost of infrastructure across many organizations, providing economies of scale. The potential threats to and vulnerabilities of this and the previous tier should be recognized. These include (Securing the Virtualized Data Center, 2011; Zissis and Lekkas, 2011; Carroll et al., 2011):

- Compromise of the virtualization layer possibly affecting all hosted workloads
- Lack of visibility and controls on internal communication between virtual machines (VMs)
- Risks from combining workloads of different trust levels on the same physical machine
- Isolation failure, untrustworthy source of images, and offline patches
- Data leakage and disaster recovery

The *cloud operating system tier* provides the cloud computing fabric as well as application virtualization, core cloud provisioning, metering, billing, load balancing, workflow, and related functionality typical of cloud platforms (Marks and Lozano, 2010). The vulnerabilities from the technical perspective in this layer are related to software breaches, service handling and operations, metering and billing evasion, Identity, Authentication, Authorization and Auditing (IAAA), and so on.

The *cloud platform tier* and the cloud operating system tier make up the cloud platform as a service (PaaS) including both preintegrated platforms offered only as a service and the middleware tools and technologies that enable platforms to be constructed from any combination of infrastructure—physical or virtual, private or public, and so forth (Marks and Lozano, 2010). Preintegrated platforms such as Microsoft Azure, Google

App Engine, and Amazon EC2 are all Web based and offer on-demand access to create scalable applications and services and deliver lucrative software as a solution (SaaS) to customers. (OWASP, 2007) has offered a list of top 10 security threats to Web applications that should be considered when dealing with PaaS and SaaS.

The *cloud business tier* deals with cloud-enabled business applications, software, data, content, knowledge, and associated analysis frameworks. These applications are offered as services across the Internet by the CSP and are predominantly subject to the same risk issues as cloud platform services.

II. The *cloud deployment model* describes the range of cloud deployment scenarios available to private, public, hybrid, and community enterprises (Marks and Lozano, 2010). The main two deployment models are private and public; the other types are a derivative of the two. Concerns depend on the deployment by which an organization chooses to address its cloud needs. The top security issues according to (Cloud Security Alliance, 2010; Srivastava et al., 2011) and the cloud adoption models are outlined in Table 15.1. Additional factors can affect the adoption of cloud computing, including the nature of an organization's business and data and the way it deals with and manages other architecture layers, laws, regulations—discussed in following paragraphs—and valid information security qualifications. Gartner (Three Strategies for Securing Public and Private Cloud Services, 2011) has offered a framework of "three styles of securing public/private cloud," indicating how to address the different security issues of cloud deployment models.

III. The *cloud governance and operations model* describes the governance, culture, and security operations as well as the support, management, and monitoring requirements for cloud computing so that all potential operational risks of adopting a cloud for an enterprise are considered (Marks and Lozano, 2010).

Cultural and cloud governance issues are vital facts facing an organization using this trend of technology. Especially when a new technology trend in business performance is implemented or such a change requires updating governance, security, and management policies, its user acceptance and cultural reform must be properly aligned.

Gartner (Ten steps to building private Cloud Computing, 2011) has considered cultural reform management as the first step toward cloud adoption.

Table 15.1 Threats Affecting Public and Private Clouds

Threats	Public Cloud	Private Cloud
Abuse and nefarious use of Cloud Computing (CC)	✓	×
Insecure interface and Application Programming Interfaces (APIs)	✓	✓
Malicious insider	✓	×
Shared technology	✓	✓
Data loss or leakage	✓	×
Account or service hijacking	✓	×
Unknown risk profile	✓	×

From the governance perspective, it is essential to balance the risk regarding customers and providers or data controllers and data processors and to ensure that services are aligned (Morrell and Chandrashekar, 2011). Can an organization consent to a provider's prior right to data stored on its equipment? Can an organization manage to deal with geographically distributed resources in terms of monitoring and audit processes? Can an organization provide a mutually acceptable policy enforcement framework to ensure the provider's transparency, liability, insurance, and so on? How can the organization integrate legacy systems and services?

Security and privacy are definitely the most critical discussions of cloud governance and could be considered the most preventive factors in cloud computing; this is, of course, obvious because of the various risks an organization takes when it stores and processes its data using a method that is beyond its control scope in order to benefit from doing so. Gartner (Heiser and Nicolett, 2008) has pointed out that if a company is considering the use of an external service of any sort, it needs foremost to assess the security, privacy, and regulatory compliance risks. Privilege user access, compliance, data location, data segregation, availability, recovery, investigative support, and viability are examples of the vast security concerns set of cloud computing. And according to (Mosher, 2011), the primary privacy-related risks and challenges are access control, internal segmentation, subcontractors, data ownership, e-discovery, data censorship, and encryption. Krutz (Krutz and Vines, 2010) believes that management initiatives should address many of these challenges including clearly delineating the ownership and responsibility roles of both the CSP (which may or may not be the organization itself) and the organization that is functioning in the role as customer. Some general management processes are required regardless of the nature of the organization's business. These include the following:

- Security policy implementation
- Computer intrusion detection and response
- Virtualization security management

IV. The *cloud ecosystem model* considers the requirements of developing and sustaining a cloud ecosystem that includes a number of cloud providers, cloud customers, and cloud intermediaries as well as the cloud network and cloud dial tone that are necessary for using the cloud (Marks and Lozano, 2010).

Note that the variety of challenges and risks of different cloud layers, especially technical weaknesses that could be sources of concerns in other domains, are not partible to exclusive security categories.

The cloud ecosystem is not all technical but includes behavioral, cultural, and trust dimensions.

From the provider viewpoint, the following concerns are pointed out (Zissis and Lekkas, 2011):

- Malicious insider
- Metering and billing evasion

- Bursting cloud (third-party trust)
- Increased management workloads

Geographical distribution of infrastructure resources mentioned previously in the section on the physical layer could be troublesome from legal and political views. Some huge cloud providers lack certain legal clarification and privacy principles (Suantesson and Clarke, 2010).

Networks and communication infrastructure that provide users on-demand and ubiquitous access to cloud resources could be vulnerable to these technical risks: data interception in transit, Internet protocol vulnerabilities, hacking and intrusion, mobile device attack, and so on (Subashini and Kavitha, 2011):

Furthermore (Cloud security alliance, 2009; Consensus Audit Guidelines, 2011) have discussed critical areas of concern that are subject to most network and Internet cyberattacks.

Certainly, the customer organization that intends to use cloud resources to benefit from its advantages must have appropriate risk tolerance and appetite to deal with possible risks. Of course, the risks may vary according to the cloud usage scenario and could be unique in their level of impact and probability.

As previously mentioned, different layers of cloud architecture consist of various risks that could be studied from the viewpoints of technical, organizational, legal, and provider aspects or ecosystems. The following section discusses confidentiality, integrity, and availability as the main information security categories that could constrain the optimized values of different cloud risk appetites

Green Cloud Computing and Risk Management

Green information technology is the set of technologies that can provide environmental sustainability. Cloud computing can be characterized as green technology because it provides many benefits for the environment such as the use of fewer machines, increased equipment efficiency, better control of temperature and humidity levels, and dynamic resource allocation. As a green IT, cloud computing includes the dimensions of environmental sustainability, the economics of energy efficiency, and the cost of disposal and recycling (Murugesan, 2008).

Cloud computing has led to the development of a new job market, and its adoption by many companies has changed the way businesses operate. In order to comprehend how businesses incorporate green technologies, the following three approaches must be examined:

- *Tactical incremental approach*: An enterprise preserves the existing infrastructure and policies and uses simple measures to achieve its moderate green goals such as the reduction of energy consumption. These short-term measures include the adoption of policies for simple operations such as the maintenance of an optimal room temperature (Murugesan, 2008).

- *Strategic approach*: The enterprise conducts an audit of its IT infrastructure and its use from an environmental perspective, develops a comprehensive plan for addressing broader aspects of greening its IT, and implements distinctive new initiatives (Murugesan, 2008).
- *Deep green approach*: This expands on the measures presented in the strategic approach, but in this case, the enterprise adopts additional measures, such as the implementation of a carbon offset policy to neutralize greenhouse gas emissions (Murugesan, 2008).

A number of issues such as authentication, the interaction of different cloud systems, the management of a cloud's huge amount of data, the identification of the appropriate virtual network topology, the provision of the appropriate quality of service (QoS), and the allocation of dynamic energy resource need to be considered. To conduct a risk-based assessment of cloud computing, a number of frameworks, such as the Committee of Sponsoring Organizations of the Treadway Commission (COSO) Enterprise Risk Management—Integrated Framework is available as well as IT domain-specific risk frameworks, and practice and process models, such as ISO 27001 and the IT Infrastructure Library (ITIL). The Cloud Security Alliance (CSA), the European Network and Information Security Agency (ENISA), and the NIST provide guidance for cloud computing. CSA has released the cloud controls matrix, which provides security guidelines for assessing the overall security risks of cloud computing. The NIST guidelines on security and privacy in public cloud computing (NIST Special Publication [SP] 800-144) contain guidelines that address public cloud security and privacy. The Risk IT framework based on COBIT® from ISACA fills the gap between generic risk management frameworks and domain-specific frameworks based on the fact that any IT risk is not purely a technical issue (Gadia, 2011).

Risk Appetite and Tolerance

An important aspect of ERM is the process of governing risk appetite, which maintains risk and potential losses within tolerances that are set in advance by the enterprise. An organization must consider its risk appetite at the same time it decides what goals and objectives to pursue.

There are several different items to measure to indicate qualitative and quantitative risk appetite. The most commons are solvency,and earning volatility with compensation adequacy (Barnes, 2006).

Risk appetite can be defined in practice in terms of the combinations of the frequency and magnitude of a risk. Risk appetite could be different among enterprises and whether to consider it acceptable or unacceptable depends on an enterprise's stakeholders. (ISACA, 2009) outlines a table of risk communication flows in which the enterprise appetite is composed of several factors. These include IT risk management scope and plan, IT risk register, IT risk analysis results, executive summary IT reports, integrated/aggregated IT risk report, Key Risk Indicators (KRIs), and risk analysis request, which the chief risk officer (CRO) and the enterprise risk committee handle. Standard & Poor's also recommends sensible tips to consider while documenting risk appetite (Barnes, 2006). See Figure 15.3 for important areas to control the risk. The extent of the discussed areas with other areas could be determined by using fuzzy values (ISACA, 2009). As is indicated in the figure, risk-taking areas take some quantitative ranges from multiplying the impact values by the probability of risk.

One advantage of presenting a statement about the risk appetite with quantitative fuzzy values is that this can give the allowed deviation of the actual risk appetite to the trapezoidal fuzzy numbers (Nasseri and Ardil, 2005). For example, the capacity of unacceptable risk taking as fuzzy places ± 1 unit allowed deviation within itself. For each risk probability, values and their impact can be expressed differently, such as in a discrete or fuzzy range of numbers. For example, the number range from 1 to 4 can be applied for unlikely risks (every two years), less likely (once a year), probable (five times per year), to very likely (more than five times a year), respectively.

However, these quantities must be defined precisely and exclusively in accordance with the valid parameters for the specific case. By definition, any risk can have destructive effects on all or part of an organization. Recognition and quantitative measurement of these effects based on standard criteria such as ISO/IEC 27004:2009 are some important tasks of the organization's CRO unit and the risk committee.

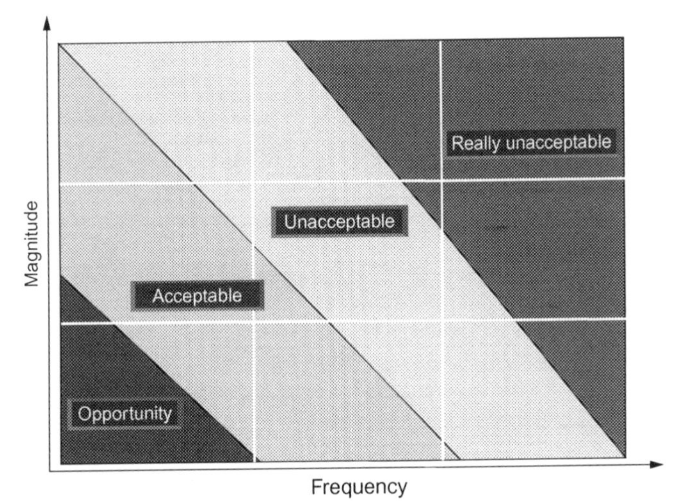

Figure 15.3

Risk map indicating risk appetite bands.

FIPSA (FIPS PUB, n.d.) developed a generalized format for expressing the security category (SC) of an information type as follows:

SC information type = {(confidentiality, impact), (integrity, impact), (availability, impact)} where the acceptable values for potential impact are low, moderate, high, or not applicable[2] or could even be numbered as {1, 2, 3, 0}.

As an example, consider an organization willing to use SaaS to expand its storage capacity and provide ubiquitous access to its data. One of the risks is a malicious insider. The consumer determines a high impact on confidentiality, high impact on integrity, and moderate impact on availability. The resulting I_e would be a three tuple for CIA, as follows:
$I_e = \{(C, 3), (I, 3), (A, 2)\}$

$$e = \text{Malicious insider}$$

The risk may cause an organization to lose its reputation, have its critical information and knowledge exposed to the public, or experience a significant reduction in the effectiveness of its core functionalities. The total impact calculation of the mentioned risk in terms of all three categories is $C + I + A = 8$.

According to the definition, the quantity is calculated for each risk as follows:

$$R_e = P_e . I_e \tag{15.1}$$

To specify the capacity range of risk taking as the range of areas in Figure 15.3, calculate the upper and lower limits of multiplying quantity ranges of final impact by the probability:

$$2 \le I_e := C + I + A \le 9 \tag{15.2}$$

$$1 \le P_e \le 4$$

$$2 \le R_e := P_e . I_e \le 36$$

From this, we use the range of numbers obtained in the numerical example in the section Risk Target and Optimization Model.

According to the numbers obtained, we can identify a quantitative continuum of risk-taking areas and, according to the experts, identify the extent of each in addition to each area's allowed deviation.

For example, we can use the following fuzzy numbers to show different areas in Figure 15.3 along with their allowed deviation.

[2] The potential impact value applies only to the security objective of confidentiality.

Colored areas cannot be identified for certain, exactly what kind of risks are involved. But it can be said that the black area includes the risks that the probability of their occurrence and intensity of its destructive effect on the business or organization's entities is high and the organization should have the most immediate and serious action to deal with it.

Davy's grey areas contain unacceptable risks and need specific policies and programs to stay in the control area. The dim grey area requires no special precautions but monitoring their situation is in order; finally, the grey area might include the risks in a financial business opportunity that results from not dealing with them.

With identifying the risk appetite of an organization in three areas—namely confidentiality, integrity, and availability—the possibility of taking the risks associated with the use of cloud services even when there is no document, log, record, or experience to accurately calculate the probability and impact of the risk, and especially in new areas of information security, the model presented in the next section can be an approach to calculate the quantity of risk appetite directly.

Risk Target and Optimization Model

So far, a variety of models and frameworks has been presented to achieve the risk appetite in information technology services from a security perspective or to study alternating and cascading effects of endangering the interests or assets of an organization faced with a threat. The QuIRC model is one of the most important models for calculating risk assessment based on the probability and impact of risks on the assets; it presents results in known security areas of confidentiality, integrity, and availability.

The Federal Information Security Management Act (FISMA) defines three security objectives for information and information systems:

Confidentiality Preserving authorized restrictions on information access and disclosure, including a means for protecting personal privacy and proprietary information.
Integrity Guarding against improper information modification or destruction; includes ensuring information nonrepudiation and authenticity.
Availability Ensuring timely and reliable access to and use of information.

Any risk associated with the use of cloud services cannot be attributed exclusively to one of the areas discussed because the risks usually influence the three areas of information security with a percentage of the intensity.

Once the organization has clarified how much risk it can tolerate in terms of CIA, it should consider the optimal amount of cloud risk appetite that it is willing to assume to achieve its strategic goals and objectives.

One goal for controlling the risk appetite in relation to the use of cloud services can be the steady increment of advantages that characterize these kinds of services. Because the risk tolerance regarding the use of cloud services make certain benefits in the areas of business, information technology, resources, and so on, it can be formulated as a quantitative goal for the linear programming model. In fact, the direct relationship between the amount of risk and reward helps to maximize the benefits.

$$\widetilde{X}_{\text{Reward}} \propto \widetilde{X}_{\text{Risk}} \text{ or } \widetilde{X}_{\text{Reward}} = \widetilde{C}^T \otimes \widetilde{X}_{\text{Risk}} \tag{15.3}$$

In the following model, the fuzzy numbers, membership function of fuzzy numbers, and operators are defined in accordance with (Gadia, 2011):

$$\begin{gathered} \text{Max } \Re\left(\widetilde{X}_{\text{Reward}}\right) = \text{Max } \Re\left(\widetilde{C}^T \otimes \widetilde{X}_{\text{Risk}}\right) \\ \text{subject to} \\ \widetilde{A} \otimes \widetilde{X} \lesssim \widetilde{B} \end{gathered} \tag{15.4}$$

$\widetilde{X}$ is a non-negative triangular fuzzy risk appetite, $\widetilde{X} = [\widetilde{x}_j]_{n\times 1}$ and $\widetilde{C}^T = [\widetilde{c}_j]_{1\times n}$, $\widetilde{A} = [\widetilde{a}_{ij}]_{m\times n}$, $\widetilde{B} = [\widetilde{b}_i]_{m\times 1}$, and $\widetilde{c}_j, \widetilde{x}_j, \widetilde{a}_{ij}, \widetilde{b}_i \in F(R)$ where $F(R)$ is a set of fuzzy numbers defined on a set of real numbers (Nasseri and Ardil, 2005).

Variables can be considered for more simple calculations such as triangle or sometimes as crisp.

$$\begin{aligned} \widetilde{x}_j &= (x_j, y_j, z_j) \quad \widetilde{c}_j = (p_j, q_j, r_j) \\ \widetilde{b}_i &= (b_i, g_i, h_i) \quad \widetilde{a}_{ij} = (a_{ij}, b_{ij}, c_{ij}) \end{aligned} \tag{15.5}$$

For every $\widetilde{x}_j$, which is the risk appetite for jth risk,

$$x_j, y_j, z_j \in [P_{\min} I_{\min}, P_{\max} I_{\max}] \tag{15.6}$$

Expanding the mentioned FFLP model has the following detailed one:

$$\text{Max } \Re\left(\widetilde{X}_{\text{Reward}}\right) = \text{Max } \Re\left(\widetilde{C}^T \otimes \widetilde{X}_{\text{Risk}}\right) \tag{15.7}$$

subject to

$$\text{Confidentiality}: \sum_{j=1}^{n} \widetilde{a}_{1j} \otimes \widetilde{x}_j \lesssim \widetilde{\mathsf{b}}_1$$

$$\text{Integrity}: \sum_{j=1}^{n} \widetilde{a}_{2j} \otimes \widetilde{x}_j \lesssim \widetilde{\mathsf{b}}_2$$

$$\text{Availability}: \sum_{j=1}^{n} \widetilde{a}_{3j} \otimes \widetilde{x}_j \lesssim \widetilde{\mathsf{b}}_3$$

$$y_j - x_j \geq 0, z_j - y_j \geq 0 \ \forall\, j = 1, 2, \ldots, n$$

And $\Re$ as a ranking function is $F(R) \rightarrow R$, where $F(R)$ is a set of fuzzy numbers defined on a set of real numbers and maps each fuzzy number into the real line where a natural order exists. Let $\widetilde{A} = (a, b, c)$ be a triangular fuzzy number, then $\Re\left(\widetilde{A}\right) = \frac{a+2b+c}{4}$ (Kaufmann and Gupta, 1985).

Identifying the issue that the risks compromising an organization's information and services in three categories of CIA at what level of risk-map reside, based on the knowledge and expertise of CRO or risk management committee, is much easier and more practical than identifying risks in each area of CIA and obtaining the exact value of the probability and effect of each risk uncompromisingly which is usually beyond the scope of the person responsible. The replacement of fuzzy quantities enables better decision making in relation to risk management and reduces the complexity of the required information.

Fuzzy coefficients of the objective function $\widetilde{c}_j$ can also be determined by the decision makers' discretion to illustrate the benefit ratio of jth risk tolerance or the importance of risk-taking concerning the organization's objectives.

The quintet fuzzy Likert range (FIPS PUB, n.d.) with triangular fuzzy numbers can be a good option to determine the importance of risk in the objective function.

As previously mentioned, risks are not certainly related to one of the information security categories of CIA and influence all three areas in different scenarios using cloud services with particular strength. So, technical coefficients $\widetilde{a}_{ij}$ can be considered as fuzzy numbers or crisp, and according to the experts, as equivalent to the impact proportion of the jth risk on the ith risk area.

Right-hand coefficient $\widetilde{b}_i$ in the presented model indicates a maximum risk tolerance of an organization in the ith category. It can be determined based on the fuzzy quantities of the risk map. The above model application based on a simulation example is explained next.

Case Study: Petrogas Jahan Co

We now provide a case study to show how the Linear Programming (LP) risk model works. Petrogas Jahan is a leading corporation in the petroleum industry whose IT department offers various services to more than 160 active users, evaluates the use of cloud mail service, and hosts software outsourcing and similar cloud-enabled added values. From the IT department's point of view after defining and describing the conditions, limitations, quantities, and variables, the risk appetite in the areas of cloud services will be calculated and the final quantitative results will be displayed according to Figure 15.4.

Suppose that the given organization's CRO has identified three samples of the most critical risks of adopting cloud computing, each of which includes security categories with different predefined impacts.

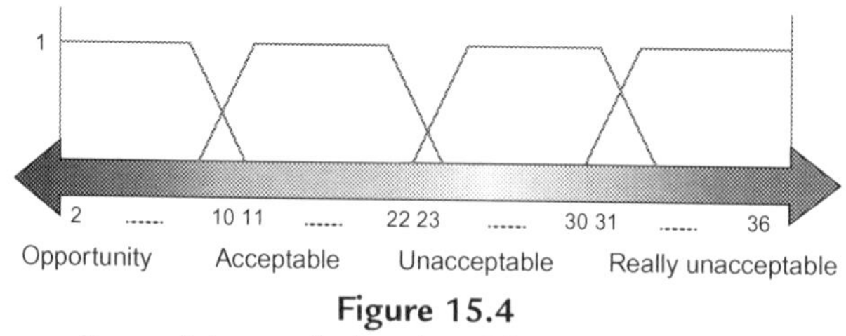

Figure 15.4
Fuzzy risk map indicating risk appetite bands.

Numerous risks and threats concerning the use of cloud computing services occur in different scenarios. The intensity of each is determined based on the circumstances of client organization, server organization, their ability to reduce risk effects, and SLA contract security standards. The obtained value and reward from the risk tolerance for the first and third risks as extracted from an information security audit are $\widetilde{c}_1 = \widetilde{c}_3 = H = (0.5, 0.7, 0.9)$ and for the second risk is $\widetilde{c}_2 = \text{VH} = (0.8, 1, 1)$ in order for the objective coefficients of linear programming to be shaped.

According to the numbers in Table 15.2, the proportion of the impact of each risk in the security categories is obtained more easily than the other values and can be applied as the technical coefficients of the model.

(Kumar et al., 2010) has described how to solve linear full-fuzzy programming.

$$\begin{gathered}\text{Max}\,\Re\left(\widetilde{X}_{\text{Reward}}\right)\text{ or}\\ \text{Max}\,\Re(H\otimes\widetilde{x}_1\oplus\text{VH}\otimes\widetilde{x}_2\oplus H\otimes\widetilde{x}_3)\end{gathered} \tag{15.8}$$

subject to

$$C:\frac{1}{2}\otimes\widetilde{x}_1\oplus\frac{1}{6}\otimes\widetilde{x}_2\oplus\frac{1}{3}\otimes\widetilde{x}_3\precsim(22, 26, 31)$$

$$I:\frac{3}{7}\otimes\widetilde{x}_1\oplus\frac{1}{7}\otimes\widetilde{x}_2\oplus\frac{3}{7}\otimes\widetilde{x}_3\precsim(10, 17, 23)$$

$$A:\frac{2}{7}\otimes\widetilde{x}_1\oplus\frac{3}{7}\otimes\widetilde{x}_2\oplus\frac{2}{7}\otimes\widetilde{x}_3\precsim(22, 26, 31)$$

Table 15.2 Cloud Risks and Their SC impacts and Appetite

SC Impacts Risk	C	I	A
R_1: Malicious insider	3	3	2
R_2: Mail system vulnerability	1	1	3
R_3: Untrusted network and internet	2	3	2
Risk appetite	Unacceptable	Acceptable	Unacceptable

$\widetilde{X}$ is a non-negative triangular fuzzy number and is bounded by risk map limits. The given LP problem may be written as:

$$\text{Max } z = \frac{1}{4}(5x_1 + 14y_1 + 9z_1 + 8x_2 + 20y_2 + 10z_2 + 5x_3 + 14y_3 + 9z_3) \tag{15.9}$$

subject to

Confidentiality	$3x_1 + x_2 + 2x_3 \leq 22$
	$3y_1 + y_2 + 2y_3 \leq 26$
	$3z_1 + z_2 + 2z_3 \leq 31$
Integrity	$3x_1 + x_2 + 3x_3 \leq 10$
	$3y_1 + y_2 + 3y_3 \leq 17$
	$3z_1 + z_2 + 3z_3 \leq 23$
Availability	$2x_1 + 3x_2 + 2x_3 \leq 22$
	$2y_1 + 3y_2 + 2y_3 \leq 26$
	$2z_1 + 3z_2 + 2z_3 \leq 31$

where $y_j - x_j \geq 0, z_j - y_j \geq 0 \forall j = 1, 2, \ldots, n$ and $x_j, y_j, z_j \in [2, 36]$.

The preceding LP is implemented by GAMS® software, and the resultant quantities of decision variables are as follows:

C110.000, C210.000, C310.000, C423.529, C523.529
C626.211, C72.000, C82.000, C913.088, C1012.340
C1116.340, C1217.225, C147.000, C157.000, C1610.275
C1710.275, C1810.275, C222.682, C2411.088, C254.000
C260.885, C277.000

Fuzzy and crisp values, and, finally, the three-risk appetite conforming to the risk map follow:

In fact, Table 15.3 can be a guide for managing identified risks. These quantities are obtained in the most efficient way because of the limitations discussed. And achieving them requires spending money, making a comprehensive plan to deal with risk, conducting a regular audit, having transparent SLAs, and having the provider's commitment to services and other requirements in order to receive the many benefits of cloud services.

Table 15.3 Cloud Risk Appetite Calculation

Risk Appetite	Crisp=$\Re$(Fuzzy)	Fuzzy	
Acceptable	10	(10,10,10)	R_1
Unacceptable	24.2	(23.5,23.5,26.2)	R_2
Opportunity	4.8	(2,2,13.1)	R_3

Conclusion

In this chapter, we have developed a quantitative linear programming model to determine the optimal amount of risk appetite for each individual organization's determined cloud risk in order to deal more easily with these risks and to compensate for dearth of data, documents, and logs to identify precisely the probability and impact quantities. The importance of cloud risk management, especially as a main part of ERM, and of risk appetite as a strategic directive regarding RTP provided orientation to the model design. There are some possible directions for future work. More mathematical models should be developed to identify and highlight clearer coefficients of risk and reward in objective function and to expand the model described here to multiobjective nonlinear programming in order to include more probable quantitative and qualitative objectives. Finally, evaluating our model's performance in real organizations is an exciting area for future work.

References

Barnes, R., 2006. Evaluating Risk Appetite: A Fundamental Process of Enterprise Risk Management. Standard and Poor's. www.standardandpoors.com.

Bhardwaj, A., Kumar, V., 2011. Cloud security assessment and identity management. In: IEEE Proceedings of 14th international Conference on Computer and Information Technology.

Carroll, M., der Merwe, A.V., Kotze, P., 2011. Secure cloud computing. In: Information Security South Africa (ISSA), Johannesburg, August 15–17, 2011, pp. 1–9.

Chorafas, D.N., 2011. Cloud Computing Strategies. CRC Press, Boca Raton, FL.

Cloud security alliance, 2009. Security Guidance for Critical Areas of Focus in Cloud Computing. http://www.cloudsecurityalliance.org/guidance/csaguide.pdf.

Cloud Security Alliance, 2010. Top Threats to Cloud Computing V1.0. http://www.cloudsecurityalliance.org/topthreats/csathreats.v1.0.pdf.

Twenty Critical Security Controls for Effective Cyber Defense: Consensus Audit Guidelines (CAG), http://www.sans.org/critical-security-controls/ Version 3.1 October 3, 2011.

Federal Information Processing Standards Publication, FIPS PUB 199: Standards for Security Categorization of Federal Information and Information Systems.

Gadia, S., 2011. Cloud computing risk assessment a case study. ISACA J. 4, 1–6.

Grobauer, B., Walloschek, T., Stocker, E., 2011. Understanding cloud computing vulnerabilities. IEEE Comput. Reliab. Soc. 9 (2), 50–57.

Heiser, J., Nicolett, M., 2008. Asssessing the Security Risks of Cloud Computing. Gartner Research G00157782, Stamford, Connecticut, USA.

ISACA, 2009. The IT Risk Framework. ISACA, USA. http://www.isaca.org/Knowledge-Center/Research/ResearchDeliverables/Pages/The-Risk-IT-Framework.aspx.

Jansen, W.A., 2011. Cloud hooks: security and privacy issues in cloud computing. In: Proceedings of the 44th Hawaii International conference on System Sciences, pp. 1–11.

Kaufmann, A., Gupta, M.M., 1985. Introduction to Fuzzy Arithmetic: Theory and Applications. Van Nostrand Reinhold, New York.

Krutz, R.L., Vines, R.D., 2010. CloudSecurit: A Comprehensive Guide to Secure Cloud Computing. Wiley Publishing Inc., Indianapolis.

Kumar, A., Kuar, J., Singh, P., 2010. Fuzzy optimal solution of fully fuzzy linear programming problems with inequality constraints. Int. J. Math. Comput. Sci. 6, 37–41.

Marks, E.A., Lozano, B., 2010. Executive's Guide to Cloud Computing. John Wiley & Sons, Inc., Hoboken, New Jersey.

Marston, S., Li, Z., Bandyopadhyay, S., Ghalsasi, A., 2011. Cloud computing—the business perspective. Decis. Support Syst. 51, 176–189.

Mell, P., Grance, T., 2009. The NIST Definition of Cloud Computing, Version 15. National Institute of Standards and Techonology, Gaithersburg, MD. http://csrc.nist.gov/groups/SNS/Cloud-computing.

Morrell, R., Chandrashekar, A., 2011. Cloud computing: new challenges and opportunities. Network Secur. 2011 (10), 18–19.

Mosher, R., 2011. Cloud computing risks. ISSA J. 4 (1), 34–38.

Mounzer, J., Alpcan, T., Bambos, N., 2010. Integrated security risk management for IT-intensive organizations. In: IEEE 6th International Conference on Information Assurance and Security, pp. 329–335.

Murugesan, S., 2008. Harnessing green it: principles and practices. IT Professional 10 (1), 24–33. http://dx.doi.org/10.1109/MITP.2008.10.

Nasseri, S.H., Ardil, E., 2005. Simplex method for fuzzy variable linear programming problems. World Acad. Sci. Eng. Technol. 8, 198–202.

OWASP top 10 web security vulnerabilities, 2007 http://www.owasp.org/index.php/top_10_2007 accessed (11.01.2010.).

Rittenberg, L., Martens, F., 2012. Enterprise risk management: understanding and communicating risk appetite. Committee of sponsoring organizations of the tread way commission, COSO. See, http://www.coso.org/documents/ERM-Understanding%20%20Communicating%20Risk%20Appetite-WEB_FINAL_r9.pdf.

Saripalli, P., Walters, B., 2010. QUIRC: a quantitative impact and risk assessment framework for cloud security. In: IEEE 3rd International Conference on Cloud, Computing, pp. 280–288.

Securing the Virtualized Data Center, Gartner Webinar, 2011.

Srivastava, P., Singh, S., Alfred Pinto, A., Verma, Sh., et al., 2011. An architecture based on proactive model for security in cloud computing. In: IEEE-International Conference on Recent Trends in Information Technology, pp. 661–667.

Suantesson, D., Clarke, R., 2010. Privacy and consumer risks in cloud computing. Comput. Law Secur. Rev. 26, 391–397.

Subashini, S., Kavitha, B., 2011. A survey on security issues in service delivery models of cloud computing. J. Network Comput. Appl. 34, 1–11.

Ten steps to building private Cloud Computing, Gartner Webinar, 2011.

Three Strategies for Securing Public and Private Cloud Services, Gartner Webinar, 2011.

Xanthopoulos, S., 2010. On Risk Appetite. University of the Aegean, Summer School 2010 - "Enterprise Risk Management for Actuaries", Karlovassi, Samos, June 29 - July 2, 2010, Samos, Greece. [online] Available at: http://www.actuaries.org.gr/samos2010/presentations/RISK%20APPETITE.pdf.

Zissis, D., Lekkas, D., 2011. Securing e-Government and e-Voting with an open cloud computing architecture. Gov. Inf. Q. 28, 239–251.

Index

Note: Page numbers followed by *f* indicate figures, and *t* indicate tables.

A

ADR technology solution providers (ATSPs)
 and DRAs, 192
 DR service, 198
AEC sector. *See* Architecture, engineering, and construction (AEC) sector
Agile manufacturing processes, 178
Analog TV signals, 19
Architecture, engineering, and construction (AEC) sector, 100–101
Architecture of prototype
 discrete efficient hardware platform, 250, 251*f*
 multiple hardware platforms, 250, 250*f*
 sample mISP deployment, 250, 252*f*
Architectures, ICT
 ecoefficiency metrics, 53–54
 energy consumption and carbon emission, 52, 52*f*
 expertise and results, 50–52
 green ICT metric pyramid, 55–56, 56*f*
 radar diagram, 54, 55*f*
 stakeholder requirements definition, 47
 system requirements, 48–49, 48*f*, 50
 systems specification process, 49*f*
 traceability matrix, 52–53, 53*t*
Arc hybrid layered manufacturing, 175, 176*f*
Article 3 TEU, 67–68
ATSPs. *See* ADR technology solution providers (ATSPs)
Automated demand response (ADR)
 architectural approach, 203
 "balancing service" programs, 198
 cyber security, 203
 demand forecasting, 207
 and demand response aggregator (DRA), 195–196
 distribution network operators (DNOs), 196–197
 and DRAS (*see* Demand response automation server (DRAS))
 DR event optimization, 207–208
 flexible method, 206
 hardware, 206
 intuitive web interface, 208
 load-shed strategies, 201–202
 market anticipation, 203
 NG, payments, 200
 OpenADR gateway, 205
 open standard, 203–204
 project goals and phases, 200–201
 STOR and DR, 198–199
 system architecture, 205, 205*f*
 TSOs, 197
 turnkey implementation, 202–203
 turnkey solution, 197, 197*f*
 value proposition, building owners, 196, 196*f*
 value streams, building owners, 196

B

Backward compatibility, 18–19
Building information model (BIM), 100–101
Building management systems (BMS), 23–24, 200–201

C

CACTI framework, 116
Carbon emission
 algorithms, 38
 embodied carbon, 38
 energy consumption and, 48–49, 52*f*
 greenhouse gas, 38
 ICT architecture, 50–51
 MoP, 44–45
 real-time measure, 52
 Smart 2020 report, 37
Casual loop diagrams (CLD), 85–86
Cellular backhaul signal strength
 configuring Minicom, 266
 Minicom, 265–267, 265*f*
 NetworkManager, 264–265
 UMTSmon, 264
Central processing unit (CPU)
 energy model, 116, 117
 execution and data processing, 116
 physical ICT system performances, 29
 scheduling, 34
Certified Energy Efficient Data Center Award (CEEDA), 228–229

Cisco promotional video, "The Storm", 157–158, 157*f*
CLD. *See* Casual loop diagrams (CLD)
CLEER. *See* Cloud Energy and Emissions Research Model (CLEER)
Cloud architecture
  and cloud-based software, 135
  and risk preferences, 297–301
CloudBIM prototype, 100–101
Cloud computing
  architecture and risk preferences, 297–301
  definition, 295
  green information technology, 301
  IT risk, 296
  mass consumer migration, 7
  Petrogas Jahan Co, 307–309
  reference model, 297*f*
  risk and reward, 296
  risk appetite and tolerance, 302–305
  risk management, 301–302
  risk target and optimization model, 305–307
  security and privacy, 295–296
Cloud Energy and Emissions Research Model (CLEER), 103
Comitology, 66–67
Command Line Interface (CLI), 276
Communitywide area network and mobile ISP
  context of application, 249
  feedback, 274–275
  Langdale pilot study, 262–274
  prototype mobile learning environment, 250–260
Connected devices, emerging technologies
  cloud storage, 14
  conventional wired communications technology, 15
  data centers, 14
  energy consumption, 14
  features, 14
  hidden connectivity, 15
  Internet, 13–14, 15
Consumers
  automated market participation, 203
  cloud enablement model, 298
  global mobile computing and environmental impact, 7
  GS1 EPC global standard, 67–68
  in-vehicle navigation, 65–66
  and market-related mechanisms, 62
  national ecolabels, 76
  voluntary labeling methods, 79
Content management system (CMS), 249, 256
Conventional wired communications technology, 15
CPU. *See* Central processing unit (CPU)
CyanogenMod, 123
Cyber security, 203

D

Dassonville-Gebhart formula, 64
Data center energy efficiency, issues
  aim and objectives, 224–225
  alternative cooling mechanisms, 237
  assumptions, 238
  cooling system, effect of change, 241–244, 243*f*, 243*t*, 244*f*, 244*t*
  DCIM, 227–228
  "digital universe", 223–224
  energy efficiency techniques, 227, 227*f*
  European Union Code of Conduct, 227
  future impact, 245
  green data center, elements, 226, 226*f*
  green ICT, 225–226
  green power usage effectiveness, 224, 229–230
  immediate impact, 244
  measurements and metrics, 229–230, 231*t*
  in mechanical and power systems, 227–228, 228*f*
  operation of experiment, 238
  PUE analysis, 224, 238–241
  RAL-UZ 161 certification, 227
  results, 238
  set point temperature, effect, 241, 241*f*, 242*f*
  target temperature "set point", 237
Data center infrastructure management (DCIM), 227–228
DECC. *See* Department of Energy and Climate Change (DECC)
Decision support systems (DSSs), 85, 88–89
Delegated legislation, 66–67
Demand forecasting, 207
Demand response (DR)
  ADR solution, 198
  demand-side services, 193
  event optimization, 207–208
  and STOR, 198–199
  strategy, 207
Demand response aggregator (DRA)
  ADR project, 192, 194, 200
  and ATSP, 198
  flexible load, 192
Demand response automation server (DRAS)
  baselining, 209–211
  morning adjustment, 210
  OpenADR-compliant software, 208–209
  OpenADR gateway and signal, 215, 215*f*
  virtual meter data, 210–211, 211*f*
Demand-side balancing reserve (DSBR), 193
Demand-side response (DSR), 193, 217
Dematerialization, product chain, 20–21
Department of Energy and Climate Change (DECC), 188, 218
DeskProto, 173
"Digital universe", 223–224

Direct metal deposition (DMD), 174–175, 176*f*
Distribution network operators (DNOs)
DR programs, 219
streams, 196–197
DMD. *See* Direct metal deposition (DMD)
3D modeling package blender, 162, 163*f*
DNOs. *See* Distribution network operators (DNOs)
3D Printing
additive, 182
Catalyst, 162–164, 164*f*
customized prosthetics, 181
design pipeline, 162–166
3D modeling package blender, 162, 163*f*
Factory 2.0 philosophy, 177–178
fused deposition modeling (FDM), 171
hybrid systems, 173–175
inkjet technology, 172
laminated object manufacturing (LOM), 170–171
made-to-measure implants, 181
media interest in, 161
open-source designs, 161
rapid prototyping (RP), 161
Red Eye3, 181
REP-RAP 3D printer, 161, 162*f*
selective laser sintering (SLS), 168
Shapeways material catalog, 164–165
solid ground curing (SGC), 168–169
stereolithography apparatus, 166–168
stereolithography (STL) process, 161
Urbee car1, 181
DR. *See* Demand response (DR)
DRA. *See* Demand response aggregator (DRA)
DRAS. *See* Demand response automation server (DRAS)
DSBR. *See* Demand-side balancing reserve (DSBR)
DSR. *See* Demand-side response (DSR)
DSSs. *See* Decision support systems (DSSs)
DVFS. *See* Dynamic voltage-frequency scaling (DVFS)
Dynamic Host Configuration Protocol (DHCP), 276
Dynamic voltage-frequency scaling (DVFS), 117, 120

E

Ecodesign directive
harmonized interpretation and application, 75–76
IT equipment and devices, 73
measures implementation, 74–75
resource efficiency and sufficiency, 73–74
responsibilities for compliance, 74
Ecoefficiency metrics, 53–54
Ecolabeling, 76–78
Ecological and ethical consideration, ICT
base transceiver station (BTS), 37–38
carbon emission, 38
contamination, 36
electromagnetic transmitters, 39
electrosmog, 39–40
embodied carbon, 39
emitted carbon, 39
energy consumption, 37–38
global initiatives, 36
GS1 EPC global standard, 37
International Organization for Standardization (ISO) 14001 standard, 35–36
key performance indicators (KPIs), 36
light pollution, 40
measure of performance (MoP), 42–45
noise pollution, 40
power usage effectiveness (PUE), 37–38
radio frequency identification (RFID) technologies, 37
resource efficiency, 40–41
Smart 2020 report, 37, 38
standardization requirement, 36
sustainable development approach, 35–36
Ecopliant, 76
EEA. *See* European Environment Agency (EEA)
Electricity market reform (EMR), 188, 192
Electronic Numerical Integrator and Computer (ENIAC), 4
Electrosmog, 39–40
Embodied carbon, 39
Emitted carbon, 39
EMR. *See* Electricity market reform (EMR)
End-of-life recycling rate (EOF-RR), 42–43
Energy consumption
applications, 115–117
compiler techniques, 118–119
MoP, 43–44
physical properties, semiconductors, 117–118
probabilistic approaches, 120–121
by retailers, 78–79
runtime approaches, 119–120
Energy harvesting, 8
applications, 153
carbon monoxide alarm, 154
components, 153
"hyped" and "hidden, 154–159
IoT, 154–159
LED and LCD, 152
Peltier module, 154
scavenging, 151
solar-powered calculator, 151
Energy labeling directive, 76–78
Energy scavenging, 151
Energy star label, 76
ENIAC. *See* Electronic Numerical Integrator and Computer (ENIAC)
Environmental impact
analog TV signals, 19
balance sheet construction, 20
building management systems (BMS), 23–24

Environmental impact *(Continued)*
- camera resolution, 15
- connected devices, 13–15
- desktop PC, 113
- distribution, 113
- greening by IT, 19
- hidden connectivity, 15
- improvement process, 12
- intelligent energy metering, 22–23
- in-vehicle navigation, 15–16
- IT resources, 24
- mode of operation, 19
- multifunction devices (MFDs), 16
- negative, 16–17
- obsolescence, 12–13, 18–19
- product chain dematerialization, 20–21
- resources, 19
- of software, 111
- speed and reliability demand, 17–18
- travel advice/road traffic control, 21–22
- videoconference, 20
- XP operating system, 18

EOF-RR. *See* End-of-life recycling rate (EOF-RR)

Equivalent isotropic radiated power (EIRP), 271

EU. *See* European Union (EU)

EU environmental law
- article 3 TEU, 67–68
- consumers and market-related mechanisms, 62
- integrated product policy (IPP), 68
- legal bases in, 65
- natural resources conservation law, 68–69
- nonstate standard setting and voluntary instruments, 63
- producers and production process, 62

European Environment Agency (EEA), 103

European legal instruments, Green computing
- ecodesign directive, 72–76
- energy consumption by retailers, 78–79
- energy labeling directive and voluntary ecolabeling, 76–78
- hazardous substances and natural resources conservation, 79–80
- public procurement, 70–72
- recycling and disposal, 80

European Union (EU)
- delegated legislation, 66–67
- directives, 66
- ecolabel and energy star label, 76
- environmental law, 61, 65
- integrated product policy (IPP), 61
- life cycle approach, 61, 69–70
- and national law, 64–67
- regulations, 65–66
- supremacy and market freedoms principles, 64–65
- voluntary ecolabel, 76

European Union Code of Conduct, 227

F

Factory 2.0 philosophy, 177–178

Factotum personal fabricator, 174, 175*f*

FDM. *See* Fused deposition modeling (FDM)

Feedback, mobile learning system
- by module leader, 274
- by tutors, 274–275

Flexible load, building
- "capacity crunch", 193
- DRA and ADR, 192
- DSBR, 193
- DSR, 193
- potential value, 194
- STOR, 192
- sustainable and carbon-neutral solution, 191–192
- UK electricity system, 191–193, 194

Free and open source software (FOSS), 249, 252, 263

Full reference (FR), multimedia service, 35

Fused deposition modeling (FDM), 171

G

General Services Administration (GSA), 101–102

German Blue Angel, national ecolabels, 76

GIF. *See* Green infrastructure framework (GIF)

Greenhouse gas, carbon emission, 38, 103, 129–130

Green ICT
- green IT, 225–226
- metric pyramid, 55–56, 56*f*
- second industrial revolution, 4–7

"Green Information and Communication Technology", 7–8

Green infrastructure framework (GIF), 99

Green IT
- consumers and market-related mechanisms, 62
- direct and indirect governance, 61
- EU and National Law, 64–67
- European legal instruments, 69–80
- nonstate standard setting and voluntary instruments, 63
- producers and production process, 62
- sustainability in EU Law, 67–69

Green technology
- Cisco, 157–158, 157*f*
- cloud computing, 98–99
- "hyped" and "hidden, 154–159
- inexpensive device, 159
- internet, 155, 156*f*
- IoT, 154
- machine-to-machine applications, 155
- Minnesota Department of Transportation, 159
- Philips intelligent and traditional lighting operations, 158, 158*f*
- smart building management, 157
- smarter trash can, 157
- smart street lights, 157
- US Department of Defense, 154

Green theme
green IT and cloud, 129–131
information and communication technology (ICT), 132
simple network management protocol (SNMP), 132
sustainability on cloud, 133–136
virtualization, 131–132
virtual machines (VMs), 132
GSA. *See* General Services Administration (GSA)
GS1 EPC global standard, 37

H

Hardware stack, 253–255, 254*t*, 275–276
Hazardous substances and natural resources conservation, 79–80
Health and Human Services (HHS), 101–102
Hybrid systems, 3D printing
agile manufacturing, 178
arc hybrid layered manufacturing, 175, 176*f*
DeskProto, 173
direct metal deposition (DMD), 174–175, 176*f*
disadvantages, 173
Factory 2.0 philosophy, 177–178
Factotum personal fabricator, 174, 175*f*
laser-engineered net shaping (LENS), 173–174, 174*f*
Makerbot Thingiverse Website, 177
material subtraction, 173
organic materials, 179–180
recycling, 178–179
short delivery time, 173
"Hyped" and "hidden", green technology, 154–159

I

Industrial revolution
ecology and humans, 3
information and communication technologies, 4–7
pollution, 3
Influence diagrams
vs. decision tree, 87–88
and knowledge representation, 86–87
Information and communication technologies (ICTs)
computer technology, 5–6
ecosystem, 7
and environmental impact, 6–7
global mobile computing, 6–7
IC revolution, 4–5
Inkjet technology, 3D printing, 172
Integrated Circuit (IC) revolution, 4–5
Integrated circuit technology (ICT)
architectures (*see* Architectures, ICT)
business, 29
data measurement, 33
data processing service, 33–34
data storage, 34
data transport, 34
delay, 33
engineers, 29
ethics in, 45–46
functional models, 32–33, 32*f*
impact on pollution, 36–40
issues (*see* Ecological and ethical consideration, ICT)
metric, 33
multimedia service, 34–35
performances (*see* Performances assessment, ICT)
products composition, 37
response time, 33
Integrated product policy (IPP), 68
Integrated sustainability assessment (ISA), 84
Intelligent energy metering, 22–23
Intel X86 family of central processing units, 256–257
International Organization for Standardization (ISO) 14001 standard, 35–36, 280
International Telecommunication Union-Telecommunication (ITU-T), 29
Internet
connected devices, emerging technologies, 13–14
electricity intensity, 6–7
green technology, 155, 156*f*
WWW and, 6
zettabyte, 15
Internet of things (IoT)
applications, 152
Cisco approach, 159
remote space probe, 155
water management, 159
In-vehicle navigation, 15–16
IoT. *See* Internet of things (IoT)
IPP. *See* Integrated product policy (IPP)
ISA. *See* Integrated sustainability assessment (ISA)
IT risk, 296
ITU-T. *See* International Telecommunication Union-Telecommunication (ITU-T)

J

JANET Wireless LAN project, 249
Jevons paradox, 11–12
JISC Data Center Efficiency project
flood at Portland, 230, 237*f*
metering topology, 230, 236*f*
Meter installation for servers, 230, 236*f*
Woodhouse building LV switch panel, 230, 236*f*
JISC thin-client project, 285
Jitter, 34

K

Karlsruhe Institute of Technology (KIT), 102
Key performance indicators (KPIs), 36
KIT. *See* Karlsruhe Institute of Technology (KIT)
KPIs. *See* Key performance indicators (KPIs)

L

"Label shopping", 78
Laminated object manufacturing (LOM), 170–171
Langdale pilot study
AT+CSQ/RSSI lookup, 268
cellular backhaul signal strength, 263–267

Langdale pilot study *(Continued)*
client equipment, 263
core equipment, 262–263
requirements, 262
squid proxy, 271–273, 274
WAN (*see* Wide area network (WAN))
WLAN (*see* Wireless local area network (WLAN))
Large combustion plant directive (LCPD), 188, 189
Laser-engineered net shaping (LENS), 173–174, 174*f*
LCPD. *See* Large combustion plant directive (LCPD)
LENS. *See* Laser-engineered net shaping (LENS)
Lighting and security systems, 23
Light pollution, ICT, 40
Linear programming, 305–306, 308
Linux terminal server project (LTSP), 253
Load-shed strategies
DR programs, 201
HVAC plant, 202
and load profile, 201–202, 202*f*
LOM. *See* Laminated object manufacturing (LOM)

M

Made-to-measure implants, 3D printing, 181
Makerbot Thingiverse Website, 177
Management information base (MIB), 30, 140
MAs. *See* Morning adjustments (MAs)
Measure of performance (MoP)
carbon emission, 44–45
end-of-life recycling rate (EOF-RR), 42–43
energy consumption, 43–44
ethics in ICT, 45–46
green ICT development, 42
notations, 42*t*
recyclability, 42–43
MFDs. *See* Multifunction devices (MFDs)
Minicom
configuration information, 266, 267*f*
connection information, 265–266, 265*f*
usage, 266–267
Minnesota Department of Transportation, 159
Mobile Internet Service Provider (mISP) platform, 249
Modeling sustainability, quantitative and systemic methods
artificial intelligence techniques, 85
complexity, 83–84
criticisms, 89–90
decision support systems (DSSs), 85, 88–89
influence diagram, 86–88
integrated assessment (IA), 84
neural networks, 85
Russian doll model, 84
system dynamics and control, 85–86
Venn diagram, 84
Moore's law, 12, 115, 124
MoP. *See* Measure of performance (MoP)
Morning adjustments (MAs), 209–210
Multifunction devices (MFDs), 16
Mutual recognition, 64
"My Site" interface, 208

N

NASA. *See* National Aeronautics and Space Administration (NASA)
National Aeronautics and Space Administration (NASA), 21, 101–102
National grid (NG)
payments, 200
STOR program, 199–200, 199*f*
Natural resources conservation law, 68–69
Network management for green cloud
analytical engine, 142
global view, 144*f*
guidelines, 141–142
hypervisors role, in traffic characterization, 139
local machine level, interaction of elements, 143*f*
and power consumption, 137–138
statistics, 143–146
traffic characterization in virtualized environments, 138–139
VASPE's role, 142
virtualization management, 136
NG. *See* National grid (NG)
Noise pollution, ICT, 40

O

Obsolescence, 12–13, 18–19
OpenADR gateway
ADR program, 203
BMS, 201
embedded controller/server platform, 205
protocols, 205
and signal DRAS, 215, 215*f*
Open Lightweight Directory Access Protocol13 (OpenLDAP), 256–257
OpenWRT Better Approach To Mobile Adhoc Networking (B.A.T.M.A.N), 256–257

P

Packet loss rate, 34
Path capacity, 34
Performances assessment, ICT
black box and its behavior, 30–31
constructive methods, 30–31
data center energy consumption, 31–32
models and, 31
pollution assessment, 31
sustainable development pillars, 31, 31*f*
toxic material rate, 31
Petrogas Jahan Co, cloud computing, 307–309
Philips intelligent and traditional lighting operations, 158, 158*f*
Powder-based printers, 172

Power consumption
hot spots, 141
infrastructure as a service (IaaS), 132–133
network management problem, green cloud as, 137–138
network processes, 133
virtualization, 131–132
Power usage effectiveness (PUE) analysis
of data center, 224
data center efficiency, 37–38
energy consuming resource, 229–230
green power usage effectiveness, 229–230
renewable energy, 229–230
Processing quality, 34
Product chain dematerialization, 20–21
Program Support Center (PSC), 101–102
Prototype mobile learning environment
architecture of prototype, 250
configuration, 260–261
electrical topology, 257–260
hardware stack, 253–255, 254*t*
infrastructure, 253
mISP infrastructure, 251–253
software stack, 256–257, 258*t*
specifications, 260, 261*f*
technical capabilities, 251–253
PSC. *See* Program Support Center (PSC)
Public procurement law
certificates and ecolabels, 72
CJEU case law, 71
German Law, implementation in, 72
market distortion, discrimination and intratransparency risk, 70–71
role, 70
Strategy Europe 2020, 71

R

Radio frequency identification (RFID) technologies, 37
RAL-UZ 161 certification, 227
Rapid prototyping (RP), 161
Rationale
infrastructure as a service (IaaS) and power, 132–133
network processes and power, 133
thermal-aware virtualization, 133
Recycling
3D printing, 178–179
MoP, 42–43
Red Eye3, 181
REP-RAP 3D printer, 161, 162*f*
Risk appetite, 296–297, 296*ge*
Risk preferences, 296*ge*
Risk target, 296*ge*
Risk tolerance, 296*ge*
Robots development, 30
Russian doll model, 84

S

Scratchpad memories (SPM), 118, 119, 120
Second industrial revolution
Computer Technology, 5–6
ENIAC, 4
global mobile computing and its environmental impact, 6–7
Integrated Circuit (IC) revolution, 4–5
Selective laser sintering (SLS), 168
Service-level agreement (SLA), 29, 30
SGC. *See* Solid ground curing (SGC)
Shapeways material catalog, 164–165
Short-term operating reserve (STOR)
and DR, 198–199
technical requirements, 199–200
Simple network management protocol (SNMP)
energy consumption, 140–141
green usage monitoring (GUM) Information, 141
joulemeter, 140–141
management information base (MIB), 140, 141
monitoring systems, 30
operation in context of green clouds, 139–140
power consumption hot spots, 141
SLA. *See* Service-level agreement (SLA); Stereolithography apparatus (SLA)
Sleep mode, 281–282
SLS. *See* Selective laser sintering (SLS)
Smart building management, 157
Smarter trash can, 157
Smart grid technology, 213
Smart meters, 22
Smart 2020 report, 37, 38
Smart street lights, 157
SNMP. *See* Simple network management protocol (SNMP)
Software stack, 256–257, 258*t*
Solar-powered calculator, 151
Solid ground curing (SGC), 168–169
Speed and reliability demand, 17–18
SPM. *See* Scratchpad memories (SPM)
Squid proxy
control measures, 271–273
policy, 274
Stakeholder requirements definition, 47
Stereolithography apparatus (SLA), 166–168
Stereolithography (STL) process, 161
STOR. *See* Short-term operating reserve (STOR)
Storage capacity, 34
Sustainability on cloud, green theme
architectures, 135
central processing unit (CPU) scheduling, 134–135
cloud-based software, 135
computer room air conditioning (CRAC), 135
data center's activity, 135
"energy-agnostic" devices, 134–135
government policies, 133–134

Sustainability on cloud, green theme *(Continued)*
Gustafson's law, 134–135
hardware-optimized servers, 135
network interfaces, 136
power-aware intelligent networking devices, 135
traffic-aware green initiatives, 136
virtual networking algorithms, 135
VM management, 136
Sustainable cloud computing
advantages, 96–97
applications, 100–105
architecture, 95, 96*f*
business models, 99
category, 95–96
characteristics, 105
environmental sustainability, 97
European Union (EU), 106
GIF, 99
globalization, 95
green ITs, 97
green technology, 98–99
platforms, 95–96, 97*t*
reflections, applications, 106–107
UNFCC, 97
Sustainable software design
Android system version, 122
categorization, 114
computing devices, 121
devices, 121
energy consumption (*see* Energy consumption)
executing software, 121
Moore's law, 124
and product life cycle, 112–113
runtime energy consumption, 115
and scope, 111
sensor nodes, 114–115
sustainability effects, 111–112

T

TCM. *See* Tightly coupled memories (TCM)
Thames Valley Vision Project, United Kingdom
ADR, 215
benefits, 216
DSR adopters, 217
load profile, 214, 214*f*
project description, 211
project partners, 211
promotion success, trial, 217–218
smart grid project, 211–218
SSEPD TVV Project SDRC report, 216
utility, 211
Thick client. *See also* Thin-client and energy efficiency
environmental effects, 164
green IT, 162–164
higher education (HE) environment, 161
PC-based, 165
variations, 172
Thin-client and energy efficiency
aim, 280
assumptions, 288
client-server methods, 292
green IT, 281–283
higher education (HE), 280
immediate impact, 288–290
implementation, 287
JISC-funded project, 280
low-power PC systems development, 291
performance, project reports, 284–285
power requirements, 290
power use breakdown, 289*t*
technical factors, 291–292
thick client, definition, 279
and thick client power consumption, 283–284
variations, 285–287
Tightly coupled memories (TCM), 118
Top-runner approach, 78
Total path bandwidth, 34
Traceability matrix, 52–53, 53*t*
Transmission system operator (TSO), 188, 197, 198
Travel advice/road traffic control, 21–22
Treaty on the European Union (TEU), 64
TSO. *See* Transmission system operator (TSO)
Turnkey
implementation, 202–203
solution, 197, 197*f*

U

UK electricity system
buildings and flexible load, 191–193, 194
DECC, 188
EMR, 188
EU and LCPD, 188
existing, construction and planned capacity, 189, 189*f*
factors, 188
forecast demand and power generation, 189, 190*f*
forecast, factors, 188–189
NG, 188
transmission and distribution networks, 191
UMTSmon, 244
UNFCC. *See* United Nations Framework Convention on Climate Change (UNFCC)
United Nations Framework Convention on Climate Change (UNFCC), 97
Universal serial bus (USB)-compliant devices, 281
Urbee car1, 181
US Utility-Driven ADR Programs
CenterPoint Energy, Houston and Texas, 219–220
Consolidated Edison and New York, 220–221
CPS energy, San Antonio and Texas, 221–222
networks and address peak demands, 219

V

Vehicular ad-hoc networks (VANETs), 104
Venn diagram, 84
Videoconference, 20